CROP PRODUCTION SCIENCE IN HORTICULTURE SERIES

Series Editors: Jeff Atherton, Senior Lecturer in Horticulture, University of Nottingham, and Alun Rees, Horticultural Consultant and Editor, *Journal of Horticultural Science*.

This series examine economically important horticultural crops selected from the major production systems in temperate, subtropical and tropical climatic areas. Systems represented range from open field and plantation sites to protected plastic and glass houses, growing rooms and laboratories. Emphasis is placed on the scientific principles underlying crop production practices rather than on providing recipes for uncritical acceptance. Scientific understanding provides both reasoned choice of practice and the solution of future problems.

Students and staff at universities and colleges throughout the world involved in courses in horticulture, as well as in agriculture, plant science, food science and applied biology at degree, diploma or certificate level, will welcome this series as a succinct and readable source of information. The books will also be invaluable to progressive growers, advisers and end-product users requiring an authoritative, but brief, scientific introduction to particular crops systems. Keen gardeners wishing to understand the scientific basis of recommended practices will also find the series very useful.

The authors are all internationally renowned experts with extensive experience of their crops, who provide authoritative and common sense coverage of all aspects of production, ranging from propagation and planting, through husbandry and harvesting and storage. Selection with the consumer and other information on s

Titles Available:
1. **Ornamental F as**
2. **Citrus** F.S. Davi
3. **Onions and Ot** er
4. **Ornamental Be**
5. **Bananas and P**
6. **Cucurbits** R W. R
7. **Tropical Fruits** H.Y. Nakasone and E. Paull

TROPICAL FRUITS

Henry Y. Nakasone
Emeritus Professor of Horticulture (Deceased)
College of Tropical Agriculture and Human Resources
University of Hawaii at Manoa
Honolulu, HI, USA

and

Robert E. Paull
Professor of Plant Physiology
College of Tropical Agriculture and Human Resources
University of Hawaii at Manoa
Honolulu, HI, USA

CAB INTERNATIONAL

CABI *Publishing* is a division of CAB *International*

CABI Publishing
CAB International
Wallingford
Oxon OX10 8DE
UK

CABI Publishing
10 E 40th Street
Suite 3203
New York, NY 10016
USA

Tel: +44 (0)1491 832111
Fax: +44 (0)1491 833508
Email: cabi@cabi.org

Tel: +1 212 481 7018
Fax: +1 212 686 7993
Email: cabi-nao@cabi.org

A catalogue record for this book is available from the British Library, London, UK.

Library of Congress Cataloging-in-Publication Data
Nakasone, Henry Y.
    Tropical fruits / Henry Y. Nakasone and Robert E. Paull.
        p.   cm.--(Crop production science in horticulture)
    Includes bibliographical references (p.   ) and index.
    ISBN 0-85199-254-4 (alk. paper)
    1. Tropical fruit. I. Paull, Robert E. II. Title. III. Series.
SB359.N275   1998
634'.6--dc21   97-44048
                                                                         CIP

ISBN 0 85199 254 4

First printed 1998
Reprinted 1999

Typeset by Solidus (Bristol) Ltd.
Printed and bound in the UK by Biddles Ltd, Guildford and King's Lynn.

CONTENTS

DEDICATION

This book is dedicated to two men who contributed significantly to tropical-fruit breeding and production throughout their lives. The first is Professor William Bicknell Storey, who was born on 31 March 1907, in Honolulu, Hawaii, and died on 22 December 1993. He was the great-great-grandson of the Revd Elias Bond, one of the original pioneering missionaries sent by the Congregational Church of Boston, Massachusetts, to the Hawaiian Islands. He graduated in general science (1935) and received his MSc degree in genetics (1937) from the University of Hawaii and PhD in genetics from Cornell (1940). He was on the faculty of the Horticulture Department at the University of Hawaii from 1940 to 1954, when he left to join the faculty at the University of California Citrus Experiment Station, Riverside, later to become the Riverside Campus. He retired from Riverside in 1975. He focused on tropical-fruit genetics, specializing in avocado and macadamia.

Henry Yoshiki Nakasone was born in Eleele, Wahiawa, Kauai, on 6 July 1920, and died of cancer in Honolulu on 13 August 1995. He began his studies at the University of Hawaii in September 1939 and received a BA in Botany in February 1943. He then served in the US Army as a medic in the much decorated 442nd Regimental Combat Team, 100th Infantry Battalion, in Europe. He returned to the University in the 1948 academic year, receiving an MSc in genetics in 1952 under Professor Storey while being employed by the University as an instructor and then assistant professor with Dr Storey. He spent the 1957 academic year at the University of California at Davis doing course work. Upon his return from Davis, at the same time as carrying out his teaching and research responsibilities at the University, he pursued a PhD in genetics, which was awarded in June 1960. He spent 2 years at the University of Ryukyu as a visiting professor, affiliated with Michigan State University foreign-mission team. This led to a position as a consultant to the US Civil Administration and Government of Ryukyu from 1959 to 1966 for fruit, ornamental and floriculture development and production, which, in turn, led to further experience via consulting positions and invitations to speak at conferences in Brazil, Taiwan, Mexico, Jamaica, Okinawa, India, Indonesia, Madeira Island, Pacific Islands, Windward Islands, Belize, Dominican Republic and Egypt.

Henry's experience overseas served him well as an active contributor to the education of undergraduate and graduate students and many trainees at the University of Hawaii. His courses on the principles of plant propagation and tropical-fruit production were greatly enhanced by his wide experience. Two dozen masters and doctoral degrees were earned under Dr Nakasone's direction. These students are now contributing to horticulture and other agricultural fields throughout the world.

His own research focused on the development of new varieties of horticultural plants, working on numerous crops that led to new varieties of papaya, guava and acerola. He also contributed to the development of many cultural practices used by the horticultural industry in Hawaii. He published over 70 papers in scientific journals, books, book chapters and station bulletins. He retired in 1981 from the University of Hawaii and was named an emeritus professor of Horticulture, but did not stop working. This book started soon after he retired, but activity was greatly slowed by his many volunteer and consultant activities to 17 countries, providing expertise on tropical-fruit production.

Henry was a kind and very humble, modest man who believed strongly in being a team player once the discussion was finished and the decision made. He took pride in accomplishments, never taking credit himself. He is survived by his wife Bessie, his son Alan and daughter Gail, and two grandchildren, Akio and Sonoe.

Henry Yoshiki Nakasone
6 July 1920 – 13 August 1995

PREFACE

The monoaxial banana, pineapple and papaya and polyaxial mango are the most well-known tropical fruits worldwide. Avocado is better known for production in subtropical areas, but considerably more production occurs in the tropical zone. Banana, pineapple and avocado are extensively grown by large companies. Banana along with plantain is the largest fruit crop in the tropics, with only a small fraction entering international commerce. Many other tropical fruit, already well known in the tropics, are now appearing in larger temperate city markets.

Henry Nakasone started this book soon after his retirement in 1981 from the University of Hawaii at Manoa. His work on the book was prolonged because of his extensive volunteer and consulting activities from his retirement to 6 months before his death. During the development of the book, he shared with me a number of the chapters to review the fruit growth and development and postharvest-handling sections. This was followed by discussion of the book focus, publisher, audience and approach. The discussion intensified during his illness and we approached CAB International for consideration of this book. Henry had completed first and, in some cases, second drafts of most of the chapters. The extensive research carried out by Henry in preparing the draft chapters made editing to a textbook style straightforward. Henry had understood the need for a book that melded equally the genetics, physiology and cultural practices with postharvest handling of each fruit crop as an interrelated whole.

The first two chapters deal with the general aspects of the tropical climate and fruit-production techniques. Subsequent chapters deal with the principal tropical-fruit crops, including those becoming more common in temperate city markets. The last two chapters deal with some miscellaneous tropical Asian and American fruit that have potential international markets. Palm fruits, such as salak, date, peach and coconut, are not covered, due to space limitations. The full development of the international markets for a number of these crops depends upon the availability of suitable varieties, production and tree-management technologies, and suitable storage and insect disinfestation procedures. The information in each chapter deals with biotic and abiotic problems, variety development and postharvest handling. The information contained should be of use to all readers and students interested in an introductory text on tropical-fruit production.

Many have contributed to this text. Encouragement and help of Henry in this passion came from many, especially as Henry battled to understand and use his personal computer. To name just a few: Sandra Lee Kunimoto, Natalie Nagai, Catherine Cavaletto, H.C. 'Skip' Bittenbender, Charles L. Murdoch, Chung Shih Tang, Roy Nishimoto, Lynne Horiuchi, Dale Evans, Joseph DeFrank and Warren Yee. Many of the same individuals also provided me with encouragement and help. A number of special individuals must be mentioned who provided technical reviews of the chapters: Goro Uehara, Dale Evans, Skip Bittenbender, Carl Campbell, Victor Galan Sauco, David Turner, George Wilson, Ken Rohrbach, Duane Bartholomew, Steve Ferreira, Francis Zee, Brian Paxton and Frank Martin. Their numerous comments and suggestions have been incorporated in most cases. All errors and omissions are my responsibility. The illustrations of each crop were done by Susan Monden, and her perseverence and skill were greatly appreciated. Gail Uruu gave up her valuable time in an attempt to correct my English; thank you. Thanks are also due to Tim Hardwick at CAB International for his assistance and patience during the book's development. Osa Tui, Jr's many hours of typing and editing are sincerely appreciated. The assistance of the UK Department For International Development (DFID) in making a contribution towards the cost of printing the colour photographs is also gratefully acknowledged.

I would greatly appreciate receiving all comments and suggestions on this text. I can be reached at the address in the front of the text or via E-mail at <paull@hawaii.edu>.

In closing, the continued support, assistance and love of my wife Nancy and daughter Wensi enabled me to persevere in this undertaking.

Robert E. Paull
Honolulu, 1997

Plate 1. Field sanitation, such as the removal of fallen fruit from the field, is essential to minimize disease build-up, especially in a fruit crop with continuous production, such as papaya.

Plate 2. Mixed plantings are not uncommon in the tropics, especially in small plantings, such as this mango and betel-nut orchard.

Plate 3. Olive-brown chilling injury symptoms on ripening papaya. Chilling injury limits the storage potential of most tropical fruits.

Plate 4. Fruit-fly larvae damage of ripe papaya. This group of insects attacks many tropical fruit and may require postharvest disinfestation treatments before export to some markets.

Plate 5. Postharvest insect disinfestation requires treatment in chambers similar to the one shown, in which the fruit is subjected to temperatures up to 50°C at high relative humidity for 6–8 h.

6

7

8

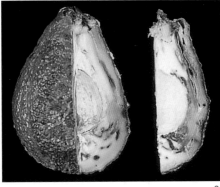

9

Plate 6. Soursop fruit are frequently borne on the main trunk and primary branches.
Plate 7. After avocado flowering, the apex continues to develop vegetatively and the fruit becomes subapical.
Plate 8. Avocado grafting on to suitable rootstocks is essential for good production.
Plate 9. Anthracnose is a serious postharvest disease of avocado and other tropical fruits.

10

11

12

14

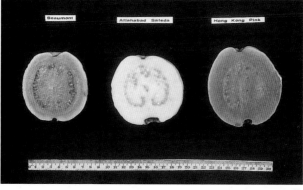

13

Plate 10. Bananas are harvested slightly immature, as fully mature fruit are subject to splitting.
Plate 11. Banana bunches are dehanded in the field or, as here, at a central site.
Plate 12. Banana packing is performed in sheds of varying size and with different levels of technology.
Plate 13. Guava flesh colour varies from white to bright red in different cultivars, as does flesh acidity and sugars.
Plate 14. Guava fruit destined for the dessert market are frequently bagged when young to protect the developing fruit from disease and insect damage.

15

16

17

18

Plate 15. Litchi cultivars differ in their vegetative flushing cycle; one tree is showing the golden-brown vegetative flush, while the other is dormant.
Plate 16. Litchi inflorescences are terminal and can have all fruit, as here, or a mixture of fruit and leaves.
Plate 17. A rambutan fruit cluster showing fruit at slightly different stages of development.
Plate 18. Litchi fruit still attached to the panicle are frequently sold tied together in bunches or in punnets of detached fruit.

19

20

21

22

23

Plate 19. Mango trees pruned and trained to less than 3 m showing uniform orchard flowering.
Plate 20. Mango fruit are bagged to prevent damage and, on these young trees, the limbs are propped because of the fruit load (courtesy of Paojen An).
Plate 21. Anthracnose on ripe mango fruit. This is a serious postharvest disease of mango.
Plate 22. Mango seed weevil causes some early fruit shedding but does not affect fruit quality; the flesh is undamaged and edible.
Plate 23. Internal breakdown 'jelly seed' indicated as water-soaked tissue around the mango seed.

24

25

26

27

Plate 24. Young papaya planting showing a double-row planting system, drip irrigation and plastic mulch.
Plate 25. *Phytophthora* can cause root rot in papaya, stem canker, as shown here, and a fruit disease.
Plate 26. Severe weed infestations in a papaya orchard can reduce yields.
Plate 27. Passion-fruit growing on north–south orientated cross-type trellis.

28

29

30

31

Plate 28. Pineapple fruit cultivars vary greatly in size, skin and flesh colour, crown size, acidity and sugar level.

Plate 29. Pineapple fruit sunburn is a serious problem in some production areas. Fruit can be protected by tying the leaves over the fruit.

Plate 30. Pineapple plantations use titratable acidity and total soluble solids to evaluate fruit quality and schedule harvesting.

Plate 31. Pineapple harvesting operations using a boom with a belt to carry fruit from the middle of the field to the access roads.

32

33

34

35

Plate 32. Durian fruit are often tied to the tree to prevent natural abscission. Fruit falling to the ground can damage the edible aril and predispose the fruit to disease.

Plate 33. Harvested durian awaiting shipment in Thailand.

Plate 34. Stages of mangosteen fruit ripening as judged by skin colour (courtesy of Sing Ching Tongdee).

Plate 35. Wax-apple tree showing a double irrigation drip line, the previous season's girdling scar and recent pruning.

THE TROPICS AND ITS SOILS

INTRODUCTION

Climate is defined as the general temperature and atmospheric conditions of an area, over an extended period of time. Atmospheric conditions include rainfall, humidity, sunshine, wind and other factors. Climates are subject to modification by various factors, such as latitude, elevation, cloudiness, whether the land mass is continental, coastal or oceanic, the direction of wind and ocean currents and the proximity to large bodies of water and mountain ranges.

The tropical region is a belt around the earth between the Tropic of Cancer at 23°30' latitude north of the equator and the Tropic of Capricorn at 23°30' latitude south of the equator (Fig. 1.1). The Tropics of Cancer and Capricorn as boundaries are rather rigid and do not take into consideration the presence of areas that do not meet the various climatic characteristics generally established to describe the tropics. Some climatologists have extended the region to 30°N and S of the equator (Henderson-Sellers and Robinson, 1986); this increases the land mass in the tropics substantially, especially in Africa, China, South America and India, and would include approximately two-thirds of Australia.

CHARACTERISTICS OF THE TROPICS

The tropical zone is generally described as possessing the following characteristics:

1. An equable warm temperature throughout the year, having no cold season at lower elevations. The average annual temperature of the true tropics is generally greater than 25°C, with no month having an average less than 18°C. Others have described the tropics as areas with a mean temperature not lower than 21°C and where the seasonal change in temperature equals or is less than the diurnal variation of temperature. The latter boundary is very

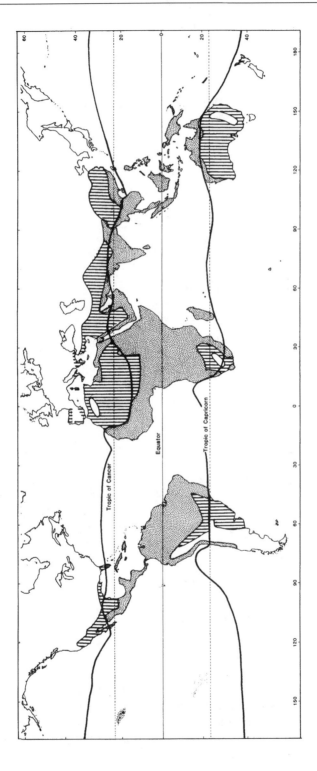

Fig. 1.1. Distribution of tropical and subtropical regions of the world separated by the absence of frost ▦; the subtropics having occasional frost ▦, and the unshaded areas being subject to frequent frost (redrawn from Bluthgen, 1966), and by the position of the 18°C sea-level isotherm for the coolest month as the boundary of the tropics (after Ayoade, 1983).

much influenced by continentality. Another boundary is the isotherm where the mean sea level temperature in the coldest months is not below 18°C (Fig. 1.1); although it can include certain errors, these are relatively small on a world scale and reliable data are available for its computation (Ayoade, 1983).
2. Rainfall is usually abundant, seldom less than the semiarid 750 mm to as high as 4300 mm, indicating considerable variation. The heaviest rainfall occurs closest to the equator. Where rainfall is marginal for agriculture, its variability takes on great significance.
3. Photoperiod varies little throughout the year at the equator, where day length is about 12 h (Table 1.1).
4. The position of the sun is more directly overhead, giving a year-round growing season (Fig. 1.2).
5. Higher potential evapotranspiration, due to rainfall, temperature and solar radiation.

These characteristics describe the true tropics in and near the equator, with latitudinal changes towards the poles producing a variety of subclimates. Even near the equator, mountain ranges and other geographical factors can produce various subclimates. Since temperature, solar radiation and photoperiod are fairly constant in the tropics, the variety of subclimates and vegetation is frequently dependent upon rainfall. A continuous succession of climates starts with long seasons of well-distributed precipitation and short dry seasons close to the wet tropics and gradually changes to short seasons of relatively low rainfall and long dry seasons as the latitude increases (Fig. 1.3). Some seasonal variation in mean daily temperature becomes apparent, with cool temperatures increasing with increasing distance from the equator.

RADIATION CONDITIONS

Day Length

The day length at the equator is about 12 h. At low latitudes in the tropics, the increase in difference between the longest and shortest days is about 7 min per degree (Table 1.1), increasing to 28 min per degree at latitudes between 50°

Table 1.1. Day-length extremes in hours and minutes at various latitudes in the tropics and subtropics.

	Latitude				
	0°	10°	20°	30°	40°
Longest day	12:07	12:35	13:13	13:56	14:51
Shortest day	12:07	11:25	10:47	10:04	9:09

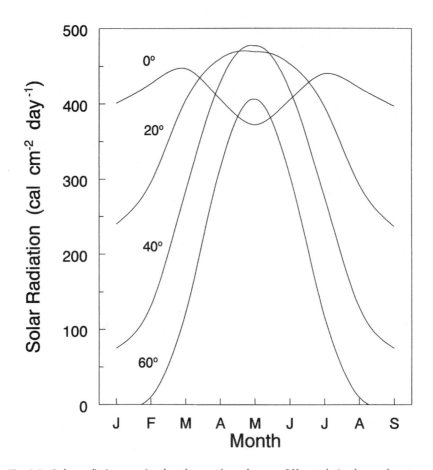

Fig. 1.2. Solar radiation received at the earth surface at different latitudes and an atmospheric coefficient 0.6 (List, 1966).

and 60°. The difference in photoperiod is associated with the earth being inclined on its axis by approximately 23°30′; hence the solar equator moves about 47° as the earth moves around the sun. The extremes are the Tropic of Cancer (23°30′N) and the Tropic of Capricorn (23°30′S); within this belt the sun's rays are perpendicular at some time during the year. At the spring and autumn equinoxes, the lengths of the day and night are equal everywhere over the earth.

Radiation

When compared with higher latitudes, the tropical latitudes have a smaller seasonal variation in solar radiation along with a higher intensity. The longer

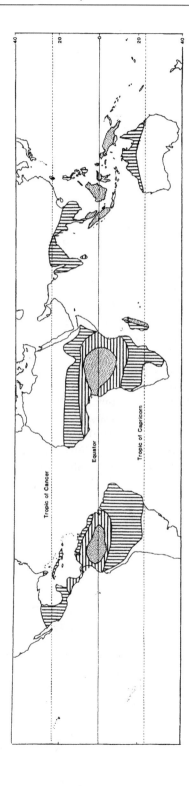

Fig. 1.3. Seasonal-rainfall distribution types of annual precipitation in the tropics (redrawn from Bluthgen, 1966; Nieuwolt, 1977). No dry season ▦, two dry and two rainy seasons ▤, one rainy season ▦ and unshaded areas dry or outside the tropics.

summer day length at the higher latitudes means that these latitudes exceed the daily amounts of solar radiation received in the tropics. The highest annual energy input on the earth's surface of about 12 MJ m^{-2} day^{-1} occurs in the more cloud-free subtropical dry belt at $20-30°$ latitude (Fig. 1.3). In the tropics, solar radiation received is reduced by clouds and water vapour in the air through reflection and absorption to a minimum of about 7 MJ m^{-2} day^{-1} at the equator. Over a large portion of the tropics, the average is 9 MJ m^{-2} day^{-1} $\pm$ 20%.

Temperature

Near the earth's surface, temperature is controlled by incoming and outgoing radiation. Surface temperatures are modified spatially and temporally throughout the year by local factors more than by radiation. The main modifying factors are continentality, the presence of large inland water bodies, elevation, topography and cloudiness. Highest diurnal temperatures occur in dry continental areas, at higher elevations and in cloud-free areas. The rate of decrease in temperature with elevation (lapse rate) varies with cloudiness (hence season) and between night and day. The normal rate is about $5°C$ per 1000 m under cloudy conditions and can range from 3.1 to $9°C$ per 1000 m.

A human's perception of temperature is modified by the rate of evaporation. Evaporation from human skin is primarily influenced by humidity, wind speed and response to sunshine. A human can endure high temperatures if the humidity is low, but feels discomfort in the tropics associated with high temperatures and humidities ($> 25°C$ and $> 80\%$ relative humidity (RH)). These conditions are also favourable for growth of micro-organisms and insects. The problem of controlling plant diseases and insect pests in the tropics is compounded by the absence of a cold winter and aridity to limit their development. Field sanitation is crucial to disease and insect control (Plate 1).

Rainfall

Temperature determines agricultural activity in the temperate regions of the midlatitudes ($30-60°$), while rainfall is the crucial factor in the tropics. The seasonal and diurnal distribution, intensity, duration and frequency of rainy days vary widely in the tropics both in space and time (Fig. 1.3). Maximum rainfall occurs near the equator, with no dry season. Surrounding this equatorial zone in Africa and South America are areas with two rainy and two dry seasons alternating; rarely are the seasons of the same duration or intensity. Further from the equator is a region of minimum rainfall at $20-30°$ latitude, associated with a subtropical high-pressure area, with one rainy season, frequently due to the monsoons (Fig. 1.3). Topography can significantly modify the generalized rainfall pattern; examples include the western coast of India and Borneo and the coastal areas of Sierra Leone, where monsoonal winds are forced to rise because of mountain ranges. Trade winds can bring considerable rainfall and are subjected to the forced rise by topographical features. Other factors influencing rainfall include changing and slowing down of wind speed as it approaches the equator and continentality, such as in South-west and Central Asia. The above factors lead to complicated rainfall patterns with generalization possible when remembering that there are numerous variations.

Strong Winds, Frost and Hail

In the equatorial zone, strong winds are associated with localized thunderstorms (diameter < 25 km) having greater intensity than those in the middle and upper latitudes and lasting from 1 to 2 h. Most occur outside the 0–10° latitude zone (Fig. 1.4) and are convectional in origin, associated with intense solar heating; others are due to sea or land breezes and unstable warm and humid air associated with a land mass. Hail occurs rarely in the tropics, but it is known to damage tea in Kenya and tobacco in Zimbabwe.

Tropical cyclones (hurricanes, typhoons) are an almost circular storm system, ranging in diameter from 160 to 650 km, with winds from 120 to 250 km h^{-1}, originating over water in the warm summer and early autumn season. Most develop within latitudes 20°N and S of the equatorial belt and may turn north-east in the northern hemisphere or south-west in the southern hemisphere to 30–35° latitude (Fig. 1.4). These systems bring violent winds and heavy rains. The Philippines are very prone to such systems. Crop damage, especially to trees, can be very severe, due to the high winds in vulnerable areas, making wind-breaks essential.

Monsoon depression is a less intense weather phenomenon. It brings 80% or more of the precipitation to the Indian subcontinent, with considerable year-to-year variation. It occurs when there is at least a 120° directional shift in prevailing wind direction between January and July. It is a characteristic of the seasonal wet and dry tropics and spreads from Asia to Africa. The intensity of rainfall can lead to considerable flooding.

In the subtropics, frost is a major limiting factor to tropical horticultural production. In isolated tropical high mountainous areas, frosts can occur frequently. Frost in the subtropics is associated with incursions of cold air masses (advection frost), while on tropical mountains it is mainly due to rapid cooling on clear nights (radiation frost).

MAJOR TROPICAL-CLIMATE TYPES

Many geographers and climatologists have classified climates into zones by temperatures (tropical, temperate and frigid zones), by vegetation or crop

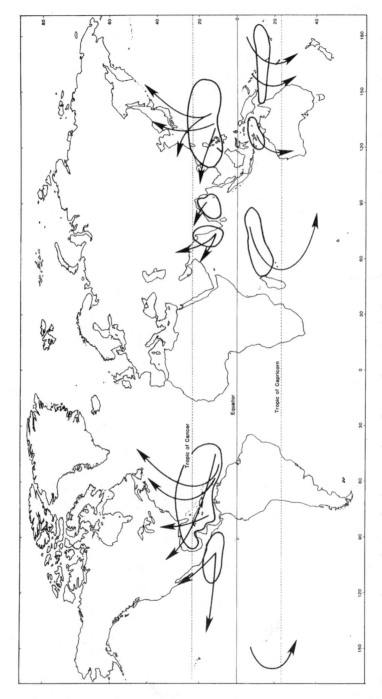

Fig. 1.4. Regions of tropical storm development that influence the tropics (redrawn from Gray, 1968) and typical storm tracks in each region (Anon., 1956).

requirements, precipitation, altitude, soils or human responses, or by combining these factors. A well-recognized classification system is the Köppen system, named after the Austrian botanist and geographer, Wladimir Köppen. This classification is based on temperature, rainfall, seasonal characteristics and the region's natural vegetation and was developed from 1870 to his death in 1940. The current system is based on a 1918 scheme of world climatic regions (Table 1.2). Another well-known modification is that of C.W. Thornthwaite (1948), who based his classification on distribution of effective precipitation (precipitation/evaporation (P/E) ratio) and temperature-efficiency evapotranspiration. This led to nine moisture and nine temperature regions. Numerous other classifications have been published based upon similar criteria (Oliver and Hidore, 1984).

The classification systems of Köppen, Thornthwaite and others focus on the major factors (temperature, precipitation and evaporation) that limit vegetation growth and hence horticultural production. The 18°C boundary of Köppen recognizes the dramatic slowing of tropical plant growth and development at lower temperature. At temperatures less than 10–12°C and above freezing, most plants that evolved in the tropics are injured, depending upon length of exposure and species; this response is called chilling injury. Precipitation and evaporation have a significant impact on natural vegetation and subsistence agriculture, although irrigation does allow horticultural production to proceed.

Several major tropical climatic types have been described.

Wet Tropics

The wet equatorial or humid tropics, equatorial zone or tropical rain forest occurs within 5–10° of the equator. It is characterized by consistently high rainfall, humidity and heat. Rainfall is well distributed (Fig. 1.3) and may range from 2000 to 5000 mm or more annually. Solar radiation is reduced, due to cloudiness. Vegetation is luxuriant on the weathered soils. Undisturbed, it supports natural vegetation very well, but under cultivation the soils lose their organic matter and porosity rapidly. Much of the land in the wet tropics is undeveloped and in some areas unpopulated.

This climatic type is common in parts of Africa within the 10°N and S latitudes and includes the Congo Basin, Gulf of Guinea in West Africa and parts of Kenya and Tanzania in East Africa; in South America the Amazon Basin of Brazil and countries bordering the Basin, such as French Guiana, Guayana, Surinam and Venezuela; and in South-east Asia most of Malaysia, Indonesia, Papua New Guinea, the Philippines and some Pacific islands (Walter, 1973). The wet tropics are not limited to the above areas and are also widespread in countries bordering the equator.

Table 1.2. Köppen's major climates based upon four major temperature regimes derived from monthly mean temperatures, monthly precipitation and mean annual temperature: one tropical, two midlatitudes and one polar. The second division is based on moisture availability. The system does not completely agree with natural vegetation and climate and frequently the boundaries are rigidly interpreted.

	Principal climatic types	Temperature	Rainfall
Tropical (A)	Rainy (Af)	Coolest month >18°C (64.4°F)	>6 cm driest month
	Wet and dry (Aw)		<6 cm driest month
Dry (B)	Monsoon (Am)		Derived formula
	Steppe (Bs)		Evaporation > precipitation
	Desert (Bw)		
Midlatitudes – mild winter (C)	Mediterranean (Cs)	Coolest month <18°C to > −3°C	3 × more summer precipitation than winter
	Wet and dry (Cw)		10 × more summer precipitation than winter
	Rainy (Cf)		≥3 cm per month
Midlatitudes – cold winter (D)	Wet and dry (Dw)	Coolest month <−3°C	As for (C)
	Rainy (Df)	Warmest month >10°C	
Polar (E)	Tundra (ET)	Average warmest month 0–10°C	
	Ice cap (EF)	Average warmest month <0°C	

Wet-and-dry Tropics

This is also known as the monsoon rain forest between 5° and 15°N and S of the equator and as far north as 25° in parts of tropical Asia. Walter (1973) extended this zone from about 10°N and S to about 30°N and S latitudes. Maximum rainfall occurs in the summer when the sun is directly overhead and the dry season occurs in the cooler months (Fig. 1.3). Tropical-fruit horticulturists will probably spend most of their time in this climatic zone.

The climate is found in wide regions of Africa, Asia, the Americas, Australia and the Pacific tropics. Many tropical-fruit species are well adapted to wet–dry climatic conditions. For fruit production, some form of irrigation is necessary, especially in areas where the wet season is relatively short. A dry period in winter may substitute for cool temperatures in crops requiring some stress prior to flowering. However, irrigation is desirable once flowering begins and during fruit development.

Dry Tropics

Also called the tropical savannah, the dry tropics occur to the north and south of the monsoon climate zones along the Tropics of Cancer and Capricorn, between 15° and 20°N and S latitudes. The climate is characterized by hot, dry, desert conditions where crop production cannot succeed without irrigation. It is found in North Africa, bordering the Tropic of Cancer, north-east India, Australia and parts of the Pacific coast of South America. The coast of Peru has arid plains and foothills and fertile river valleys. For example, average annual rainfall for the town of Piura on the north-western coast is 51 mm and average minimum and maximum temperatures are 17.6 and 30.5°C, respectively. Thousands of hectares of land remained unproductive due to lack of irrigation until large dams were built to produce hydroelectric power as well as to provide irrigation.

ALTITUDINAL CLIMATES

Climates change with altitude at the same latitude, and the change is related to temperature. The altitudinal environments can be divided into three temperature zones at the equator: the hot zone from sea level to 1000 m; the temperate zone from 1000 to 2000 m; and the cool zone above 2000 m, with frost occurring at approximately 4000 m at the equator. Temperatures in these zones differ with changes in latitude, prevailing wind patterns, precipitation and other factors.

In the Selva region of eastern Peru with large rivers, principally the Amazon, two subclimates are recognized in terms of altitude and rainfall. The low jungle, or humid tropics, extends from sea level to 800 m and has rainfall throughout the year. The high jungle, or central selva, located between 800 and 1500 m above sea level, has a wet and dry climate with about 4 months of wet season and the remainder with little or no rain. In the low jungle, subsistence agriculture prevails, with crops such as coffee, cacao, banana, mango, papaya, pineapple, soursop, citrus, black pepper, cassava and poma rosa (*Syzygium malaccensis*). Under large-scale commercial cultivation, some of the traditional crops, such as mango, papaya, pineapple and citrus, would do better at latitudes somewhat removed from the equatorial region, having better soils and reduced disease problems associated with high rainfall. The high jungle (wet-and-dry tropics) in the central selva area of Peru is better developed, with farms of large commercial size. Citrus (valencias, mandarins and limes) and coffee do well.

SUBTROPICS

Strict separation of tropical, subtropical and temperate climates is not practical because of the many factors that influence climate. Even within the geographical limits of the tropics, there are areas that are subtropical and temperate or even frigid because of altitude, topography, ocean and air currents. The subtropics occur between the two tropics and about 40° latitude, with summers being hotter and winters cooler than in the tropics; humidity is generally lower and the difference in day length becomes greater with higher latitude (Table 1.1). The limit for the subtropics is the isotherm of 10°C average temperature for the coldest month. This 10°C isotherm excludes the large land masses, whose climates are temperate, including almost half of China, three-quarters of Japan, all of South Korea, the southern half of the USA, all of the southern half of Australia, the North Island of New Zealand and more than half of Argentina and Chile between 23°30' and 40° latitude. Horticulturists who have spent their professional careers in regions where the temperatures during the coldest month at sea level are rarely lower than 15–18°C find it difficult to accept the tropical classification of regions with winter temperatures down to 4–7°C with the potential for frost.

SOILS

Soils using the US classification system are separated into ten groups, based on parent material, soil age and the climatic and vegetative regime during formation. Tropical soils are diverse (Fig. 1.5), having formed from different parent material and climatic conditions. These soils have formed in areas where the soil temperature at 50 cm differs by less than 5°C between the warm and cool seasons. The parent rock materials are as different in the temperate zone as in the tropics, erosion and deposition are similar, and soil formation could have been from recent volcanic or alluvial flood plains to 1 million years

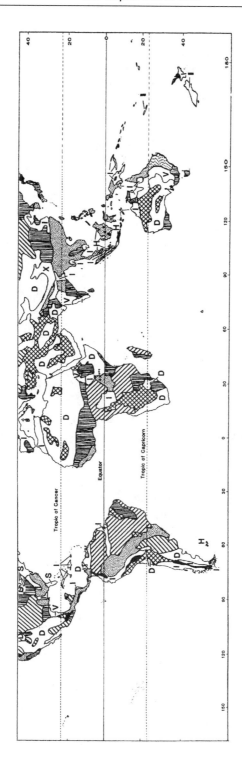

Fig. 1.5. Soil-type distribution in the tropics (redrawn from US Department of Agriculture Soil Conservation Services, 1971; Sanchez and Uehara, 1980). The major groups are indicated: U, Ultisols ▦; O, Oxisols ▨; An, Andisols ▨; M, Mollisols ▨; E, Entisols ▨; A, Alfisols ▨; I, Inceptisols; V, Vertisols; D, Aridisols; H, Histosols; S, Spodosols. X, mountain areas.

old. The differences in temperate regions lie in soil-forming factors such as glaciation and movement of loess that have not occurred in the tropics.

The majority of tropical soils come under the US orders Oxisols, Aridisols, Alfisols, Ultisols and Vertisols and these are spread widely throughout the tropics (Fig. 1.5). The soil orders are separated on the presence or absence of diagnostic horizons or features that indicate the degree and kind of the dominant soil-forming process (Soil Survey Staff, 1990). It is very difficult to make generalizations about tropical soils other than that they have less silt than temperate soils and that surface erosion and deposition have been more significant. There are greater volcanic deposits in the tropics while the temperate region has a larger proportion of younger soils. Only a small proportion (2–15%) of the tropics has so-called lateritic soils (Oxisols and Ultisols), defined as soils that have high sesquioxide content and harden on exposure. The red colour of tropical soils does not mean that they have low organic matter. For example, the average organic carbon content in the top 1 m of a black North American Mollisol is 1.11%, while a red, highly weathered tropical Oxisol may have 1.05%, the reddish temperate Ultisols 0.4% and tropical Ultisols 0.66% (Sanchez and Buol, 1975).

The more intensively farmed, fertile soils of the tropics cover about 18% of the area and are Alfisols, Vertisols, Mollisols and some Entisols and Inceptisols (Table 1.3). These soils generally developed from alluvium and sediment (Table 1.4) and are high in calcium, magnesium and potassium. This gives them a high base status with no acidity problem. Phosphorus deficiency can be readily corrected. A larger group of tropical soils (Oxisols, Ultisols and others) are of low base status, are highly leached and cover 51% of the tropics (Table 1.3). Phosphorus deficiency can be significant as it is fixed by the iron and aluminium oxide in these soils, which also often have aluminium toxicity problems, with sulphur and micronutrients (zinc (Zn), boron (B), and molybdenum (Mo)) deficiencies. However, they have good physical properties. The high-base soils (Aridisols) in tropical deserts (14%) can be very productive with irrigation. Nitrogen deficiency and sometimes salinity can be problems. The 17% covered by dry sands and shallow soils are greatly limited in agricultural productivity. The most important aspect is the development of management strategies for the different tropical soils, taking into consideration their unique properties. Management practices developed in temperate regions may not be directly transferable.

SUMMARY

Whereas temperature is the major limitation to plant growth in temperate areas, in the tropics rainfall plays that role. Year-round plant growth is generally only limited in the tropics by moisture availability. There is considerable variation in tropical climates, caused by altitude, continentality and the presence of large bodies of water. Disease and insect problems are

Table 1.3. Distribution of major soils in the tropical regions with differences in rainfall distribution (after Sanchez, 1976; Sanchez and Salinas, 1981).

	% Humid tropics[*]	% Seasonal	% Dry and arid	% Total tropics
Highly weathered, leached, red or yellow soils (Oxisols, Ultisols, Alfisols)	36	61	2	51
Dry sands and shallow soils (Entisol – Psamments and lithic group)	10	33	58	17
Light-coloured, base-rich soils (Aridisols and aridic groups)	0	15	85	14
Alluvial soils (Inceptisol – Aquepts, Entisol – Fluvents)	40	52	8	8
Dark-coloured, base-rich soils (Vertisols, Mollisols)	8	60	32	6
Moderately weathered and leached soils (Inceptisol – Andepts and Tropepts)	7	59	34	4
% of tropics	24	49	27	10[*]

[*]Classified on number of rainy months: humid tropics, 9.5–12 months; seasonal, 4.5–9.5 months; dry and arid, 0–4.5 months.

Table 1.4. Soil classification and main characteristics of tropical soils and orders.

Order	Characteristic	Agricultural productivity
Entisol	Recent alluvium, also on barren sands and near deserts. Lacks significant profile	Some very productive alluvial soils
Vertisol	Swelling clay. Cracks in dry weather	Productive, difficult to till
Inceptisol	Horizons forming, little accumulation of Al of Fe oxides. Tropic with long, wet season, acid	Very prone to erosion, good if well drained
Aridisol	Deserts, little organic matter, $CaCO_3$, $CaSO_4$ and soluble salts	Needs irrigation to be agriculturally productive
Mollisol	Granular or crumbly soils. High in silt and organic matter	Most productive worldwide
Alfisol	Humid-regions soil with grey or brown upper horizon and silicate clay below	Good soils, prone to erosion, easily compacted
Ultisol	Highly weathered, acidic in moist, warm tropical areas, moderate fertility	Very productive, easily compacted, prone to erosion, needs fertilizer
Oxisol	Highly weathered, in hot, heavy-rainfall areas deep clay of hydrous oxides of Al and Fe, can be very acidic, leached	Needs nutrients to be productive, susceptible to erosion if left bare
Andisol	Surface deposits of volcanic ash, free-draining, low bulk density	Erosion serious problem, high phosphorus fixation, high Al and Fe fixation, excellent structure, productive

Al, aluminium; Fe, iron; $CaCO_3$, calcium carbonate; $CaSO_4$, calcium sulphate.

more severe in high-rainfall warm tropics. Good soils are available in the tropics, and the high value of horticultural crops means that they can command the use of more favoured areas.

FURTHER READING

Alvim, P. de T. and Kozlowski, T.T. (1977) *Ecophysiology of Tropical Crops*. Academic Press, New York, 502 pp.

Ayoade, J.O. (1983) *Introduction to Climatology for the Tropics*. John Wiley & Sons, Chichester, 258 pp.

Bennema, J. (1977) Soils. In: Alvim, P. de T. and Kozlowski, T.T. (eds) *Ecophysiology of Tropical Crops*. Academic Press, New York, pp. 29–55.

Sanchez, P.A. and Buol, S.W. (1975) Soils of the tropics and the world food crisis. *Science* 188, 598–603.

CULTIVATION AND POSTHARVEST HANDLING

INTRODUCTION

Environmental variables directly and indirectly influence every aspect of plant growth and development. These variables determine the final plant phenotype, while cultural practices (fertilization, irrigation, pruning, etc.) aim to reduce negative impacts of these variables to achieve maximum productivity and product quality. The focus of much research on tropical fruit crops has been on the impact of environmental variables on whole-plant growth and development. There have been extensive observations on the impacts of various environmental and cultural variables on the components of fruit production: germination, establishment, growth, flowering, fruit set, fruit growth and quality and yield.

Results from studies on temperate-fruit crops have relevance and provide direction as to possible management practices worthy of testing in the tropics. This chapter will consider the various environmental factors that influence fruit production and quality and common cultural practices. Requirements specific to a particular crop are treated in the separate chapters.

ENVIRONMENTAL FACTORS

Temperature

Tropical fruit crops such as mango, guava, acerola, papaya, pineapple, some annonas and others originated in the warm, lowland tropics. Others such as the litchi, Mexican and Guatemalan races of avocado, cherimoya and purple passion-fruit are subtropical fruits by virtue of their origin in the subtropics or at higher elevations in the tropics. Humans, in their attempt to commercialize tropical fruit crops, have extended production into subtropical regions beyond the Tropics of Cancer and Capricorn and have generated considerable knowledge on the range of temperature adaptability of these crops. However, data on threshold temperatures and durations of exposure for various stages of plant development of tropical fruit trees are often unavailable, so minimum temperatures in the coolest month that support either survival, commercial or best production have been approximated (Table 2.1). The minimum temperature criteria take into account differences in elevation and latitude. In regions subjected to marginal winter temperatures, site selection becomes a paramount consideration, such as southern-facing slopes in the northern hemisphere and northern-facing slopes in the southern hemisphere. Young plant growth may also be inhibited when soil temperature exceeds 35°C, a condition not uncommon in the tropics. Leaf temperature can exceed air temperatures by 20°C; for example, pineapple fruit and leaf temperatures have been recorded in excess of 50°C in the field. Hence, maximum temperatures in the orchard microclimate need to be considered in site selection.

Three crops (mango, litchi and avocado) illustrate this temperature adaptability and the varietal and race adaptation at various stages of development. Mature mango trees have been found to withstand temperatures

Table 2.1. Guidelines for survival, commercial and best production of tropical fruit based upon mean minimum temperature in the coldest month (after Watson and Moncur, 1985).

Crop	Botanical name	Mean minimum temperature (°C) for coolest month		
		Survival	Commercial	Best
Acerola	Malpighia glabra	10–12	>12	>14
Banana	Musa spp.	6–8	>8	>16
Breadfruit	Artocarpus altilis	14–16	>16	>16
Carambola	Averrhoa carambola	6–8	>8	>14
Chempedak	Artocarpus polyphema	12–14	>14	>16
Cherimoya	Annona cherimoya	4–6?	5–12?	6–10?
Durian	Durio zibenthinus	14–16	>16	>18
Duku and langsat	Lansium domesticum	12–14	>14	>18
Guava	Psidium guajava	4–8	>8	>14
Jackfruit	Artocarpus heterophyllus	6–10	>10	>14
Longan	Euphoria longana	4–8	8–18	8–14
Litchi	Litchi chinensis	4–8	>14	>16
Mango	Mangifera indica	6–8	>8	>12
Mangosteen	Garcinia mangostana	10–14	>14	>16
Papaya	Carica papaya	6–8	>8	>14
Pineapple	Ananas comosus	6–8	>8	>10
Rambutan	Nephelium lappaceum	8–12	>12	>14
Sapodilla	Manilkara zepota	6–10	>10	>14
Soursop	Annona muricata	6–10	>10	>16

as low as −16°C for a few hours with some injury to leaves. Flowers and small fruits may be killed when temperatures less than 4.5°C occur for a few hours during the night. Mango varietal differences for cold resistance have not been observed. Mango responds to cool night temperatures (10–14°C) with profuse flowering. Ideally, day temperatures during this period should be warm (21–27°C). When winter night temperatures are mild (16–18°C), flowering is more erratic.

The vegetative and flowering behaviour of litchi is very similar to that of mango except that it is adapted to even lower minimum temperatures than mango. The total duration of relatively low temperatures seems to be the determining factor, rather than the frequency or time of occurrence during a critical period. There are considerable cultivar differences in the temperature exposure necessary to induce flowers; for example, Brewster litchi flowers better in seasons with 200 or more hours of temperature below 7.2°C. Other cultivars flower profusely when night temperatures of 14–15°C occur.

The three races of avocado originated under different ecological conditions. The West Indian race is best adapted to humid, warm climates, with the optimum around 25–28°C. It is susceptible to frost, and the minimum temperature tolerated by the foliage is recorded at 1.5°C. Mature Mexican race trees show tolerance to as low as −4 to −5°C without damage to foliage, although flowers are damaged. The Guatemalan race is adapted to a cool tropical climate but is less tolerant of low temperatures than the Mexican race. Guatemalan cultivars have shown tolerance down to −2°C, with flower damage by even light frost. The Mexican–Guatemalan hybrids, such as Fuerte, have shown wider tolerance of cold than the Guatemalan cultivars.

Rainfall and Evaporation

Tropical-fruit production is normally limited by available soil moisture. The stage of growth or development at which water stress occurs greatly affects the final yield. Many factors influence the amount of rainfall available to plants, including evaporation and transpiration rates, surface runoff, soil water-holding capacity and percolation through the soil profile beyond the rooting area. Using the average tropical daily net radiation of 9 MJ m^{-2} and the latent heat of vaporization for water (2.45 MJ kg^{-1}) gives an evaporation rate of *c.* 4 mm day^{-1}, similar to evaporation in a temperate summer. Higher rates of 10–15 mm day^{-1} occur for irrigated crops in the semiarid tropics due to advection of hot dry air. Rainfall and irrigation need to make up this evaporative loss, and a mean monthly rainfall of 120 mm (*c.* 4 mm day^{-1}) would be required.

Excessive rainfall causes major problems with flowering, pests, diseases and fruit quality. Many trees, such as mango and litchi, require a dry (or cold) period to stop vegetative growth and induce flowering. Mango and litchi originated in areas with a monsoon climate, which provides distinct wet and dry seasons. Dry conditions, preferably accompanied by cool temperatures, during the preflowering period promote flowering, while cool, wet conditions reduce flowering in both crops. When mango flower buds begin to emerge, some soil moisture is needed, preferably from irrigation rather than from rainfall. Light drizzling rains coinciding with the mango flowering lead to severe anthracnose (*Colletotrichum* spp.), which can destroy most of the inflorescences. Too much rain during litchi anthesis reduces either flower opening or the insect activities needed for pollination, or both.

Total rainfall is frequently less important than its distribution throughout the year. In Loma Bonita and Acayucan, Mexico, two large pineapple-producing areas, mean rainfall over a 5-year period is 1600 mm and 1500 mm, respectively. These approximate the upper limits of the optimal range for pineapple; however, periods of serious drought are encountered, as 89 and 82%, respectively, of the rainfall occurs from June to November. In tropical rain forest, 85% of a 1 mm shower may be intercepted by plants, but only 12% of a 20 mm fall, indicating that fall intensity, duration and frequency are significant factors (Norman *et al.*, 1995). The interception by plants is significantly influenced by species; for example, pineapple, with its upright leaves, funnels water to the centre. Plant density similarly affects rainfall interception.

Orchards located on flat lowland areas can experience flooding during the rainy season, particularly if drainage is poor. This is important for avocado, papaya, litchi and pineapple, for which waterlogging causes severe root-rot problems. Mango is slightly flood-tolerant, as indicated by reductions in leaf gas exchange and vegetative growth and variations in tree mortality. Mangosteen, in contrast, grows well under conditions of flooding and a high water-table. A high water-table may prevent trees from experiencing the moisture stress needed to induce flowering.

Water deficit and therefore irrigation demand are determined by evaluating rainfall, evaporation and soil-water storage. The two most common approaches are the water balance (rainfall, evapotranspiration, water storage, change in the root range, surface runoff) and actual soil-moisture measurement. The crop needs, along with the soil type, determine the frequency of irrigation; this determination should be made on at least a weekly basis for fruit crops. The use of drip (trickle) and microsprinkler irrigation enables a grower to match the needs of the crop to irrigation requirements at different stages of growth and development, avoiding the necessity of relying on rainfall. These irrigation methods allow precise placement of the water, reduce surface evaporation and seepage, and increase water use efficiency. Other irrigation methods used are basin, furrow, overhead sprinklers and cannons.

Radiation Conditions

Photoperiod

Fruit trees such as mango, papaya, bananas, the annonas, avocado, acerola and guava show no response to photoperiod and are capable of flowering at any season of the year. In equatorial Colombia, with approximately 12 h day length, it is not uncommon to find mango trees flowering during February–March and again in August. Flowering is more precise in the subtropics, occurring in the spring as a function of temperature and moisture. For guava, Choudhury (1963) reported that seedlings grown under 15 h day length from germination to 140 days and field-transplanted produced fruit within 376 days from sowing; the control seedlings under 10 h day length did not flower. Photoperiod is useful in predicting days to guava harvest after pruning to induce flowering.

Pineapple cv. Smooth Cayenne flowers naturally at any time of the year, depending upon the size of the planting material and time of planting. However, it is a quantitative, but not an obligatory, short-day plant. Interruption of the dark period by illumination suppresses flowering. Although pineapple does not require low temperatures or diurnal variations in temperature to flower, there is an interaction with temperature. No flowers are produced on yellow passion-fruit (*Passiflora edulis* f. *flavicarpa*) vines under artificially induced short days (8 h). Long days promote passion-fruit vine growth and flowering, while short days promote vine growth only. This observation, however, could be due to amount of solar radiation received and not photoperiod.

Radiation

In the tropics, atmospheric radiation transmissivity varies from 0.4 to 0.7, due largely to clouds and seasonal variation. The maximum recorded irradiance under cloudless skies at noon in the tropics is 1.1 kW m^{-2}, with a daily total received of from 7 to 12 MJ m^{-2}. About 50% of this energy is in the 0.4–0.7 μm waveband, which is known as photosynthetically active radiation (PAR). In full sunlight, C$_3$ plants, including all fruit crops discussed in this book except pineapple, are 'light'-saturated. This saturation is due to ambient carbon dioxide (CO$_2$) availability limiting the rate of photosynthesis.

High shade (3–5 MJ m^{-2} day^{-1}) does not influence litchi flowering, although it does increase early fruit drop. Flowering in passion-fruit is reduced once irradiance falls below full sun. Radiation, however, is normally not the factor limiting plant growth in the tropics, except under circumstances of heavy mist and cloud or shade from vegetation and mountains.

Soils

Soil physical characteristics are of primary concern for tropical-fruit production, with soil nutrients being secondary because they can normally be readily corrected. Soil texture and structure, soil-water storage and drainage are crucial. Deficiencies in these characteristics are major constraints to production because they are difficult and expensive to correct. Under natural conditions, most soils considered for fruit crops in the tropics have a good topsoil structure. This includes the highly weathered Oxisols and Ultisols. Loss of organic matter may lead to loss of structure and crusting of these soils after heavy rains. Some soils, however, do not favour root development, due to a dense subsoil layer, which needs to be broken during soil preparation to avoid shallow root systems. Heavy machinery may also cause the formation of a compact subsurface layer in medium-textured Oxisols with low iron and in fine-textured Oxisols. Low calcium and phosphorus and high aluminium contents in the subsoils can also restrict root growth.

Tropical fruit crops have shown a wide range of adaptability and have been observed to grow and produce well in a wide variety of soil types, provided other factors are favourable. In some cases, considerable management skill is required to maintain the crops in good growth and production. Soil pH can be corrected by liming during field preparation, with most trees preferring pH 5.5–6.5. Papaya is one of the few fruit crops that is adapted to a wide range of soil pH, growing and producing well in soil pH ranging from 5.0 to well into the alkaline range. Deficiencies in phosphorus associated with adsorption and excess aluminium need to be addressed in the Oxisols, Ultisols and some Inceptisols. Soil organic matter can be maintained by use of manure, ground-cover crops and mulches to preserve soil moisture and structure and improve the rhizosphere. Magnesium, zinc and boron deficiencies may also be encountered in some tropical soils, but these are relatively easy to correct in a management programme. Saline and alkaline soils, along with deep peat soils, should be avoided for fruit production because of their difficult nature. Acid sulphate soils require specialized management strategies, such as raised beds, to be productive.

A prime soil requirement for all crops is good drainage to prevent waterlogging, which leads to root diseases. Drainage is crucial for crops that are susceptible to *Phytophthora* root rot, such as avocado, papaya, passion-fruit and pineapple. Mango and avocado have been observed to show branch dieback in parts of fields in western Mexico with a water-table around 50–60 cm below the soil surface.

Nutrient Requirement and Fertilization

Plant tissue analysis is less influenced by site and situation than soil tests. Plant analysis data need to be standardized to the plant part, its age at sampling, the crop phenology and in some cases the cultivar. Soil analysis is more widely used for annual and short-term crops, while plant analysis is used to monitor the nutrition of a permanent crop having more extensive root systems and to compute nutrient uptake and fertilizer need. The analysis of a standardized tissue integrates the effects of soil, plant, climate and management. Use of 'crop logging' to monitor nutrition over seasons helps to maintain the tissue level in the desired range. Deficiency symptoms and low levels in early growth are indicative of some nutrient deficiencies. However, by the time plant analysis indicates a problem, it is often too late to correct without yield loss. Tissue levels that are toxic, causing poor growth and reducing yield, are frequently not known.

Adequate levels of nutrients need to be maintained for growth and production and to replace the nutrients lost in production. Non-seasonal monoaxial-plant fruits (banana, papaya) remove more nutrients than seasonal polyaxial-tree fruits (durian, mango, rambutan, litchi). Pruning, leaf fall, leaching and runoff also need to be taken into account. The fertilizer programme should be designed to meet nutrient needs for vegetative growth and fruiting (Table 2.2). Application times are adjusted to meet the different seasonal needs at the various phenological stages. These nutrient needs can be met from organic and inorganic fertilizer sources.

Wind Protection

There are inherent differences among trees in the degree of resistance to winds, but all fruit species benefit from wind protection. Mango, acerola and guava exhibit greater resistance than other tropical tree crops and banana, to the extent that they survive strong gusts of wind without losing limbs or being blown down. Leaves, flowers and fruits are often completely blown away. Pineapple, by virtue of being low-growing, gives the appearance of resistance, but wind can damage leaves and, during the fruiting period, the peduncles may be broken, resulting in loss of fruit. However, wind breaks are almost never found on pineapple plantations. The annona species, avocado and litchi are known for their brittle branches and show limb splitting even under moderate gusts of 65–80 km h^{-1}. Limb braces are occasionally used to prevent splitting of large limbs forming 'Y' crotches on litchi trees. Guava trees propagated by grafting have tap roots that provide substantial anchorage. However, guava trees propagated by rooting cuttings or by air-layering are subject to uprooting during the first 3 years, due probably to faster top growth than root growth. Papaya plants and passion-fruit vines are vulnerable to even moderate winds. Papaya trees are easily blown over, especially if the soil is softened by heavy rains. Passion-fruit vines on trellises can be tangled and broken, or the entire trellis may be blown down. Developing carambola fruits are easily bruised and marked by rubbing on branches and adjacent fruits due to wind, reducing fruit appearance and grade.

Table 2.2. Nutrient levels for various selected tropical fruit crops that are sufficient for growth (after Reuter and Robinson, 1986; Jones et al., 1991).

	Tissue	Nutrient levels (%) adequate for growth					
		Nitrogen	Phosphorus	Potassium	Calcium	Magnesium	Sulphur
Annona sp.	Leaves	2.5–3.0	0.16–0.2	1.0–1.5	0.6–1.0	0.35–0.5	
Avocado	Leaves	1.6–2.0	0.08–0.25	0.75–2.0	1.0–3.0	0.25–0.8	0.2–0.6
Banana	Leaves	3.5–4.5	0.2–0.4	3.5–5.0	0.8–1.5	0.25–0.8	0.25–0.8
Guava	Leaves	1.3–1.6	0.14–0.16	1.3–1.6	0.9–1.5	0.25–0.4	
Litchi	Leaves	1.3–1.5	0.15–0.2	0.8–1.2	0.56	0.21	0.1–0.16
Mango	Leaves	1.0–1.5	0.08–0.25	0.4–0.9	2.0–5.0	0.2–0.5	
Papaya	Petioles	1.01–2.5	0.22–0.4	3.3–5.5	1.0–3.0	0.4–1.2	
Passion-fruit	Leaves	4.75–5.25	0.25–0.35	2.0–2.5	0.5–1.5	0.25–0.35	0.2–0.4
Pineapple	Leaves	1.5–1.7	< 0.1	2.2–3.0	0.8–1.2	< 0.3	

Mangosteen trees may also require protection from full sun during early establishment, as well as from wind (Fig. 2.1). Use of windbreaks for these crops is standard practice.

Windbreaks of appropriate tree species closely planted across prevailing wind directions protect orchard trees from severe damage and excessive drying, especially during flowering and fruiting periods. Windbreaks are especially helpful when normal prevailing winds become gusty with velocities exceeding 65 km h^{-1}. Occasional high-velocity winds are damaging to orchard trees but equally harmful in the long run are the mild prevailing winds of 40–50 km h^{-1}. Trees exposed constantly to such prevailing winds gradually develop a lopsided shape with all branches growing away from the winds. The effectiveness of the windbreak depends upon the height and lateral extent of the barrier, its permeability and the angle of incidence of the wind to the barrier. Wind velocity is reduced to 35% within a distance of four times the height of the windbreak (Fig. 2.2). The windbreak similarly reduces crop evaporation on the leeward side to 40% within twice the height of the windbreak, being more effective at the higher wind speeds. The reduction in evaporation allows the trees in the orchard to develop higher humidity in the canopy, which is important for crops such as rambutan, mangosteen and durian.

Various tree species are used to protect fruit-tree crops (Table 2.3). Few windbreak trees can provide crop protection under hurricane (typhoon) wind velocities. One of the most substantial windbreak trees is *Garcinia spicata*, called 'Fukugi' in Okinawa, Japan. Windbreaks composed of this species have

Fig. 2.1. Young mangosteen trees are very sensitive to wind and direct sunlight during establishment. Protection is provided by windbreak trees (in the background) and temporary plastic shade around the seedling trees.

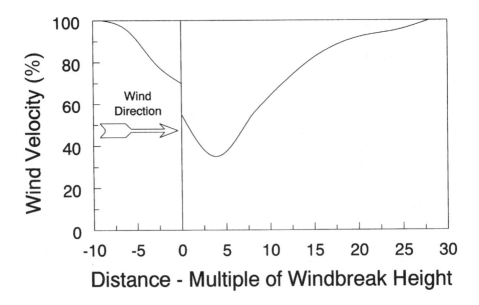

Fig. 2.2. Influence of windbreaks on the wind velocity on the windward and leeward side of the shelter belt.

been observed to withstand hurricane winds of approximately 210 km h^{-1} in northern Okinawa, providing crop protection without itself being damaged. This species is related to mangosteen and grows well in subtropical and tropical climates. Its only drawback is the very slow growth rate.

FIELD MANAGEMENT

Preparation

Field preparation is determined by the terrain and the equipment available. In hilly areas, terracing is preferred for erosion control. On gently sloping or level lands, conventional methods of clearing and ploughing and planting rows on the contour are suitable. If the area to be cleared is wooded, wind-rows should be created by leaving lines of existing tree vegetation for windbreaks. For grassland or shrub land, only planting rows are opened, and the wild vegetation between rows is left undisturbed to prevent erosion and conserve moisture. Planting holes are dug manually, trenched by use of a backhoe, or with a 46–61 cm diameter auger attached to a tractor. In moist, heavy, clay soils, an almost impervious glazed surface is formed by the auger blade on the walls of the holes. This glazed surface must be broken to re-establish permeability.

Table 2.3. Some suggested windbreak trees for tree-crop orchards.

Common name	Scientific name	Elevation range	Growth habit
Norfolk Is. pine, Australian pine	*Araucaria heterophylla* (Salisb.) Franco (*Araucaria excelsa*)	Low to 915 m	Tall to 46 m; cone-shaped, drought-resistant, salt-tolerant
Small-cone ironwood	*Casuarina cunninghamiana* Miq.	Low to 610 m	Fast-growing to 25 m tall; drooping branches, wind-resistant, drought- and salt-tolerant; free from pests, does not propagate from root-runners
Short-leaf ironwood	*Casuarina equisetifolia* Frost	Low to 610 m	Fast-growing to 30 m tall; characteristics similar to small-cone ironwood
Portuguese cypress	*Cupressus lusitanica* Mill.	Low to 900 m	Spreading evergreen to 15 m tall; shallow, surface roots; drought-susceptible; pest-free
Eucalyptus, swamp mahogany	*Eucalyptus robusta* Sm.	Low to 1200 m	Medium height; symmetrical, wide adaptation, good regrowth
Red ironbark	*Eucalyptus sideroxylon* Cunn., Cunn. ex Wools	Low to 1200 m	Medium height; slow grower, drought-resistant, grows in poor, dry soils
Fukugi	*Garcinia spicata* Hook, f.	Low to 400 m	Medium height to 15 m; slow-growing evergreen; upright, leaves thick, wind-resistant
Paperbark	*Melaleuca leucodendron* L.	Low to 760 m	Medium height to 25 m; surface roots, not drought-resistant; can be pruned to form dense growth at ground level
Madagascar olive	*Noronhia emarginata* (Lam) Poir.	Lowlands	Height to 18 m; dense; good salt barrier; resembles the 'Kamani' (*Calophyllum inophyllum* L.)
Panax	*Polyscias guilfoylei* (Bull) L.H. Bailey (syn. *Nothopanax guilfoylei*)	Lowlands	Shrub 2–6 m; upright branches, used as hedge plants; good in-field windbreak; easily propagated by cuttings
Turpentine tree	*Syncarpia laurifolia* Ten.	Low to 610 m	Dense tree to 15 m; low branches persist; good grower on exposed areas
Java plum	*Syzygium cumini* (L.) Skeels	Low to 460 m	Dense, medium to 20 m; drought-resistant, good regrowth
Brush box, Brisbane boxwood	*Tristania conforta* R. Br.	Low to 460 m	Medium to 20 m; dense, low-branching; drought-tolerant; good regrowth

In fields where the soil has been compacted, subsoiling or ripping to 50–200 cm is desirable to break up the hardpan in order to improve drainage and soil aeration. In the pineapple plantations of Hawaii, soils are ploughed below 60 cm. Subsoiling provides better subterranean drainage if done parallel to contour lines. Discing, levelling and furrowing follow standard practice for all crops. Discing to break up clumps of soil and improve soil texture is an important phase of land preparation, especially if polyethylene sheets are to be used for mulching. Soil fumigation is also more efficient when soil clumps are broken down.

When root-rot-susceptible crops, such as papaya, passion-fruit, avocado and pineapple, are grown on level land with excessive seasonal rains, raised beds or berms are recommended. The deep furrows between the beds drain water from the field, if properly constructed. These furrows may be used for ground irrigation during dry seasons.

Weed Control

Weed control is a year-round problem in the tropics, except for regions with well-defined dry seasons. If irrigation is practised during the dry period, some form of weed control is necessary. Weed control is troublesome during the first 2–3 years of orchard establishment when the trees have not yet developed substantial canopies to shade weed growth. For example, neglect of weeding can result in a 20–40% decrease in pineapple yields. Weed growth can be reduced in young orchards by heavy mulching with organic materials, such as straw, dried grass, wood shavings or whatever materials are easily available at low cost. Black polyethylene sheets cut to about 1 m square and placed around the newly transplanted trees reduce weed growth and minimize weeding. Planting low-growing ground-cover crops is useful.

Methods of weeding vary from manual weeding to tractor-operated discing, mowing or application of herbicides, depending upon the orchard size and financial status of the grower. Growers of small orchards control weeds manually or with herbicides. Mechanical discing is rapid, with minimal cost, but is not considered to be a good practice inasmuch as it destroys feeder roots that are close to the soil surface.

Chemical weed control is efficient and rapid if done correctly at appropriate times. Pre-emergence application, either before or immediately after planting, is effective, least costly and a desirable management practice. Spray apparatus ranges from simple knapsack sprayers to the more sophisticated boom-type sprayers. Herbicides are generally not recommended in young orchards due to the possibility of damage from spray drift or direct contact with the stems, unless the crop is resistant to the rate used. Young avocado and papaya trees are especially sensitive to herbicides, so weeding around young transplants is usually done manually. For young seedlings and newly transplanted orchards, glyphosate, a systemic herbicide, may be applied by rope wick or weed wipers saturated with the herbicide solution and wiped on the weed leaves. These applicators eliminate spray drift and can be applied close to the stem. Application of herbicides through the irrigation system, known as herbigation, has been shown to be effective. There are restrictions on herbicide use imposed by manufacturers and government agencies, and all applicators must know the proper uses. Growers who intend to export fruit should become acquainted with regulations of the consuming countries governing the use of registered chemicals, residue levels and other precautions.

Intercropping in orchards with vegetables, leguminous crops and cucurbits is a common practice in some areas. In Mexico, fruit orchards have been observed with excellent intercrops of beans (*Phaseolus vulgaris*), water melons (*Citrullus vulgaris*), cantaloups (*Citrullus melo* var. *reticulatus*) and chili (*Capsicum annuum*). Good weed control is obtained under this intercropping system.

CULTURAL PRACTICES

Tropical fruit crops either display continuous growth (papaya, banana) or grow in cycles (avocado, mango, litchi) of vigorous vegetative growth, followed by periods of low vegetative growth during which flowering is frequently initiated. Vegetative growth and flowering of tropical fruit crops are often determined by rainfall if the crop is not irrigated. Temperature influences the rate of vegetative, flower and fruit development and, in litchi and longan, determines the extent of flowering and fruiting. Although vegetative growth can be controlled by withholding water and nutrients, controlling water is often difficult in the tropics with frequent rainfall, and withholding nutrients can limit subsequent development if applied or withheld at the wrong stage. Vegetative growth can be manipulated by training, shoot and root pruning, girdling, root restriction and applying plant-growth regulators. Many of these practices and combinations of these have been used for centuries to initiate flowering and control vegetative growth in fruit crops. The application of these techniques to manipulating growth and development, however, requires an understanding of the growth and development characteristics of the crop being grown.

Single-stemmed (monoaxial) plants grow continuously and either flower concurrently (e.g. papaya) or culminate growth by flowering (e.g. pineapple, banana) (Table 2.4). After initial growth, the leaf number and root growth stabilize, leading to a constant top : root ratio (Verheij, 1986). Good management to maintain growth leads to improved flowering and fruiting. Optimization of fertilization, irrigation, equal spacing and disease and insect control of these crops is, therefore, essential. Cultivar improvement also offers scope for yield improvement.

Table 2.4. Tropical seedling fruit-tree growth is normally continuous and the trees develop different growth phenomena and branching habits. Flowering then becomes synchronized to either or both vegetative growth or more responsive to the environment (Verheij, 1986).

Branching	Vegetative growth	Flowering	Example		Growth characteristics
Single-stemmed (monoaxial)	Continuous	Concurrent	Papaya	Banana	Constant leaf area and top : root ratio
		Terminal	Pineapple		Improvement via altered growing techniques to maintain high growth rate, equidistant planting patterns frequently used in single or double row
Branched (polyaxial)	Flushes	Concurrent	Passion-flower		Rapid shoot growth
		Separate spatially	Durian Jackfruit Lanson	Soursop Abiu	Rhythmic growth → competition major between vegetative and flowering
		Separate temporally	Atemoya Mango Litchi Rambutan Longan	Avocado Mangosteen Carambola Guava Chiku	Feedback control top : root, improvement in yield via manipulation of the tree using pruning, fertilization and irrigation

The branching of polyaxial species increases their ability to fill in open areas with a canopy. Examples of these include vines (e.g. passion-flower), which have concurrent flowering and vegetative growth, and trees with either spatial (e.g. durian) or temporal (e.g. mango) separation of flowering from vegetative growth (Table 2.4). The trees with spatial or temporal separation of vegetative (root and shoot) growth from flowering can have their rhythmic development altered by tree manipulation. Fertilization and irrigation need to be timed to the tree's growth rhythm, supplementing cultural practices such as shoot and root pruning, girdling, defoliation, fruit thinning and branch bending to promote flowering. These tree-management practices can also reduce the asynchrony found in one tree where one part of the tree flowers and another section will remain vegetative (e.g. mango). The objective is to balance vegetative growth and flowering, improving vegetative growth alone, as in a monoaxial plant, is not desired.

Propagation and Transplanting

Propagation

The objectives of propagation are to either increase the number of plants or perpetuate a species or cultivar. Besides seed propagation, plants can be propagated vegetatively, either by division, cutting or grafting (Gardner and Chandhri, 1976). Division includes all processes by which a part of the plant, usually the stem, is induced to grow roots and shoots before separation from the parent (e.g. stooling and air layering, or marcotting). Propagation by cutting uses a piece of root, stem, leaf, leaf and bud or meristem. Normally, the propagule is treated with growth-promoting chemicals to induce root and shoot development, and the planted propagule is held in misting chambers (e.g. passion-fruit, guava and breadfruit).

Grafting is used to perpetuate and increase clonal cultivars, with the upper part referred to as the scion and the lower part the stock (rootstock). It is necessary for the cambial layers of the scion and rootstock to be in contact to allow regeneration of the union, which includes the xylem, phloem and cambium. There are two types of grafting: approach grafting, involving two plants, and detached scion, which uses either a stem piece or a bud. All types of grafting have been tested on tropical fruit trees; general agreement is not always possible as to the most effective method, as operator skill plays a major role, and some methods are preferred for particular tropical fruit crops (Table 2.5). Stooling is used if numerous root-bearing shoots occur when pruned and covered. Layering (marcotting, gooteeing) is used for plants that readily root from the stem and on trees whose bark readily peels off.

Many crops are propagated by tissue culture, in order to ensure freedom from disease and trueness to type. Somaclonal variation is a major problem in some crops, such as pineapple. Tissue culture is possible with soft-tissue

Table 2.5. Propagation methods commonly used for selected tropical fruit crops (Gardner and Chandri, 1976).

Crop	Seed	Sucker stooling	Layering (marcotting)	Cuttings	Grafting/budding
Annona			Air layers	Stem	Budded/grafted
Avocado					Grafting
Banana		Rhizome			
Breadfruit				Root	
Carambola					Budded
Durian	Seed		Air layers	Stem	Budded
Guava			Air layers	Stem and root	Budded
Jackfruit			Air layers	Stem	Budded
Lanson, duku, langsat	Seed				Budded
Litchi and longan			Air layers		Budded
Mango	Polyembryony		Air layers	Stem	Budded
Mangosteen	Apomixis			Stem	Grafting
Papaya	Seed			Stem	Cleft grafting
Passion-fruit	Seed		Air layers	Stem	
Pineapple		Crown and shoot		Leaf and bud	
Rambutan			Air layers		Budded
Sapodilla			Air layers	Stem	Budded approach

plants; pineapple, banana and papaya. Some success has also been reported for mango, avocado, jackfruit and *Annona* spp. It is essential to develop tissue-culture techniques for other tropical fruit trees if gene transfer is to be accomplished using molecular biology techniques.

Transplanting

In areas with seasonal dry periods and without irrigation facilities, field transplanting is best done at the beginning of the wet season. In the cool subtropics, where low winter temperatures can be expected, field transplanting is delayed until the coldest period has passed. Fields should be prepared in advance, planting holes dug and the irrigation system installed so that actual transplanting can be done with minimum loss of time. If the plant is grafted or budded, the graft union should be kept 15 cm above the ground surface, although this may be influenced by the height at which the graft or bud union is made.

The use of container-grown trees has dramatically minimized transplanting mortality. Polyethylene bags are removed, leaving the soil intact around the roots. However, plants kept in containers too long become root-bound, which is not conducive to good root development. Root-bound mango and litchi plants have been observed to remain almost dormant with little vegetative growth for over a year after field transplanting, and some have eventually died. After removing the polyethylene bags, the roots spread out to the extent possible, and roots that are severely bent and twisted should be pruned. Mangosteen is extremely sensitive to disruption of its root system and high failure rates are not uncommon; during transplanting the tap root should remain intact.

Spacing and Orchard Layout

Spacing

Tree spacing is influenced by such factors as terrain, species and cultivar, degree of mechanization, climate, soil and method of irrigation. The primary objective is to plant as many trees as possible per unit area without adverse effects upon yield, quality and orchard management. Spacing schemes are adjusted to induce the placement of pollinizer trees in the orchard, if necessary, so that an adequate ratio can be maintained after thinning. Interplanting with other fruit crops can lead to further adjustment in spacing (Plate 2).

Recent trends promote spacing as close as possible initially, with later thinning, to obtain maximum benefits per unit area (Table 2.6). Close spacing justifies the additional cost if precocious cultivars are available. Even for non-precocious cultivars, if sufficient yields can be obtained before thinning

Table 2.6. Recommended between-tree and row spacings in tropical orchards for some fruit crops.

Crop	With fillers (m)	Trees ha^{-1}	Permanent spacing (m)	Trees ha^{-1}
Acerola	2.7 × 5.5	673	5.5 × 5.5	332
			4.6 × 5.5	395
Atemoya	4.5 × 9.0	247	9.0 × 9.0	124
			8.0 × 12.0	104
Avocado	4.5 × 6.0	370	9.0 × 12.0	93
	6.0 × 7.5	222	12.0 × 15.2	56
Fuerte	7.5 × 7.5	177	15.2 × 15.2	44
Hass	7.0 × 7.0	204	14.0 × 14.0	51
Cherimoya			5.0 × 6.0	333
			5.0 × 7.6	263
Durian			10.0 × 10.0	100
Guava	3.1 × 7.6	424	6.2 × 7.6	212
			4.6 × 7.6	286
Litchi	5.0 × 5.0	200	10.0 × 10.0	100
	6.0 × 12.0	139	12.0 × 12.0	69
Mango	5.0 × 10.0	200	10.0 × 10.0	100
	12.0 × 12.0 (+ 1 tree in centre)	138	12.0 × 12.0	69
Papaya			2.4 × 3.1	1344
Passion-fruit			3.3 × 4.0	1324
Sapodilla			6.6 × 6.6	400
Soursop	3.7 × 7.6	356	7.4 × 7.6	178
			6.0 × 9.0	185

becomes necessary, the added cost of grafted trees, planting, maintenance and thinning can be justified. The greatest drawback in a close-spacing scheme is the strong reluctance on the part of the grower to thin the excess trees when overcrowding begins to lower yields. Removing producing filler trees temporarily reduces overall yields, with production recovering as the permanent trees find unrestricted space to expand. Mango trees are relatively rapid growers, bearing commercial-size crops in 4–5 years. Significant early cash flow can be obtained before tree thinning. In most instances, even when close planting begins to overlap, some pruning of the filler trees can be done to delay tree thinning for several years. Thinning can also be done in increments, removing only the largest filler trees each year over several years.

Orchard Layout

There are a number of orchard layout systems (Fig. 2.3). Four of the more commonly used ones include: (i) the square system with a filler tree (hexagon

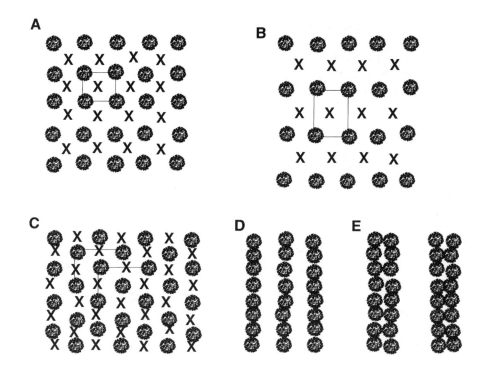

Fig. 2.3. Orchard layout systems. A. Square systems. B. Rectangular system with wider spacing between rows. C. A quincunx system. D. Modified rectangle system with close spacing of trees. E. Double-row system frequently used for pineapple and suitable for shrubby tree crops such as acerola and papaya. The original position of the filler trees is marked by X.

design) in the centre; when the filler tree is removed in later years, the orchard will show a square pattern; (ii) close planting in the square system with alternate trees being the filler trees; when the trees begin to overlap, alternate filler trees are removed on the diagonals, resulting in a diagonal or quincunx system; (iii) permanently designed quincunx system without filler trees, which gives the most effective use of space; and (iv) a rectangular system in which spacing within rows is less than between rows. The last is most common in highly mechanized orchards.

The selection of layout and spacing varies with crop, soil and management practices, such as pruning and harvest methods (Table 2.6).

A 9 × 9 m square system provides each tree with 81 m² of space and results in about 124 trees ha⁻¹. The same spacing of 9 m between trees in a quincunx system has a distance between rows of approximately 8 m and provides 72 m² of space per tree with about 139 trees ha⁻¹. Papaya planting at 2.4 m × 3.1 m is a good example of the rectangle system. This can be modified to a double-row rectangle system, which is often used for papaya. The quincunx system can be used at initial planting when the most suitable cultivar has not been decided. Later, the less suited cultivar is removed, leaving a square pattern. Sufficient space for tractor operations and truck access when trees are full size is essential.

Pruning

Training and pruning are designed to shape a tree, or control its size to optimize fruit yield and quality. Many tree models have been defined (Tomlinson, 1987), such as monopodial rhythmic growth for avocado, mango and litchi. In avocado, the branches are morphologically identical to the main trunk with flowers borne laterally with only minor effect on the vegetative shoot. These models provide a basis for comparing pruning and training protocols for different tropical fruit trees. Pruning to shape a tree aims to control the number, orientation, size and angle of branches and thereby improve a tree's structural strength to carry a fruit load, reduce wind damage, increase light penetration, stimulate flower-shoot development and enhance fruit yield and quality. Too much pruning can delay flowering and lead to vegetative growth. Pruning is not practised on monoaxial crops with short life cycles (banana, pineapple and papaya), except that excess suckers are removed in bananas. Broken, weak, drooping and diseased branches should be removed as a matter of course. The shape of the tree to be developed depends upon the species and the specific management objective.

Pruning of many tropical trees may reduce yield in the following season (e.g. mango, litchi), but some form of tree-size control is necessary to maintain a manageable tree. The trend is to maintain trees by yearly or more frequent pruning at 2–3 m for ease of flower cycling, pruning and harvesting. Carambola can be extensively pruned to remove overlapping branches and topped to maintain 2–4 m height without reducing yield significantly; it can also be trellised.

The rhythmic flushing of vegetative growth and flowering in guava, mango and rambutan influence when and how pruning is carried out. Flowers develop on shoots that are not flushing and the objective of pruning should be to increase the number of shoots that can flower simultaneously.

POSTHARVEST HANDLING

Estimates of tropical-fruit postharvest loss vary widely (10–80%) in both developed and developing countries. The losses given in published reports highlight the total postharvest losses of products, but do not consider the loss of quality or downgrading that may reduce the price received. Losses may be due to mechanical injury, physiological damage or pathogens. The reduction of the losses in a systematic way requires a knowledge of postharvest physiology, its applied technical aspect, handling, and the appreciation of its biological limitation represented as storage potential (Paull, 1994).

Tropical-fruit physiology does not differ from the basic knowledge gained from studies of temperate and subtropical fruits. There are differences in the major substrates involved in ripening, the rate of ripening and senescence and, in some cases, variation in the order in which various components of ripening occur. The aspects of tropical-fruit physiology that make these fruits unique are the chilling sensitivity of most tropical fruits (Plate 3), the generally more rapid ripening of climacteric tropical fruits when compared with temperate fruit and the frequent need in postharvest handling of tropical fruit to expose them to high temperatures or other stresses during insect disinfestation prior to export.

Storage potential is crucial if tropical fruit are not processed but shipped fresh from production areas to consumers. Storage potential needs to include all the time spent in the various marketing steps (Fig. 2.4). The data needed for decision-making should come from simulated handling schemes that approximate what happens in a commercial marketing system. This necessitates an understanding of the current handling system and the constraints that limit changes.

Tropical-fruit Physiology

Climacteric and non-climacteric

Like temperate fruits, tropical fruits can be divided into climacteric and non-climacteric (Table 2.7). This division is based on the respiratory pattern after harvest. In climacteric fruit, there is generally a dramatic and rapid change in respiration during ripening. In commercial handling, ethylene can lead to earlier ripening of climacteric fruit but not of non-climacteric fruit.

Respiration and ethylene

Tropical fruits vary widely in their respiration rate and ethylene production (Table 2.8), varying with stage of ripening and senescence as well as variety, preharvest environment and culture. However, knowing the rates of respiration is essential for determining heat loads in refrigerated cold rooms and containers. The ethylene-production rate is also essential information, because it relates to mixed loads and the effect of one commodity on another; for example, ethylene is used to ripen banana fruit.

Respiration rate and storage life are related (Fig. 2.5). Fruit with high respiration rates have shorter postharvest lives – hence the need to reduce respiration and thereby increase postharvest life. Temperature management is

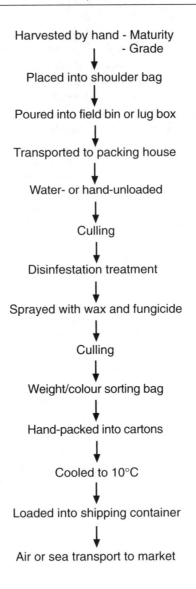

Fig. 2.4. Generalized handling scheme for tropical fruit.

the major method of controlling respiration rate, although it is limited in most tropical fruit by their chilling sensitivity.

Chilling injury

The ability to use temperature to extend the postharvest life of tropical fruits is limited by chilling injury. The symptoms of chilling injury are similar for most

Table 2.7. Classification of selected tropical fleshy fruits according to their respiratory pattern.

Climacteric	Non-climacteric
Avocado (*Persea americana*, Mill.)	Carambola (*Averrhoa carambola*, L.)
Banana/Plantain (*Musa* spp.)	Litchi (*Litchi chinensis*, Sonn.)
Breadfruit (*Artocarpus altilis*, Parkins, Fosb.)	Mangosteen (*Garcinia mangostana*, L.)
Cherimoya (*Annona cherimola*, Mill.)	Mountain apple (*Syzygium malaconse* (L.) Merril & Perry)
Durian (*Durio zibethinus*, J. Murr.)	Pineapple (*Ananas comosus* (L.), Merrill)
Guava (*Psidium guajava*, L.)	Rambutan (*Nephelium lappaceum*, L.)
Mango (*Mangifera indica*, L.)	Rose apple (*Syzygium jambos* (L.), Alston)
Papaya (*Carica papaya*, L.)	Star apple (*Chrysophyllum cainito*, L.)
Passion-fruit (*Passiflora edulis*, Sims)	Surinam cherry (*Eugenia uniflora*, L.)
Sapote (*Casimiroa edulis*, LLave.)	
Soursop (*Annona muricata*, L.)	
Chiku (*Achras sapota*, L.)	

Table. 2.8. Respiration and ethylene-production rate of selected tropical fruits at 20°C.

	Respiration			Ethylene	
Class	Range (mg kg^{-1} h^{-1})	Commodity		Range (µl kg^{-1} h^{-1})	Commodity
Very low	< 35	Pineapple, carambola			
Low	35–70	Banana (green), litchi, papaya, jackfruit, passion-fruit, mangosteen		0.1–1.0	Pineapple, carambola
Moderate	70–150	Mango, rambutan, chiku, guava, durian, lanzone		1.0–10.0	Banana, guava, mango, plantain, mangosteen, litchi, breadfruit, sugar apple, durian, rambutan
High	150–300	Avocado, banana (ripe), sugar apple, atemoya		10–100	Avocado, papaya, atemoya, chiku
Very high	> 300	Soursop		>100	Cherimoya, passion-fruit, sapote, soursop

commodities and include pitting, skin darkening, failure to ripen completely and increased susceptibility to decay (Plate 4). Carambola and the subtropical fruits longan and litchi are crops that are somewhat resistant to chilling injury, requiring a considerable time (>14 days) at 1°C before injury occurs.

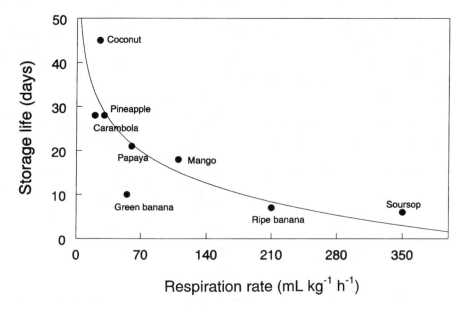

Fig. 2.5. Relationship between respiration rate and postharvest life of tropical fruit held at their optimum storage conditions (Paull, 1994).

In any discussion of chilling, two aspects must be considered: the temperature and the time spent at that temperature. Although there is not an exact, fruit-specific, reciprocal relationship between temperature and time, it takes a longer time at a higher temperature to develop injury than at a lower temperature. As the storage temperature is lowered from 30°C, duration of storage life increases, the limitation being fruit ripening and senescence. Storage life reaches a maximum at 15°C for Brazilian banana, 10–12°C for papaya, 8°C for rambutan and 5°C for carambola (Fig. 2.6). Lowering the temperature further leads to a shorter storage life, but the limitation changes from ripening to the development of chilling injury, ripening being completely inhibited. Recommendations for optimum storage of most tropical fruits are just inside the chilling range (8–12°C), as this allows ripening to be controlled and, if removed before the chilling-stress threshold is exceeded, the fruit still has a number of days of useful marketing life as it ripens. Unfortunately, similar data are not available or are fragmentary for many tropical fruits. The actual relationship between storage temperature and duration can vary with cultivar, preharvest conditions, stage of ripeness and postharvest treatments.

Moisture loss

Frequently, injury induced by water loss is confused with chilling injury. Loss of mass in tropical fruits postharvest, mainly a loss of water, is dependent upon the commodity, cultivar, preharvest conditions, water-vapour pressure deficit (WVPD), wounds, postharvest heat treatments and the presence of

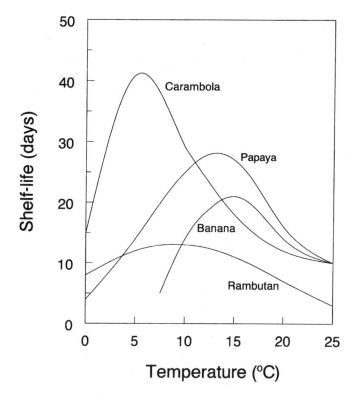

Fig. 2.6. The shelf-life at various storage temperatures for carambola, rambutan, Brazilian banana and papaya fruit (Chen and Paull, 1986; O'Hare *et al.*, 1993; Paull, 1994).

coatings or wraps. Tropical fruit can be grouped from low moisture-loss rate (such as from a coconut) to medium (avocado, banana, pummelo) to high (guava, litchi, mango, papaya, pineapple) (Table 2.9). On a per-unit-area basis, the stem scar is frequently the site of highest water loss, although most water is lost through lenticels, stomata and skin cuticle. Tropical fruit have water-loss rates between about 0.1 and 0.3% day^{-1} mbar WVPD^{-1}. Fruit that have lost 6–8% of their fully turgid initial weight begin to show signs of mass loss. Usually the initial sign is skin wrinkling, although skin discoloration is the first symptom in some fruits. Loss of mass, besides affecting overall appearance, is also an economic loss if fruit is sold by weight.

Storage Potential

Storage potential is normally defined as the postharvest life of a commodity held at its optimum storage temperature. This potential is dependent upon cultivar, preharvest environment and culture, maturity at harvest and storage

Table 2.9. Comparison of water loss from different tropical fruits expressed as loss per day per water-vapour pressure deficit (WVPD).

Tropical fruit	Water loss	
	% loss day^{-1}	% loss day^{-1} mbar WVPD^{-1}
Avocado	0.5–1.0	
	1.0	
Banana	0.4	0.3
Guava		0.3
Mango		0.1
Papaya ('Sunset')		0.08
Plantains	0.5	
Rambutan		0.32
Chiku		0.3

conditions (Arpaia, 1993). Postharvest life is terminated because of physiological, mechanical and pathological stress with associated symptoms, such as excessive water loss, bruising, skin scald, failure to ripen and decay. The limitation imposed by chilling injury on tropical fruits puts a different constraint on storage potential, because storage temperatures higher than 8–12°C lead to ripening and senescence, with lower temperature leading to chilling injury, as discussed above.

Commercial-shipping publications present storage potentials of many fruits, vegetables and flowers, including many tropical fruits (Table 2.10). Such values for postharvest life should be taken as maximums. The ranges of storage life given indicate the variability in the available information. These estimates are frequently based upon laboratory studies that do not allow for the vagaries of commercial handling. Subjective eating-quality criteria, such as aroma, flavour and texture, are frequently not considered in these estimates.

Postharvest Insect Disinfestation

Postharvest treatments are required to disinfest economically important host fruit, vegetables, dried fruits and vegetables, nuts, flowers and ornamentals of insect pests before they are moved through marketing channels to areas where the pests do not occur (e.g. various tephritid fruit flies and mango seed weevil), or the importing country has a 'zero tolerance' for all live insects, whether or not they are economically important. While there are many insects and arthropods of quarantine importance, fruit flies represent a major group of destructive pests that attack a wide range of fruit (Plate 4). Disinfestation-treatment technologies and strategies include pest-free zones, non-host status,

Table 2.10. Recommended storage temperature and storage potential as postharvest life of tropical fruits. Data extracted from commercial publications (Paull, 1994).

Fruit	Optimum storage temperature (°C)	Postharvest life at optimum temperature (days)
Acerola	0	50–58
Atemoya	13	28–42
Avocado		
Mexican	5	14–28
West Indian	10	14–28
Banana	14	7–28
Breadfruit	13	14–40
Carambola	1	21–28
Cherimoya	13	14–28
Durian	4	42–56
Granadilla	10	21–28
Guava	10	14–21
Jackfruit	13	14–45
Langsat	11	10–15
Litchi	1	21–35
Longan	2	21–35
Mango	10–12	14–25
Mangosteen	13	14–25
Papaya	8–12	7–21
Passion-fruit	12	14–21
Pineapple	10	14–36
Rambutan	12	7–21
Sapote	12	14–21
Star fruit	9	21–28
Sugar apple	7	28
White sapote	19	14–21

fumigation, irradiation, insecticides, heat and cold treatments, controlled or modified atmospheres and combinations of treatment methods. Either treatment monitoring or commodity inspection by regulatory personnel or both are required to ensure that quarantine treatment procedures and regulations, including handling requirements to preclude reinfestation after treatment, are followed. Failure to control the spread of insect pests can result in expensive quarantine and eradication procedures, product losses due to infestation and costly new disinfestation-treatment requirements.

Many of these disinfestation treatments cause damage to the commodity, which limits postharvest life and reduces quality. Current research efforts are directed to the use of physical disinfestation treatments, such as heat (Plate 5), cold and irradiation. Such physical disinfestation treatments are non-polluting and they leave no toxic residues. No single quarantine treatment or system can be expected to work equally against all insects or for all host fruits, and the response to any treatment can vary greatly (Paull and Armstrong, 1994).

FURTHER READING

Arpaia, M.L. (1993) Preharvest factors influence the postharvest fruit quality of tropical and subtropical fruit. *HortScience* 29, 982–985.

Gardner, R.J. and Chandhri, S.A. (1976) *The Propagation of Tropical Fruit Trees.* Horticultural Reviews No. 4, Commonwealth Agricultural Bureau. East Malling, UK, 566 pp.

Norman, M.J.T., Pearson, C.J. and Searle, P.G.E. (1995) *The Ecology of Tropical Food Crops.* Cambridge University Press, Cambridge, 430 pp.

Paull, R.E. (1994) Tropical fruit physiology and storage potential. In: Champ, B.R. Highley, E. and Johnson, G.I. (eds). *Postharvest Handling of Tropical Fruits.* ACIAR Proceedings No. 50, Canberra, pp. 198–204.

ANNONAS

BOTANY

Family

Annonaceae, commonly referred to as the custard-apple family, consists of about 75 genera, widely distributed. Some are grown as ornamentals, while others are known for their edible fruit and perfume.

Important Genera and Species

The genus *Annona* is the most important since, among its 100 or more species, seven species and one hybrid are grown commercially. All are native to the American tropics. Leaves are alternate, simple and entire and flowers may be solitary or in clusters, with two series of three thick and fleshy petals (Fig. 3.1). The other closely related genus with some commercial fruit is *Rollinia*.

The important commercial species are:

Annona cherimola Mill.	cherimoya; chirimoya (Spanish); cherimolier (French); anona (Mexican); noina ostrelia (Thai)
Annona diversifolia Saff.	ilama (English, Spanish)
Annona glabra L.	pond apple; mamon (Philippine)
Annona montana Magfady	mountain soursop (English), guanabana cimarrona (Spanish)
Annona muricata L.	soursop; guanabana (Philippine); catoche (Spanish); durian belanda (Malay); nangka belanda (Indonesian); thurian-khaak (Thai); sitaphal (Indian); atio (Philippine); fruta de conde (Brazilian); nona sri kaya (Malaysian);

Fig. 3.1. Leaves and flower of soursop.

	zapote agrio (Mexican); graviola (Brazilian); corossol epineux (French)
Annona reticulata L.	custard-apple, bullock's heart (English); anon; anona; corazon (Spanish)
Annona squamosa L.	sugar apple, sweetsop (English); anon, rinon (Spanish); noina (Thai); nona seri kaya (Malay); custard-apple (Indian)
A. squamosa × *A. cherimola*	atemoya, custard-apple
Rollinia orthopetala R.DC.	birriba (Brazilian)

Area of Origin and Distribution

A. cherimola Mill. (cherimoya) is speculated to have originated in the highlands of Peru and Ecuador. The antiquity of the fruit is attested to by ancient artefacts shaped in the form of the fruit in Peru. Distribution through Central America and Mexico probably occurred at an early date, as it has become naturalized in the cool highland areas. Distribution continued from Mexico to the Caribbean islands and then to the African coast and the Mediterranean. Introduction to Africa and the Far East is attributed to early Spanish navigators.

The cherimoya is considered the best of the annonas and is cultivated in subtropical regions and in the tropical highlands. In most areas it is grown as a garden tree or as part of a subsistence farming system at appropriate elevations. Commercial production occurs in Spain, Bolivia, Chile, Peru and New Zealand. Experience has shown that the California coastal regions are more conducive to cherimoya production, having higher relative humidity (RH) (70–80%) in spring and summer than in the interior valleys where the RH can drop to 40% and below, during the hotter part of the day in summer.

The tree grows to about 7.3 m; it is vigorous when young but tends to decline in growth with age. New buds cannot sprout until the leaves have been shed, as the leaf bases grow over the axillary buds, as in sweetsop (Fig. 3.2). Leaves are 10–25.4 cm long, light green and arranged alternately.

A. diversifolia Saff. (ilama) produces fruit considered to be comparable with the cherimoya. The tree is more adapted to warmer tropical lowland conditions, being a native of the foothills of western Mexico and Central America. However, its distribution is very limited because of its poor adaptability. The tree has a slender habit and grows to a height of 6–7 m. Fruit size ranges from 455 to 680 g and they mature at midsummer.

A. muricata L. (soursop, guanabana) is the most tropical and produces the largest fruit among the *Annona* species (Plate 6). It is considered to be best suited to processing. It is native to the American tropics, with the Caribbean being the area of origin. It was distributed very early to the warm lowlands of eastern and western Africa and to south-east China. It is commonly found on

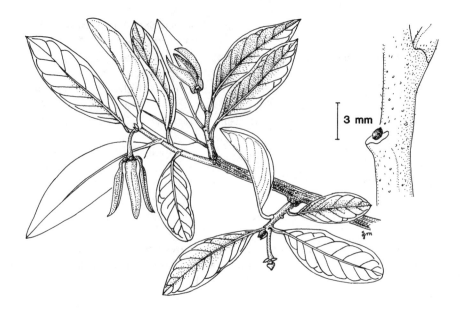

Fig. 3.2. Leaves, axillary buds and flower of sweetsop. The figure shows the buried nature of the axillary buds, which can be vegetative or a mixed flower and vegetative bud. The leaf must abscise before the bud can develop.

subsistence farms in South-east Asia and was established very early in the Pacific islands. It is grown extensively in Mexico, from Culiacan to Chiapas and from Veracruz to the Yucatan Peninsula in the Gulf region. Orchards as large as 20 ha occur.

The soursop is a small, evergreen, slender and upright or low-branching and bushy tree, growing to heights of 4.5–9 m. The leaves are glossy, dark green, obovate to elliptic and 12.7–20 cm long (Fig. 3.1). When crushed, the leaves will emit a strong odour. Flowers are solitary, yellow, 2.5–4 cm long and with three thick, fleshy petals and three minute inner petals alternating with the outer petals (Fig. 3.1). Fruit size varies from less than 0.45 kg to more than 4.5 kg, largely depending upon extent of pollination and fertilization. A normal fruit is generally heart-shaped to oval (Fig. 3.3), but, if there is poor pollination, unfertilized ovules fail to develop and the resulting fruit assumes distorted irregular shapes and is usually undersized. The skin is dark green with many recurved, soft spines. The flesh is juicy and white, with a cottony texture, and contains many dark brown seeds about 2 cm long. The pulp has an agreeable subacid flavour with a distinct aroma. The soursop produces fruit throughout the year but peak production in most areas comes during summer and early autumn, sometimes with a secondary peak during early spring (Fig. 3.4).

A. reticulata L. (bullock's heart) is a semideciduous species indigenous to the American tropics, found most widely in the lowlands. It followed the usual

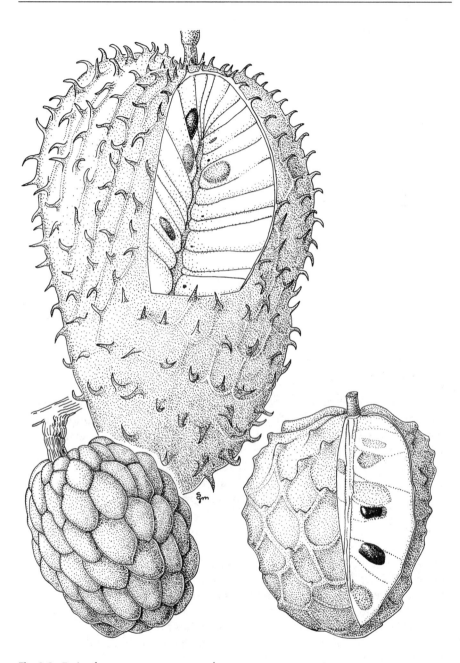

Fig. 3.3. Fruit of soursop, sweetsop and atemoya.

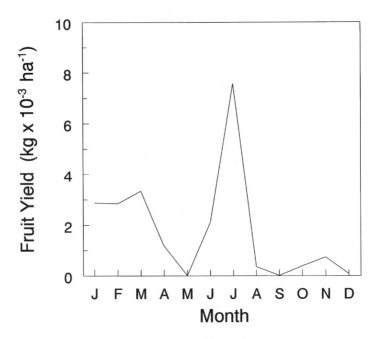

Fig. 3.4. Soursop yield in Hawaii, mean of 3 years (after Nakasone, 1972).

distribution route of the early traders from the American tropics to the coastal areas of Africa and the Pacific region. The fruit is not as favoured as the cherimoya or sweetsop, due to generally inferior quality, but some trees bear excellent fruit. The fruit is usually heart-shaped but may be oval or even conical and grows in size from 0.45 to 1 kg. The skin is smoother than that of cherimoya and is reddish yellow to reddish brown when ripe. Fruit matures during the winter and spring, at times when few other annonas are available.

Annona squamosa L. (sweetsop) probably originated in the Caribbean region and is the most widely distributed of the *Annona* species. This species is the most commonly grown *Annona* sp. in the tropical regions of the Americas, Africa, Asia and the Pacific. It is less tolerant to cold temperatures than the cherimoya but more tolerant than the soursop, as it is found thriving rather well in some subtropical areas, such as the coastal areas of south Florida. The fruit is frequently found in village markets but has not shown any potential for large commercial cultivation, due to the small fruit size, frequent cracking at maturity and poor shelf-life.

The tree is normally smaller than the cherimoya, attaining heights of 3.0–4.6 m, with slender branches. The leaves are oblong–lanceolate, narrower than those of the cherimoya and 10–15 cm long (Fig. 3.2). All leaves are shed before new shoots appear. Flowers are axillary, in clusters of two to four on leafy shoots (Fig. 3.2). Fruit set is better than in the cherimoya.

The fruit is more nearly heart-shaped, 5–10 cm in diameter. It is yellowish green in colour but a purple-fruited variant is also known. The exterior parts of adjacent carpels are not completely fused and these rounded protuberances frequently separate, exposing the white flesh upon ripening (Fig. 3.3). The fruit matures during the summer but can extend for several months.

A. glabra L. (pond apple) has greater potential as rootstock than for fruit. It grows wild in southern Florida around lakes and rivers and is widely distributed in the lowlands of tropical America. It is a vigorous grower, reaching heights of 12 m and growing well in swampy areas. The flower is fragrant, creamy-yellow and reddish inside. The fruit is about the size of the sweetsop and heart-shaped; its yellow pulp is described as insipid.

A. montana Macfady (mountain soursop) is considered a wild species, native to the Caribbean islands, and is not well known outside the area of origin. Fruit quality is inferior to that of soursop. The tree is larger than the soursop, with foliage and flowers resembling those of soursop. The fruit is somewhat spherical, about the size of a baseball to a softball. The pulp colour is yellowish when ripe. This species is a possible rootstock.

Atemoya is a hybrid between *A. squamosa* and *A. cherimola*. P.J. Wester of Florida produced the first hybrids in 1908 and called it the 'atemoya', using the Brazilian name 'ate' for sweetsop and 'moya' from cherimoya. In 1927, hybrids were also developed in Poona, India. Numerous cultivars have been selected in Israel (Gazit and Eistenstein, 1985). There is considerable variation among seedlings. Leaves are glabrous and larger than those of the parents, with flowering and fruiting seasons resembling those of *A. squamosa*. Seedlings also differ widely in the external and internal structure of the fruit. Favourable characteristics inherited from the cherimoya parent include many seedless carpels, and carpels that adhere together instead of breaking apart into individual tubercles, as in the sweetsop (Fig. 3.3). The atemoya is more tropical in its requirements than cherimoya. While young trees require some protection from frost, older trees have shown tolerance in Australia. The atemoya is currently grown commercially in Israel, Australia, California, Florida and Hawaii and in other countries for domestic consumption.

ECOLOGY

Soil

All species are capable of growing in a wide range of soil types, from sandy soil to clay loams. Nevertheless, they differ in their soil preferences. Higher yields occur on more well-drained sandy to sandy loam soils, except in the case of *A. glabra*, which can succeed in shallow ponds. Drainage is essential to avoid root-rot diseases; hence the interest in *A. glabra* as a rootstock, related to its tolerance of wet soils.

Climate

Rainfall

Atemoya benefits from uniform soil moisture for good production, with extremes of moisture lowering production. Rainfall and high humidity during the peak flowering season greatly enhance fruit production by preventing desiccation of stigmas, prolonging their receptive period and increasing fruit set and early fruit growth. Similar conditions are also ideal for other annonas. The sweetsop is probably the most drought-tolerant species and it grows and produces poorly where rains are frequent. This is supported by the fact that sweetsop does much better in northern Malaysia, where dry periods occur, than in the southern part, which has year-round high moisture.

Temperature

Temperature is the limiting factor, with frost killing young trees but older trees showing some tolerance. Cherimoya (7–18°C mean minimum) is more tolerant to low temperatures (Fig. 3.5), followed by atemoya (10°–20°C, mean minimum), and soursop is the least tolerant (15°–25°C mean minimum). Except for cherimoya, the annonas do not require chilling periods and do well under lowland conditions (George and Nissen, 1987a). All benefit from dry periods, except for soursop and pond apple. Cherimoya is susceptible to high temperatures, with a growing temperature of 21–30°C (George and Nissen, 1986b). Poor pollination is a frequent problem with all species and occurs under high temperature (30°C) and low humidity (30% RH), even with hand-pollination. Lower temperature (25°C) and high humidity (80% RH) greatly improve pollination. Hand-pollination is recommended for cherimoya and atemoya to achieve more uniform fruit shape (Fig. 3.6).

Light and photoperiod

Heavy shading of vigorous trees can reduce fruit set in atemoya. Light penetration to the base of vigorous trees with a dense canopy in close spacing can be 2% of full sunlight and there is very little fruit set. Pruning practices and spacing need to be adjusted for this growth aspect. No photoperiod responses have been reported.

Wind

The soft wood of the trees makes them susceptible to wind damage and limb breakage. Tree shaking may also be partially responsible for collar-rot organism penetration. The fruit skin is easily damaged by rubbing and exposure to drying winds (Marler et al., 1994). Annona leaf stomata are very

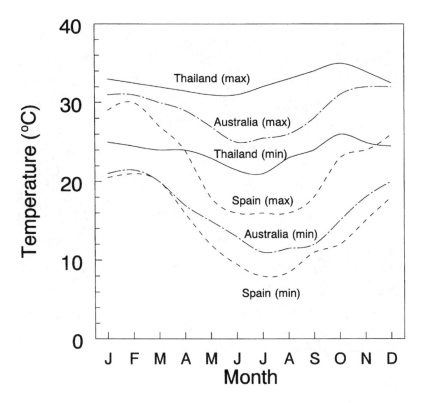

Fig. 3.5. Seasonal changes in mean monthly minimum temperature for areas suited to sweetsop (Bangkok, Thailand), atemoya (Mareeba, Australia) and cherimoya (Malaga, Spain) in southern-hemisphere equivalents (after Marler *et al.*, 1994).

sensitive to changes in RH. Productivity can be improved by windbreaks and under-tree sprinkling to raise the RH above 60%.

GENERAL CHARACTERISTICS

Flowers

The flowers of *Annona* species are hermaphroditic and are produced singly or in small clusters on the current season's growth, although flowers arising from old wood are not uncommon, especially in soursop (Plate 6). New flowers continue to appear towards the apex of the shoot as flowers produced earlier at the basal portions mature (Fig. 3.7). Defoliation of *A. muricata* manually or by using ethephon spray promotes lateral branch growth and induces additional flower formation near the apex of the branches.

    Annona species generally require 27–35 days for flower-bud development from initiation to anthesis. Differences in floral behaviour in the various areas

Fig. 3.6. Hand-pollination of atemoya using a hand-blower to disperse the pollen collected the day before on to the open flower (courtesy R. Manshardt).

may be attributed to both genetic variability and climatic differences (Kshirsaga *et al.*, 1976). Flowering can extend from 3 to 6 months or even longer, with heavy peaks. Two major flowering periods occur after periods of vegetative flushes, with the second peak coinciding with the onset of monsoon in India (Kumar *et al.*, 1977). Atemoyas in Australia produce the first flowers during spring and have a second flowering during the summer, the latter being the more productive (Sanewski, 1991).

Pollination and Fruit Set

Natural pollination

The flowers exhibit both dichogamy and a protogynous nature. This poses a serious problem in obtaining high yields. The atemoya female parts are receptive between 4 p.m. and 8 a.m. and appear moist and sticky (Thakur and Singh, 1964). The pollen is discharged in the afternoon of the same day from 3 to 6 p.m. if the RH is above 80% and the temperature > 22°C. At lower temperatures, pollen is released on the afternoon of the second day. The pollen

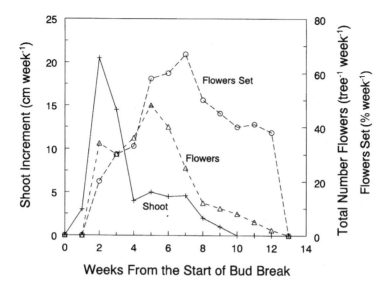

Fig. 3.7. Atemoya: newly emerging vegetative-flush shoot increment, total number of flowers produced per week and percentage flowers set per week. Peak fruit set occurs after completion of vegetative flush (after George and Nissen, 1988).

sacs turn a greyish colour as pollen is discharged. Upon opening, flowers are receptive for about 24 h. *Annona muricata* floral anthesis takes place mostly between noon and 8 p.m. and from 4 a.m. to 8 a.m., with pollen release occurring between 4 a.m. and 8 a.m.

The flowering seasons of *A. squamosa* and *A. cherimola* coincide. When sweetsop pollen is shed at about 2 a.m., cherimoya flowers are receptive, opening around 7–9 a.m. and, when cherimoya pollen is shed at 3–4 p.m., sweetsop flowers are receptive. This flower synchrony, together with complementary functional sexes, favours cross-pollination, leading to natural hybridization. This is attested to by the frequent appearance of hybrid seedlings under the trees of sweetsop and cherimoya when grown in close proximity.

Nitidulid beetles (*Carpophilus* and *Uroporus* spp.) are the important pollinators of annona flowers, with wind and self-pollination being low (1.5%). Fruit set of 'African Pride' atemoya increases linearly with increasing numbers of nitidulid beetles per flower. Three or more beetles per flower increased fruit set to nearly 25%. Studies also showed that these beetles breed rapidly in rotting fruit media and that populations of these beetles can be increased by maintaining the rotting-fruit attractant.

Pollen grains of flowers appearing early in a flowering season have thick walls, are high in starch, germinate poorly and give poor fruit set. Pollen of later flowers shows a high proportion of individual pollen grains without starch grains and which germinate well. Hand-pollination is frequently

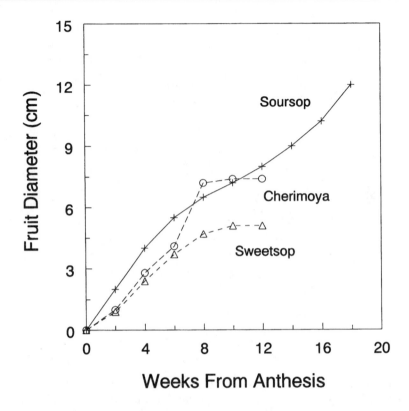

Fig. 3.8. Increase in fruit diameter from anthesis for soursop, sweetsop and cherimoyas (after Thakur and Singh, 1965; Worrell *et al.*, 1994).

practised to ensure pollination and good fruit shape. Pollen must be collected in the evening from fully open flowers, when the sacs have turned from white to cream. The flowers are held in a paper bag, not a closed container, and should discharge that afternoon. The flowers are shaken over a shallow tray or paper to collect the pollen, which is transferred to a small container and held in the refrigerator for use the next morning. Twenty to thirty flowers can give enough pollen to pollinate 50–60 flowers. Pollination is done before 7 a.m. every week, using a small brush or puffer (Fig. 3.6). Hand-pollination has shown some variable results and is less successful on very humid, overcast days and on young vigorous trees. About 150 flowers can be pollinated in 1 h and a success rate of 80–100% can be achieved.

Growth regulators for fruit setting

Hand-pollination in commercial orchards is tedious, time-consuming and a costly practice. Attempts have been made to use growth regulators, with considerable variation in the results obtained. Auxin-induced (indole-acetic acid (IAA), naphthalene acetic acid (NAA)) fruit grow very slowly with less fruit drop, while gibberellic acid (GA$_3$) promotes fruit set and growth rate; however, there are no postset retention properties (Yang, 1988). Application of the two substances separately at appropriate times have produced seedless fruit of 200–300 g (Saavedra, 1979). Gibberellin (1000 p.p.m.) is more effective than hand-pollination and produced seedless fruit in atemoya cultivars 'African Pride' and 'Gefner'. Repeated spraying is necessary to prevent fruit abscission during the first 2 months. Seedless fruit are smaller, with less flavour and less fruit splitting than occurs in the seedy fruit that result from pollination (Yang, 1988). Similar results have been obtained for cherimoya and sweetsop. GA spraying is not recommended as a general management practice for atemoya, because of variable results, although it could be used in areas with poor natural pollination.

Fruit

Fruit growth shows the typical sigmoidal curve, with maturation occurring in 16–24 weeks, depending upon species and growing conditions (Fig. 3.8). Low humidity (<60% RH) and temperature (<13°C) near fruit maturity can increase the severity of fruit-skin russeting, as well as delaying fruit maturation. High temperature can cause premature fruit ripening and fermentation of the fruit.

CULTIVAR DEVELOPMENT

Genetics, Cytogenetics and Breeding

The chromosome numbers of *A. cherimola*, *A. reticulata*, *A. squamosa* and atemoya are 2n = 14. *Annona glabra* has been shown to be a tetraploid (2n = 28), and the occurrence of high chiasma frequency and the absence of multivalents during meiosis suggest that it is of amphidiploid origin.

Cross-pollination between species is conducted primarily to determine compatibility for increasing fruit set. Atemoya is the only hybrid that has gained importance and it has inherited the glabrate leaf character of *A. squamosa* and a leaf size almost as large as that of *A. cherimola*. Flowering and fruiting seasons are similar to those of sweetsop. Skin, pulp and seed characters of both parents are inherited in varying degrees by each plant. A desirable hybrid would be between the cherimoya and soursop. This would combine the larger fruit size and acidity of the soursop and the sweetness, flavour and texture of cherimoya. Attempts to cross the soursop with cherimoya, ilama, bullock's heart or sweetsop have not been successful and may reflect a considerable genetic distance of soursop from the other species (Samuel *et al.*, 1991).

Problems in Breeding

Existing commercial cultivars show considerable variation in growth, fruit set, fruit size and quality. No single variety has all the desirable characteristics. The length of the juvenile period varies, with earliest production occurring in 2 years and full production in 5–6 years. This juvenile period is extremely variable with scions on seedling rootstocks. The seedling rootstocks are derived from extremely heterogeneous openly pollinated seeds; hence it is difficult to fix specific characters in a short period. Breeding programmes have focused on selections from seedling populations. Early maturity, better fruit appearance and, in the subtropics, greater cold tolerance are the most frequent objectives.

Cultivar Development

Except for cherimoya and atemoya, very few named clonal cultivars have been developed among the annonas (Table 3.1). Most of the plantings have been of seedlings. In California, some old cultivars of cherimoya include 'McPherson', 'Deliciosa' and 'Bays'. Considerable work in Peru has been done on the development of cultivars, but they are not widely known outside Peru. Chile, Spain and New Zealand grow the cherimoya, as it is more tolerant of cold temperatures with more successful self-pollination than the atemoya. New Zealand's principal cultivars are 'Reretai', 'Burton's Wonder' and 'Burton's Favourite'. Chilean cultivars 'Bronceada' and 'Concha Lisa' have performed well in Australia, the former having been reported to possess a postharvest cold-storage life of 3 weeks. Pascual *et al.* (1993) report seven cherimoya cultivars in Spain, of which 'Fino de Jete' and 'Campa' are the most extensively cultivated, due to their superior yield and quality. Isozyme studies indicated that these two cultivars showed identical banding patterns for 15 enzymes, indicating that they were most probably the same cultivar. A cluster analysis of isozyme patterns showed that Spanish cherimoya cultivars were distinctly different from cultivars in California (Pascual *et al.*, 1993) and atemoya (Ellstrand and Lee, 1987).

Six atemoya cultivars have been described in Australia (Sanewski, 1991). 'Pink's Mammoth' is reported to have been introduced from British Guiana to Australia. It takes 6–7 years to begin producing commercial-size yields of fruit, which are large, weighing 800 g to as much as 2 kg, and it is a less precocious bearer than 'African Pride'. 'African Pride' was probably introduced into Australia from South Africa, although its origin may have been Israel. It produces good yields of small to medium-sized fruit. 'Bullock's Heart' (an

Table 3.1. Selected cultivars of cherimoya and atemoya.

Cherimoya		Atemoya	
Name	Origin	Name	Origin
Andrews	Australia	African Pride	Southern Africa or Israel
Bays	USA – California	Bradley	USA – Florida
Booth	USA – California	Gefner	Israel
Bronceada	Chile	Island Gem	Australia
Burton's Wonder	New Zealand	Kabri	Israel
Campa	Spain	Malalai	Israel
Concha Lisa	Chile	Nielsen	Australia
Cristalino	Spain	Page	USA – Florida
E-8	Ecuador	Pink's Mammoth	Australia
Fino de Jete	Spain		
Kempsey	Australia		
Libby	USA – California		
Lisa	USA – California		
Mossman	Australia		
Negrito	Spain		
Reretai	New Zealand		
White	USA – California		

atemoya) is described as being similar to 'Pink's Mammoth'. 'Island Gem' is an early-maturing, small-fruited, heavy yielder. Atemoya cultivars 'Bradley' and 'Page' are Florida cultivars. The former produces small-sized fruit with relatively smooth and thin skin. The latter produces medium-sized, well-shaped fruit with prominent skin segments. 'Gefner', an Israeli cultivar, is the main cultivar grown in Florida. It produces small to medium-sized fruit resembling 'Page'.

In order to develop cultivars adapted to cooler environments, Australia has concentrated on self-progenies and interspecific crosses of *A. cherimola* with *A. reticulata* and *A. diversifolia.* Progenies of *A. cherimoya—A. reticulata* crosses are late-maturing, showing flowering and fruiting characteristics of *A. reticulata*, which flowers in the autumn and has mature fruit in late spring. Four promising selections, possessing most of the fruit qualities of commercial cultivars, have been established in various areas for further evaluation (George *et al.*, 1992).

India and Taiwan have produced a few named cultivars of *A. squamosa* that are propagated vegetatively. 'Cuban Seedless' is a seedless cultivar with medium-sized fruit developed in Cuba; another Cuban cultivar is low in fibre content. The soursop is largely planted by seed. Soursop clones are separated into groups, such as acid and low acid or juicy and non-juicy types. Seedling populations have been established in Mexico and Malaysia to select superior clones with increased yields and improved processing qualities.

CULTURAL PRACTICES

Propagation

The annonas are usually propagated by seed (George and Nissen, 1987b). There is a rapid loss of seed viability (6 months) and seeds should be planted as soon as possible after removal from the fruit. Seeds can take up to 30 days to germinate and GA (10,000 p.p.m.) can significantly increase germination and enhance seedling growth. Atemoya, as a hybrid, has to be propagated by cuttings from selected trees. Seedlings require at least 3–4 years to bear fruit (Sanewski, 1991).

Clonal propagation by cuttings, layering, inarching, grafting and budding have been tried (Table 3.2). Inconsistent results have been obtained with cherimoya when 1 year cuttings are treated with rooting hormones. Cuttings from mature trees have not been difficult to root. There are cultivar differences in the rooting ability of atemoya, with 'African Pride' having a higher rooting response (60–80%) than 'Pink's Mammoth' and cherimoya (< 20%). Atemoya tip cuttings are superior to stem cuttings, with rooting percentages between 50 and 60% as compared with about 25% for stem cuttings (Sanewski, 1991). Time of cutting removal is crucial for success, cuttings taken at the end of the cool season showing the greater success. Roots should occur in 8–12 weeks and the cuttings are ready to pot in 16–20 weeks. Air layering can be used with some cultivars, although cherimoya is not propagated easily. A modification where the new shoot is clamped and only the shoot tip is exposed is successful. Inarching of *A. squamosa, A. cherimola, A. glabra* and atemoya to *A. reticulata* rootstock has been successful, with only *A. glabra* giving less than 70% success. Although inarching has given good results, it is time-consuming and costly for large-scale propagation (George and Nissen, 1987b).

Grafting is superior to budding in percentage takes and subsequent growth, with side-whip graft and cleft-graft techniques giving the best results (Duarte *et al.*, 1974). The branches should be defoliated 1–2 weeks before scion wood is cut to induce bud swelling. T-budding and chip-budding methods are successful. There are considerable graft incompatibilities among *Annona* (Table 3.3) and *Rollinia* species and types. Cherimoya has been found to be a vigorous rootstock for 'Pink's Mammoth' (atemoya), although atemoya is not compatible with *A. glabra, A. montana, A. muricata* and *A. reticulata* as rootstocks (Sanewski, 1991). This is complicated by cultivar differences in compatibility with common rootstocks. Atemoya cultivars 'Bradley' and 'Page' are compatible with custard-apple rootstocks but 'Gefner' shows partial incompatibility with the same rootstock (Table 3.3).

Table 3.2. Propagation methods and success for *Annona* spp. (after George and Nissen, 1987b). Root-cutting success rate is unknown.

Method	Species			
	Atemoya	Sweetsop	Cherimoya	Soursop
Seedling				
Genetically	Variable	Uniform	Variable	Uniform
Use	Not recommended	Good	Rootstock	High
Tip and stem cutting	Some cvs only	Some cvs only	Not successful	Successful
Micropropagation	Possibly high	Unknown	Unknown	Unknown
Layering	Unknown	High if modified techniques used	Unknown	Unknown
Air layering	< 5%	< 8.3%	< 5%	Unknown
Budding	> 70%	> 70%	> 70%	> 70%
Grafting	> 70%	> 70%	> 70%	> 70%

Table 3.3. Rootstock and scion compatibility of different annonas (after Sanewski, 1991).

Rootstock	Scion							
	Atemoya	Gefner	Pink's Mammoth	Page/Bradley	A. cherimola	A. glabra	A. muricata	A. squamosa
Atemoya	C				C	–	–	C
Gefner		C						
Pink's Mammoth			C					
Page/Bradley				C				
A. cherimola	C		C		C	C	–	C
A. glabra	N				–	C	C	–
A. muricata	N				–	C	C	N
A. palustris	–				–	–	–	N
A. reticulata	N	P	C		C	P	–	C
A. squamosa	C				C	–	–	C

C, compatible; P, partial; N, not compatible; –, unknown.

Field Preparation

A soil sample should be taken 4–6 months before planting to determine lime requirements and soil-nutrient levels. Soil phosphorus can also be adjusted at this time or in the planting hole. Minimal tillage can be achieved with a 2-m-wide band cultivated where the trees are to be planted. Drainage should be installed at this time to avoid flooding, with either contour or subsurface drains. Windbreaks should be established prior to transplanting. Bana grass can be used when shelter is required in the first 12 months.

Transplanting and Spacing

Transplanting should be done at the beginning of the wet season if there are seasonal dry periods and no irrigation facilities. In the subtropics, planting should not occur if there is a risk of frost. Plants should have attained a height of 30–46 cm at transplanting time, with the union of grafted or budded plants placed 15 cm or so above the ground. Trees should be irrigated as soon as possible after transplanting, with wind and sun guards sometimes required.

Transplanting distance depends upon the species of *Annona.* The periodically pruned small sweetsop can be spaced at 3.7 m × 4.6 m. In dry areas with less luxuriant growth, closer within-row spacing than 3.7 m can be considered. A closer spacing would increase humidity and benefit stigma-receptivity longevity. Cherimoya plantings at 5 m × 6 m within and between rows are adequate with appropriate pruning practices. Soursop trials in Hawaii showed that the spacing should be 4.6 m × 6 m without affecting growth or interfering with cultural practices. Row spacing of 5–10 m and between-row spacing of 7–12 m are recommended for atemoya in Australia, depending on cultivar rootstock, and pruning to an open goblet. Narrower spacing is used for 'African Pride' on *A. squamosa* rootstock and the widest spacing is for 'Pink's Mammoth' on *A. cherimola.* In Florida, narrow plant spacing of 4–6 m and row spacing of 6–7 m are used (Campbell, 1985). A triangular layout is recommended whatever planting distance is selected, with the rows running north–south to avoid shading.

Irrigation Practices

The annonas are grown in many areas without irrigation when rainfall is well distributed. Except for pond apple (*A. glabra*), most annonas can stand periods of drought and prefer rather dry conditions. There must be adequate soil moisture to encourage vegetative growth, since flowering occurs on new growth. Bearing atemoya trees may need up to 1440 L tree^{-1} every 4 weeks during the low-growth phase and from 500 to 750 L tree^{-1} every 3–5 days during flowering, fruit set and fruit growth (Sanewski, 1991). Reducing irrigation in late winter to force atemoya and cherimoya trees into dormancy for 1–2 months in spring is recommended in Australia and California, respectively. The amount and frequency of irrigation must be determined by experience in any particular location and soil type. Water stress should be prevented during flowering, fruit set and fruit development, as fruit are more sensitive than leaves (Fig. 3.9).

High soil moisture to increase humidity during the flowering season may prolong stigma receptivity and fruit set and growth. Low-rise sprinklers beneath the tree canopy during flowering can increase humidity. The stomata of annona respond to RH not water stress, and will continue to lose water if the humidity is greater than 80%, making maintenance of soil moisture crucial (Marler *et al.*, 1994).

Pruning

Training of trees should begin in the nursery and pruning should continue after transplanting. It is desirable to train the tree to a single trunk up to a

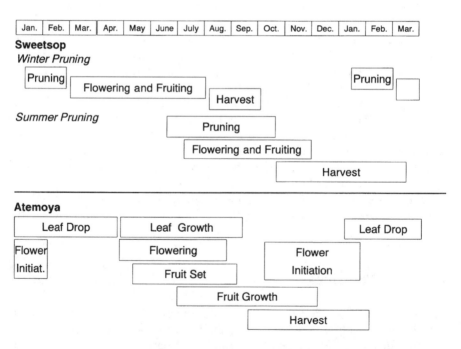

Fig. 3.9. Growth and management practices for sweetsop in Taiwan (Anon., 1995) and atemoya (Sanewski, 1991). In Taiwan, normal sweetsop pruning occurs in January and February and there are extended periods of pruning of selected trees in summer. The atemoya cycle has been converted to northern-hemisphere seasons for comparison purposes.

height of about 90 cm and then it should be headed back to produce lateral branches. The lateral branches should be spaced 15–25 cm above each other and be allowed to grow in different directions to develop a good scaffold. After about 2 m, they can be left to natural growth. Pruning is carried out when the trees are dormant and, in heavy trees, involves removal of lower limbs touching the ground and branches in the centre that may be rubbing against each other. The objective is to allow sunlight access to the centre of the tree (George and Nissen, 1986a).

All lateral buds can have up to two vegetative buds and three flower buds. Soursop and *A. reticulata* lateral buds are exposed in the leaf axil (Fig. 3.1). The lateral buds of atemoya, cherimoya and sweetsop are normally 'buried' (subpetiolar) in the base of the swollen leaf petiole (Fig. 3.2). Leaf shed must occur prior to the elongation of 'buried' buds (George and Nissen, 1987). Removal of leaves mechanically by stripping or chemically with urea or ethephon releases these buds. These buds are more suppressed in 'Pink's Mammoth' and 'African Pride' and may require leaf stripping in the summer. Adventitious buds can arise at any point on a trunk.

Sweetsop is pruned so that 1-year-old branches are cut back to 10 cm. The pruning leaves 120–150 branches per tree, with root pruning not being recommended. Flower initiation begins at the basal end of the growing branch (Lo, 1987). In Taiwan, normal pruning occurs in January/February with fruit harvest from July to September (Fig. 3.9). A summer pruning with fruit thinning (June–October) of selected trees can lead to harvesting fruit from October to the next March (Yang, 1987). Highest winter fruiting occurs when summer sproutings are pruned, compared with pruning unfruiting shoots or pruning in late May.

Atemoya produces the first flowers of a season and the most flowers on 1-year-old wood. The second and third flushes of flowers occur on new growth and produce most of the crop (Fig. 3.7). The flowers of the first flush may be removed, as they show poor fruit set and have greater problems with fruit splitting upon ripening. In south-east Queensland, each year the 1-year-old fruiting wood is shortened to about 10–15 cm with four to six buds (Sanewski, 1991). These pruned laterals are retained after producing fruit and allowed to become part of the framework, thus extending the canopy by 30–50 cm each year. However, dense shading within the canopy is a major problem. The open-goblet system is used to produce an open, spreading tree that will produce fruit early and avoid dense shading. The strategy is to produce one or two fruit on each shortened 1-year-old lateral. These laterals are removed completely after 2–3 years and replaced by new growth closer to the leader. The ideal seasonal growth of laterals producing fruit is about 60 cm long and about 10 mm in diameter at the base. Annual lateral growth over 100 cm is considered excessive for mature fruiting trees. This system allows the development of fruiting laterals along all the main limbs, with a moderately open centre although trees under this system rapidly grow tall. Pruning is normally carried out in spring before large amounts of bud growth have occurred, with a moderate summer pruning. Defoliation during the cool season to allow early development of the buried buds and more uniform flowering is sometimes carried out with an ethephon–urea mixture (Sanewski, 1991). Rejuvenation by heavy pruning is occasionally needed.

The soursop will, by nature, usually produce a symmetrically conical tree and is well adapted to the central-leader system. An alternative is to develop a mushroom-shaped tree that is topped at 2–2.6 m. The fruit in this system is borne on the lateral branches and hangs down for ease of harvesting. When properly trained, little pruning is required except to thin out poorly placed and weak branches. To contain trees within a certain space allocation and height limitation, the longest branches extending horizontally and vertically may be pruned annually, preferably immediately after harvest. Very severe pruning reduces subsequent fruiting.

Fertilization

The annonas have an indeterminate growth habit (axillary flowering) and applying nitrogen (N) in somewhat excessive amount does not interfere greatly with floral initiation, as is the case with plants having determinate growth habit. However, excessive tree vigour is usually associated with reduced flowering and yields in many tree crops and the atemoyas are no exception (George *et al.*, 1989).

In Australia, continued research and field observations of atemoya nutrition (Sanewski, 1991) have led to greater refinements in terms of quantity of fertilizer and times of incremental applications during the annual growth and fruiting cycles (Table 3.4). After 10 years of age, the annual amounts of N, phosphorus (P) and potassium (K) remain the same, as tree size is kept relatively constant by annual pruning and competition from adjacent trees. The annual requirements of N and K are split into four increments (Table 3.4). In the cool subtropical areas, greatest vegetative growth takes place during the warmer months from spring to autumn. Reduction in N during the winter minimizes new vegetative growth in young trees that are vulnerable to cold temperatures. This adjustment is not necessary in the warm tropics. Phosphate is given once per year during the early autumn application.

The use of foliar nutrient analysis has become a useful management tool in determining atemoya fertilizer programmes (Sanewski, 1991). Sampling for foliar analysis consists of obtaining the most recently matured leaf – the fourth or fifth leaf below the growing point. Sample leaves are selected from non-bearing shoots without a leaf flush, during late summer or early autumn (Table 3.5).

The primary sink for K in the atemoya is the fruit, rather than the leaves and thus there is a high requirement, with deficiency likely. About 60% of the K requirement is applied during the fruit-development period. Atemoyas also have a fairly high requirement for magnesium (Mg) and calcium (Ca). Heavy

Table 3.4. A guide to annual application of NPK for 'Pink's Mammoth' atemoya trees of different ages using straight fertilizers (g tree^{-1} year^{-1}) and percentage distribution of application of annual amounts (Sanewski, 1991).

Tree age (years)	Urea (g tree^{-1} year^{-1})	Superphosphate (g tree^{-1} year^{-1})	Potassium chloride (g tree^{-1} year^{-1})
2	400	500	360
4	860	550	930
6	1300	780	1170
8	1600	880	1500
10	1750	880	1650

Fertilizer	Early spring	Early summer	Early autumn	Late autumn
Urea	20%	30%	40%	10%
Superphosphate			100%	
Potassium chloride	10%	30%	40%	20%

NPK, nitrogen, phosphorus, potassium.

Table 3.5. Tentative leaf nutrient standards for atemoya (Pink's Mammoth) in Queensland, Australia presented as a guide (Sanewski, 1991).

Nutrient	Acceptable range
Nitrogen	2.5–3.0%
Phosphorus	0.16–0.2%
Potassium	1.0–1.5%
Calcium	0.6–1.0%
Magnesium	0.35–0.5%
Sodium	< 0.02%
Chloride	< 0.3%
Manganese	30–90 p.p.m.
Copper	10–20 p.p.m.
Zinc	15–30 p.p.m.
Iron	50–70 p.p.m.
Boron	15–40 p.p.m.

vegetative growth during the fruit-development period competes for nutrients, such as boron and Ca, resulting in boron-deficient fruit. Boron-deficient fruit develop hard, brown lumpy tissues around the central core. Deficiency in boron and, to some degree, Ca is considered a causal factor for these lumps (Cresswell and Sanewski, 1991). Applying boron at a slight excess can be phytotoxic, especially in sandy soils. A desirable practice is to use organic fertilizers, with inorganic fertilizer as a supplement to maintain a balance and to control cropping (Sanewski, 1991).

Observations in Hawaii and Mexico of soursops have indicated the desirability of providing 1.3 kg of a triple-15 fertilizer formulation during the first year of production, split into two applications. In Hawaii, the first increment should be given around February for the primary crop in July and the second increment applied in August for the December–January secondary crop. Each year thereafter, up to approximately the sixth bearing year, the total amount can be increased by approximately 0.45 kg tree^{-1} year^{-1}.

Pest Management

Diseases

A number of diseases have been reported in the literature (Table 3.6). Anthracnose, caused by *Colletotrichum gloeosporioides* (*Glomerella cingulata*), is the most serious on soursop, particularly in areas of high rainfall and atmospheric humidity and during the wet season in dry areas (Alvarez-Garcia, 1949; Dhingra *et al.*, 1980). This disease causes twig dieback, defoliation and dropping of flowers and fruit. On mature fruit, the infection causes black lesions. Another severe fruit rot disease is attributed to *Gliocladium roseum* on sweetsop only and affects 20–90% of the fruit in India. Symptoms consist of water-soaked spots, which turn soft and brown.

Black canker (*Phomopsis anonacearum*) and diplodia rot (*Botryodiplodia theobromae*) occur mostly on neglected trees and cause similar symptoms of

Table 3.6. Major diseases of annonas.

Common name	Organism	Parts affected, symptoms	Region or country
Anthracnose	*Colletotrichum gloeosporioides* (*Glomerella*)	Flowers, fruit, leaves, dieback, seedling damping off	Universal
Armillaria root rot	*Armillaria leuteobubalina*	Roots, base of trees, decline	Australia
Bacterial wilt	*Pseudomonas solanacearum*	Tree wilt	Australia
Black canker (diplodia rot)	*Botryodiplodia theobromae*	Leaf scorch, twig dieback	Universal
Black canker	*Phomopsis*	Same effects	Australia
Purple blotch	*Phytophthora palmivora*	Spots on immature fruit, fruit drop, twig dieback	Australia
Rust fungus	*Phakopsora cherimoliae*	Leaves	Florida
Fruit rot	*Glioclacium roseum*	Fruit	India

purplish to black lesions, resulting in mummified fruit. Marginal leaf scorch is also caused by *P. anonacearum* and *B. theobromae* and causes twig dieback. Diplodia rot has darker internal discoloration and deeper, more extensive corky rot in fruit. *Cylindrocladium* fruit and leaf spot is caused by a soil-borne fungus, *Cylindrocladium colhounii*. It can cause almost total loss of fruit during years of persistent heavy rains. Symptoms begin with small dark spots, primarily on the shoulders of the fruit, which spread along the sides, enlarge, become dry and crack. Infection is skin-deep but fruit becomes unmarketable. Control measures recommended are good orchard maintenance with heavy mulching and lower-branch pruning to prevent splashing of soil during heavy rainfall (Sanewski, 1991).

Bacterial wilt of atemoya is caused by *Pseudomonas solanacearum* and is characterized by rapid wilting and death of young trees and slow decline of old trees. There is a general decline of vigour and defoliation on affected limbs. Vascular discoloration of woody tissues occurs in the roots and up to the trunk at ground level. It has caused up to 70% tree death in 12 years in orchards using *A. squamosa* rootstocks in Queensland.

Insect pests

Some insect pests occur in numerous growing areas (Table 3.7). One of the most serious insects in Trinidad is the *Cerconota* moth, which lays its eggs on young fruit. The emerging larvae tunnel into the pulp, causing blackened, necrotic areas. It is not uncommon to find every fruit larger than 7.5 cm infested. Bagging the fruit is sometimes done. This moth has been reported in the American tropics as far south as Brazil and is a major limiting factor in Surinam.

The *Bephrata* wasp is widely distributed throughout the Caribbean, Mexico and Central and northern South America. This wasp is considered to be the most important pest in Florida (Campbell, 1985). Considerable damage to the soursop fruit has been observed in Mexico by the senior author. The larvae infest the seeds and cause damage to the pulp, as they bore through the flesh to emerge when the fruit matures. The *Thecla* moth is widespread through parts of the Caribbean and in the American tropics, but it is not considered to be as serious as the *Cerconota* moth and *Bephrata* wasp. Primary damage is to the flowers. The larvae feed on flower parts, such as the perianth, stamen and stigmas, with the flowers failing to set fruit.

The banana-spotting bug (*Amblyphelta lutescens*) and the fruit-spotting bug (*Amblyphelta nitida*) are considered to be serious atemoya pests. The banana-spotting bug is reported to be confined to northern Queensland, with both being found in southern Queensland. The bugs cause small black 2–10 mm spots on the shoulders of young fruit and penetrate about 1.0 cm into the fruit. The damage resembles the symptoms of diplodia rot (black canker) (Sanewski, 1991).

Mature green annonaceous fruits have been shown to be rarely infested

Table 3.7. Major insect pests of annonas.

Common name	Organism	Parts affected	Country/region
Bephrata wasp (soursop wasp)	*Bephrata meculicollis*	Fruit	Mexico, Americas, Trinidad, Surinam
Wasp	*Bephratelloides paraguayensis*	Fruit	Americas, Barbados
Cerconota moth (soursop moth)	*Cerconota anonella*	Fruit	Americans, Trinidad, Surinam
Thecla moth	*Thecla ortygnus*	Flower, young fruit	Americas, Caribbean
Banana spotting	*Amblyphelta lutescens*	Young fruit	Queensland
Mealy bug	*Dysmicoccus* spp.	Stem, leaves	Universal
Citrus mealy bug	*Planocuccus citri*	Fruit	Queensland
Southern stink bug	*Nezara viridula*	Fruit	Caribbean
Caribbean fruit fly	*Anastrepha suspensa*	Fruit	Caribbean, Mexico
Queensland fruit fly	*Dacus tryoni*	Fruit	Australia
Potato leaf hopper	*Empoasca fabae*	Leaves	Caribbean
Red spider mite	Several genera, species	Leaves, flowers	American tropics
Scale insects	*Saissetia coffeae*	Leaves, stem	Universal
Coconut scale	*Aspidiotus destructor,* other genera and species	Leaves, stem	Caribbean

by the Mediterranean fruit fly (*Ceratitis capitata*) and Oriental fruit fly (*Dacus dorsalis*), but they are found on occasion in tree-ripened fruit. In Australia, the Queensland fruit fly (*Batcrocera tryoni*) infests ripening atemoya fruit. 'African Pride' appears to be more susceptible than 'Pink's Mammoth'. Use of bait sprays and field sanitation are recommended measures to minimize fruit-fly infestation (Smith, 1991). Fruit bagging also provides protection.

Mealy bugs and various species of scale insects are found universally and usually become a serious pest on neglected trees. The former is reported to be a major pest on marketable fruit in some areas of Australia (Sanewski, 1991). Red spider mites can become a serious problem in dry areas or during dry seasons. Heavy infestations have been observed on soursop flowers and leaves in the Tecoman area of Mexico during the prevailing dry period, with trees showing heavy flower drop.

Weed Management

Problem weeds, especially grasses and twining weeds, should be controlled before planting by cultivation and herbicides. Young trees should be protected from weed competition by hand-weeding, mulching or contact herbicides. The shallow root systems limit the use of cultivation under the tree. A translocated herbicide may be needed for perennial weeds and is applied as a spot spray.

HARVESTING AND POSTHARVEST HANDLING

Harvesting Season, Yield and Harvesting

The harvesting season is quite similar in most areas, especially for soursop and sweetsop, differing only in range (Table 3.8). Mean monthly soursop production in Hawaii shows a year-round production (Fig. 3.4) with two peaks. Sweetsop is usually harvested during the summer months, from July to November, and can be extended to March (Fig. 3.9). Atemoya is harvested in autumn in the southern hemisphere and winter in Australia (Fig. 3.9) and from October to January in Poona, India.

A major problem in *Annona* cultivation is obtaining commercial yields. In order to increase yield, hand-pollination (Fig. 3.6) has become an important aspect of cultivation practices in some areas, especially for atemoya. Use of pollen stored under poor conditions, naturally defective pollen and pollinating during the hottest period of the day result in very poor fruit set.

Rootstocks have been shown to greatly influence yield (Sanewski, 1991). Trees of the same age and cultivar on cherimoya rootstock yield almost double that of trees grafted on sweetsop. Cultivar differences in yield have also been noted (Fig. 3.10). Estimated yield per tree of 6-year-old 'African Pride' on cherimoya rootstock is about 90 kg, while that of 'Pink's Mammoth' of the same age and on the same rootstock is about 30 kg.

Table 3.8. Peak harvesting seasons for some *Annona* species.

Countries	Soursop	Sweetsop	Cherimoya	Reticulata	Atemoya
Australia					Mar.–Sept.
Argentina			Feb.–July		
California			Mar.–Apr.		
Caribbean	Year-round	June–Sept.		Feb.–Apr.	
Chile			Aug.–Dec.		
Florida	June–Nov.	July–Sept.	July–Oct.	Feb.–Apr.	July–Sept.
Hawaii	Jan.–Oct.				Aug.–Nov.
India (Poona)		Aug.–Nov.	Nov.–Feb.		Oct.–Jan.
Indonesia	Year-round				
Mexico	June–Sept.				
The Philippines	June–Aug.				
Portugal (Madeira)			Nov.–Feb.		
Puerto Rico	Mar.–Sept.				

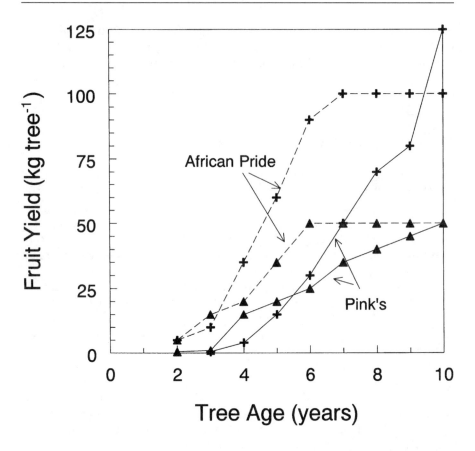

Fig. 3.10. Approximate yields of two atemoya cultivars, 'Pink's Mammoth' and 'African Pride' on cherimoya (+) and sweetsop (▲) rootstocks with tree age (after Sanewski, 1991).

Soursop yields in Hawaii from trees grown in a marginal field have shown approximately 43 kg tree^{-1} on 4-year-old trees, increasing to 83 kg tree^{-1} on 6-year-old trees. At Paramaribo, Surinam, soursop yields of 54 kg tree^{-1} at 278 trees ha^{-1} are reported.

Fruit is harvested when fully mature and firm. The skin colour changes as the fruit approaches maturity. The immature soursop is dark green and shiny, losing its sheen and becoming slightly yellowish-green upon reaching maturity. Skin of immature cherimoya and sweetsop is greyish green but turns to yellow-green at maturity. In the latter fruit, adjacent carpels towards the blossom end commonly separate, exposing the white pulp ('creaming'). 'Creaming' between segments of the atemoya fruit is considered to be an indication of maturity. Fruit are harvested when 40% of the surface shows 'creaming' (Sanewski, 1991). Determining harvest time by dating floral anthesis is impractical as flowering occurs over many months. If a rigid hand-pollination protocol is used, with removal of naturally pollinated fruit, days from anthesis can be used.

Fruit is hand-harvested and put into lug boxes or baskets. Harvesting is more difficult and time-consuming for soursop, because the trees are generally taller than other annonas and the fruit is much larger. In large soursop orchards, mechanical harvesting aids are feasible and accelerate handling. Atemoyas are harvested every 3–7 days, with experienced pickers harvesting from 150 to 180 kg of fruit h^{-1}.

Postharvest Handling

Harvested fruit should be handled with care to prevent bruising of the skin. This is especially important for fruit that are marketed for fresh consumption, such as cherimoya, sweetsop and atemoya. Firm soursop fruit needs to be held after harvest for 4–7 days at room temperature before softening begins, optimum quality processing occurring 5–6 days after softening begins (Paull, 1983). The skin of ripening soursop will gradually turn dark brown to black, but the flesh is unspoiled. Fruit of some cherimoya cultivars can be kept for 7–10 days at 17°C. Normal ripening occurs at temperatures between 15 and 30°C, and storage temperatures below 15°C cause chilling injuries and a failure to develop full flavour. Storage conditions recommended for sweetsop are 15°C, with an RH of 85–90%. Atemoya fruit can be stored up to 2 weeks at 15–16°C or 1 week at 10–12°C; fruit stored at 0–5°C turn black. At lower temperatures, skin discoloration occurs rapidly. Precooling of fruit is essential to help extend shelf-life.

Except for the atemoya in Australia, there are no reports of established grade standards. Australian standards specify mature atemoya fruit 75 mm in diameter, firm with 'creaming' between segments on the skin. Containers are 0.5 bushel (8–10 kg) in size, made of wood, fibreboard or polystyrene (450 mm × 215 mm × 180 mm), well ventilated and marked with 'custard-apple' and the number of fruit. The presence of soft fruit and even one fruit-fly-damaged fruit can lead to rejection of the consignment.

At processing plants, soursop fruit are stored on racks in the shade and inspected daily. All fruit found to yield to finger pressure are removed for processing. Slightly immature fruit will ripen but they lack the full flavour and aroma and nectars prepared from the purée of such fruit have a flat taste. Pulp extraction is an important economic factor for processed fruit, with a high recovery being desirable. In soursop, pulp-recovery percentages ranged from 62 to 85.5% (Paull, 1982). Differences in recovery percentages are due to differences in equipment, extraction methods, cultivar and cultural practices, including environmental influences. The number of seeds per fruit also influences pulp recovery.

Compositional Changes During Fruit Ripening

All annonas are climacteric fruit. Soursop respiration begins to increase within a day after harvest and reaches its peak at the sixth to eighth day. Ethylene production is initiated approximately 48 h after initiation of respiration rise and reaches its peak at about the same time as the respiration peak reaches a plateau (Paull, 1983). Total soluble solids increase from around 10° to 16° Brix during the 3 days of ripening. The major titratable acids are malic and citric acids. After day 5–6, titratable acidity, ethylene production and total phenols decline, changes that produce a bland flavour and even a slightly objectionable odour. Days 6 and 7 are considered to be the optimum edible stage and coincide with the peak of ethylene production (Paull, 1982).

'African Pride' atemoya respiration reaches a peak about 3 days after harvest and the eating stage is reached in another 2 days. Total time from harvest to eating ripeness is about 5 days at 20°C.

UTILIZATION

Cherimoya is a fair to good source of vitamins, while sweetsop is a good source of P, thiamine and ascorbic acid (Table 3.9). Atemoya is also a good source of ascorbic acid, while soursop is a good source of K, riboflavin and niacin.

The annonas are usually consumed as dessert fruit. The perishable nature and supply shortage make marketing localized or air shipment essential. The soursop is marketed fresh in local markets. This fruit, of all the annonas, has the best processing potential because of the excellent flavour characteristic of the pulp and high recovery from large fruit. Unfortunately, soursop has to be hand-peeled and cored, an expensive and time-consuming operation. The fragility of the skin and the fruit's irregular shape and softness limit machine processing.

Soursop pulp is too viscous and requires proper dilution to produce a desirable nectar; however, this product is flat and weak. To produce a desirable nectar, the pH needs to be adjusted to 3.7 by addition of citric acid and sugar to 15° Brix to create a desirable balance between acidity, sweetness and flavour. Unsweetened and sweetened soursop pulp processed below 93°C shows no changes in organoleptic properties, although freeze preservation produces a higher-quality product. Enriched pulp, sweetened or unsweetened, can be processed and stored frozen for remanufacture as various products or reconstituted directly by the consumer. Purée can be used to prepare iced soursop drink or mixed with other juices, or it can be made into sherbets and gelatin dishes.

Table 3.9. Proximate fruit composition of cherimoya, sweetsop, soursop and atemoya in 100 g edible portion (after Wenkam, 1990).

Nutrients	Cherimoya	Sweetsop	Soursop	Atemoya
Proximate				
Water	68.71 g	75.97 g	80.1 g	78.7 g
Energy (kJ)	460	360	247	310
Protein	1.54 g	1.89 g	0.69 g	1.4 g
Lipid (fat)	0.13 g	0.57 g	0.39 g	0.6 g
Carbohydrate	28.95 g	20.82 g	18.23 g	15.8 g
Fibre	–	1.41 g	0.95 g	2.5 g
Ash	0.67 g	0.75 g	0.58 g	0.4 g
Minerals				
Calcium	9 mg	17 mg	9 mg	17 mg
Iron	0.25 mg	0.30 mg	0.82 mg	0.3 mg
Magnesium	–	22 mg	22 mg	32 mg
Phosphorus	24 mg	54 mg	29 mg	–
Potassium	–	142 mg	320 mg	250 mg
Sodium	–	2 mg	22 mg	4 mg
Vitamins				
Ascorbic acid	12.20 mg	35.9 mg	16.40 mg	43 mg
Thiamine	0.11 mg	0.10 mg	0.07 mg	0.05 mg
Riboflavin	0.11 mg	0.06 mg	0.120 mg	0.08 mg
Niacin	1.0 mg	0.89 mg	1.520 mg	0.8 mg
Vitamin A	0	0	0	–
Seed/skin %	35.0	45.0	34.0	28.0

FURTHER READING

Anon. (1995) *Production and Marketing of Sweetsop*. Special Publication No. 1, Taidong Agricultural Improvement Station, Taiwan, 48 pp. (in Chinese).

George, A.P. and Nissen, R.J. (1987) Propagation of *Annona* species: a review. *Scientia Horticulturae* 33, 75–85.

George, A.P., Nissen, R.J. and Brown, B.T. (1987) The custard apple. *Queensland Agricultural Journal* 113, 287–297.

Marler, J.E., George, A.P., Nissen, R.J. and Andersen, P.J. (1994) Miscellaneous tropical fruits – annonas. In: Schaffer, B.C. and Andersen, P.C. (eds) *Handbook of Environmental Physiology of Fruit Crops*, Vol. II, *Subtropical and Tropical Crops*. CRC Press, Boca Raton, Florida, pp. 200–206.

Merodio, C. and De La Plaza, J.L. (1997) Cherimoya. In: Mitra, S.K. (ed.) *Postharvest Physiology and Storage of Tropical and Subtropical Fruits*. CAB International, Wallingford, UK.

Sanewski, G.M. (ed.) (1991) *Custard Apples – Cultivation and Crop Protection*. Information Series QI90031, Queensland Department of Primary Industry, Brisbane.

4

AVOCADO

Introduction

The laurel family (*Lauraceae*) is composed of about 47 genera with 2000–2500 species. They are mostly evergreen trees and shrubs, occasionally aromatic, and native mostly to tropical and subtropical regions. There are about 150 species of tropical evergreen trees, many of which are cultivated as ornamentals for their laurel-like leaves.

Important Genera and Species

The genus *Persea* Mill. is the most well known for the fruit called avocado or aguacate, *Persea americana* Mills. (*Persea gratissima* C. F. Gaertn.). The fruit of most *Persea* species are small and worthless. Only *P. americana* and *Persea schiedeana* Ness. bear large fruit, the latter species being watery and fibrous but pleasant in flavour and eaten by people in its native habitat in Mexico and Central America. Other species of commercial ornamental value include *Cinnamomum camphora* (L.) Nees and Eberm., camphor tree; *Cinnamomum zeylanicum* Breyn., cinnamon tree; *Laurus nobilis* L., bay or sweet bay; and *Umbellularia californica* Nutt., California bay tree.

Origin and Distribution

There is general agreement that the centre of origin of the *Persea*, including the avocado, is in the highlands of central and east central Mexico and in the adjacent highland areas of Guatemala. Early European travellers during the sixteenth century found avocado in cultivation and distributed throughout Central America and northern South America. This is evidenced by the native

76

names given to avocado in many languages and by archaeological findings. Carbon datings indicate that Mexican avocados were used as food as early as 9000–10,000 years ago (Williams, 1976). Early accounts indicate that it was not cultivated in the Caribbean islands during the pre-Columbian period and was introduced to Jamaica by the Spaniards in about 1650. Distribution to the African and Asian tropics occurred during the 1700s and 1800s. The first recorded importation into Florida was in 1833, into California in 1848 and into Hawaii during the early nineteenth century. By 1855, avocado trees were common in gardens of Oahu and were distributed to the other islands of the Hawaiian chain (Yee, 1978). It is now widely distributed throughout the tropics and subtropics but the use of the fruit differs in different areas. Although the avocado has been available in most South-east Asian countries, it has not been cultivated widely due to a preference for many other fruit.

Three horticultural races of avocado are recognized and reflect geographical areas of origin, based on abundance of each race in cultivation (Popenoe, 1920). The West Indian race is not native to the West Indies but rather to the lowlands of Central America (Fig. 4.1) and possibly to northern South America. It is tropical in ecological requirements and characterized by producing small fruit with medium thin, leathery skin, low oil content, loose seed and maturing 160–240 days after flowering (Table 4.1). The closely related Guatemalan race is postulated to have originated at somewhat higher elevations in Guatemala and adjacent areas, based on abundance of wild populations (Fig. 4.1). Leaves of cultivated types of this race are not anise-scented; Popenoe (1920) reported anise-scented wild forms. The

Table 4.1. Comparison of selected characteristics of three horticultural races of avocado (Bergh, 1975; Bergh and Ellstrand, 1986).

Trait	West Indian	Guatemalan	Mexican
Climate	Tropical	Subtropical	Semitropical
Cold tolerance	Least	Intermediate	Most
Salt tolerance	Most	Intermediate	Least
Leaf anise	Absent	Absent	Present
Leaf colour	Pale yellow	Green with red tinge	Green
Fruit bloom to maturity	5 months	12 months or more	6 months
Size	Variable	Variable	Small
Colour of fruit	Green or reddish	Green	Often dark
Skin thickness	Medium	Thick	Very thin
Skin surface	Shiny	Rough	Waxy bloom
Seed size	Variable	Small	Large
Seed cavity	Variable	Tight	Loose
Oil content	Low	High	Highest
Pulp fibre	Less common	Less common	Common
Pulp flavour	Sweeter, milder	Rich	Anise-like, rich

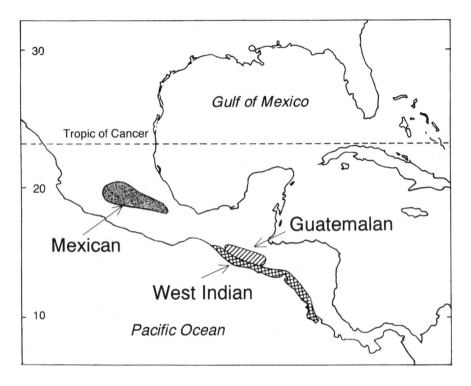

Fig. 4.1. Possible evolutionary centres of the three cultivated avocado races (redrawn from Scora and Bergh, 1990).

Mexican race is predominantly found in the higher elevations of Mexico (Fig. 4.1). Leaves of the Mexican race are anise-scented and 60% of the essential oil is the monoterpene estragol (Scora and Bergh, 1992). These races are recognized as subspecies *P. americana* var. *americana* (West Indian), var. *drymifolia* (Mexican) and var. *guatemalenis* (Guatemalan) and are based upon biochemical and isozyme data (Bergh and Ellstrand, 1987). Deoxyribonucleic acid (DNA) restriction fragment variation adds to the complexity of the taxonomy of this genera (Furnier *et al.*, 1990). Due to the outbreeding nature of these taxa and human selection and cultivation, there are many interracial hybrids and some of the principal commercial cultivars are of hybrid origin.

ECOLOGY

Ecological requirements of avocado should be viewed in terms of the areas of geographical origin. It is generally considered to be a subtropical plant, except for the tropical West Indian race. Regardless of origin, the avocado has shown adaptability to a wide range of ecological conditions, from the tropics to approximate latitudes of 30° north and south. This wide distribution may be due to the broad genetic differences of the three horticultural races.

Soil

Avocado is grown in a wide variety of soil types. Deep soils of volcanic origin, sandy loam soils, calcareous soils and other soil types have supported good growth. Soil pH may range from 5 to around 7. Since avocado is highly susceptible to root rots, good drainage is crucial and a high water-table undesirable. Trees show dieback in parts of fields when the water table is less than 1 m.

Avocado has little tolerance for saline conditions. Kadman and Ben-Ya'acov (1970a) produced various degrees of leaf burn by irrigating an avocado orchard with water containing 150–170 mg chlorine L^{-1}. It has also been shown that West Indian seedlings are more tolerant to salt than Mexican seedlings, with Guatemalan having an intermediate tolerance (Table 4.1).

Climate

Rainfall

Most cultivars are sensitive to water stress and to excess moisture caused by poor drainage. In Hawaii, avocado trees have grown well under annual rainfall of 3125 mm, largely due to the excellent drainage provided by the 'a'a' soil (crushed lava rock). Generally, a moderate rainfall range between 1250 mm and 1750 mm per annum with good distribution is desirable. Almost all avocado-growing areas have wet and dry periods, necessitating some form of supplemental irrigation.

Avocado inflorescences are not damaged by moderate amounts of rain for short periods, although relatively dry conditions are preferred during flowering. Avocado roots are shallow and prolonged dry conditions during the critical periods of flowering and fruit set can cause flower and young-fruit drop. Among the three races, West Indian cultivars are more adapted to high summer rains, while the Mexican races possess greater tolerance to water stress and lower humidity.

Temperature

Humans have attempted to broaden the range of the three races for economic reasons by cultivation in areas with seasonally adverse conditions. The West Indian race is best adapted to a humid, warm climate and monsoon rains with optimum temperatures around 25–28°C (Table 4.1). Higher temperatures depress photosynthesis, thus lowering yields. This race is susceptible to frost, and has a foliage tolerance to a minimum temperature of 1.5°C. Mature trees of Mexican cultivars have been observed to tolerate temperatures as low as

−4 to −5°C without damage to the foliage and wood, although flowers are damaged. The Guatemalan race is adapted to a cool tropical climate but is less tolerant of low temperatures than the Mexican race (McKellar *et al.*, 1992). Cultivars have shown tolerance to light frost down to −2°C, but flowers are damaged by even light frost (Bower, 1981a).

Night temperatures of 15–20°C and day temperatures of *c.* 20°C are the most suitable temperatures for floral development, pollen-tube growth and embryo development (Table 4.2). Generally, high humidity, exceeding 50% is desirable, especially during flowering and early fruit set. The Mexican–Guatemalan hybrids, such as the 'Fuerte', have shown wider temperature tolerance to cold than those of the Guatemalan race. Vegetatively, the 'Fuerte' survives temperatures above −4°C, although the flowers are damaged and fruit set can be very poor at 17/12°C (day/night). Temperatures of 17/12°C and 33/28°C during flowering time can prevent pollen-tube and embryo growth, resulting in the production of unfertilized, underdeveloped fruit. This disruption of fruit set, with endosperm and embryo development not being observed at low temperatures (17/12°C) is more marked in type B flowers (cv. 'Fuerte') where no female flower opening occurs (Sedgley and Grant, 1983). The flowering cycle is extended by cool temperatures from the usual 36 and 20 h for types A and B, respectively. Growing temperatures of 25–30°C during the day and night temperatures of 15–20°C are considered optimum (Table 4.2). A much cooler temperature range during winter will stimulate flowering (Whiley, 1984).

Solar radiation and photoperiod

Day length is apparently of little importance, as there have not been any published studies on avocado responding to photoperiod.

Wind

The avocado tree is easily damaged by winds, due to its brittle branches. Moderately high winds can cause severe damage. If orchards are not located in naturally sheltered areas, windbreaks are advised.

Table 4.2. Effect of growing temperature (day/night) on vegetative growth and flowering of 'Hass' avocado (after Chaikiattiyos *et al.*, 1994).

Temperature	Percentage of terminal shoots		
	Panicle	Vegetative	Dormant
15/10°C	35	0	65
20/15°C	20	0	80
25/20°C	0	100	0
30/25°C	0	100	0

GENERAL CHARACTERISTICS

Tree

The avocado tree is variable in shape, from tall, upright trees to widely spreading forms with multiple branches. Trees can attain heights of 15–18 m, with manageable height being controlled by pruning. Although classified as an evergreen, some cultivars shed leaves during flowering, which are replaced rapidly from terminal shoots (Plate 7). Others shed their leaves gradually, so that they are never without leaves completely. The dark green leaves are spirally arranged and variable in size from 10 to 13 cm by 20 to 25 cm long, entire, elliptic or ovate to lanceolate (Fig. 4.2).

New growth occurs in flushes (Fig. 4.3). Flushes of shoot and root tend to alternate on a 30–60 day cycle (Ploetz *et al.*, 1991). Root growth continues throughout the year in subtropical areas, even at a low rate between the flushes, while shoot growth may stop. The period between flushes varies with location (Fig. 4.3) and cultivar. Growth flushes in summer tend to be asynchronous, with only a portion of a tree canopy being involved. Trunk starch concentration ranged between 4.5 and 6.3% during active tree and fruit growth. In trees where the fruit are not harvested until full maturity with 30% dry matter, a trunk starch concentration of 5.3% occurs prior to flowering versus 7.7% in trees harvested earlier (Kaiser and Wolstenholme, 1994).

The juvenile period in avocado can be from 5 to 15 years. Girdling in early autumn 3 years after planting can significantly increase flowering and fruit set. 'Pinkerton' and 'Gwen' have precocious offspring, implying a genetic basis for juvenility modified by environment (Lavi *et al.*, 1992).

Inflorescence and Flower

The small, pale green or yellowish green flowers are borne on multibranched axillary panicles terminating in a shoot bud (Fig. 4.2). One or two million flowers may be produced in a single flowering period, although only about 200–300 fruit mature (Whiley *et al.*, 1988b). This flowering leads to considerable water loss and the recommendation is to irrigate during this period. The flower is bisexual with nine stamens, six of which form the outer circle and three in the inner circle. At the base of the inner circle are located a pair of nectaries, alternating with three staminodes, which also secrete nectar. Each stamen has four pollen sacs, which release cohesive pollen. The single pistil contains one carpel and one ovule (Bergh, 1976).

The avocado flower has a unique flowering behaviour (Fig. 4.4) and all avocado cultivars and seedlings, irrespective of race, fall into one of two complementary groups, designated A and B (Table 4.3). Flowers of the A class open in the morning for 2–3 h functioning as females with a white stigma,

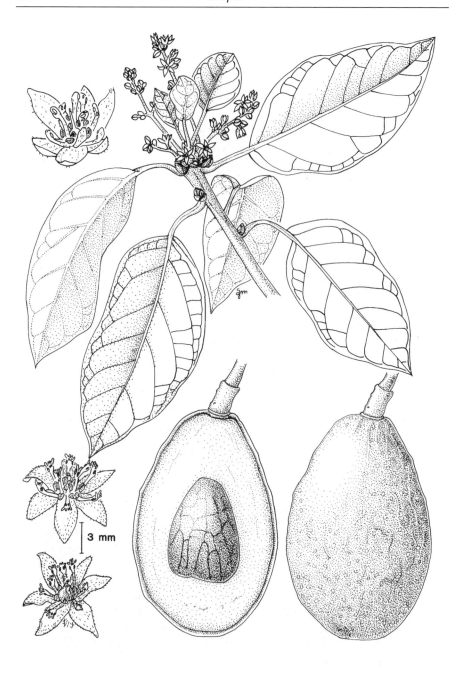

Fig. 4.2. Flower, leaf and fruit of avocado.

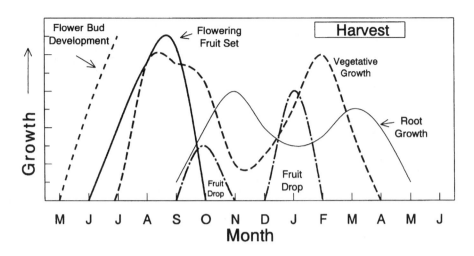

Fig. 4.3. Phenological cycle for bearing cv. 'Fuerte' in Queensland, Australia, showing the development and interaction between root, shoot, flower and fruit growth. The cycle will vary with the cultivar and needs to be determined for each location (redrawn from Whiley *et al.*, 1988a).

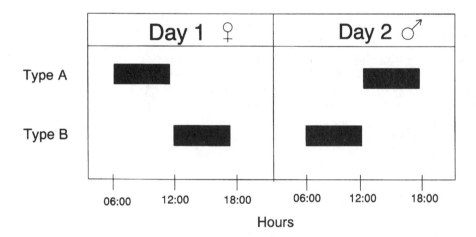

Fig. 4.4. The protogynous dichogamy patterns of flower opening involves two types, A and B, with both types found in the three ecological races. This behaviour enhances cross-pollination. These patterns may be disrupted by low temperatures and some overlap occurs.

while the stamens remain closed. These flowers close at approximately noon and reopen the following day during the afternoon hours for 3–4 h, functioning now as males. The stigmas are no longer functional. Flowers of the B class open in the afternoon as females, the stamens remaining closed. These flowers close in the evening and reopen the next morning as male

Table 4.3. Classification of selected avocado cultivars according to flower types.

Type A	Type B
Hass	Bacon
Simmonds	Fuerte
Guatemala	Zutano
Ruehle	Gwen
Nadir	Nabal
Lula	Regina
Rincon	Ruehle
Choquette	Hardee
Rodrigues	Sharwil

flowers. This phenomenon is called protogynous, diurnally synchronous dichogamy (Bergh, 1969). The dichogamy is protogynous, as the pistil matures before the stamens in both classes, and all flowers of a class are synchronized to be functionally female at one time of the day and functionally male at another time. Type B cultivars ('Fuerte') are generally less productive than type A ('Hass') in cool flowering conditions. This is due to greater disruption of the flowering cycle by low temperature during flowering of type B, where most flowers would be male with no female stage (Sedgley and Grant, 1983).

Flowering occurs from late autumn to spring, depending upon the cultivar and prevailing temperatures (Fig. 4.3). West Indian and Mexican cultivars generally may start to flower early, in October or November and October–December, respectively, north of the equator. The West Indian avocado is called the summer avocado in Hawaii. Guatemalan cultivars and many Guatemalan × Mexican hybrids begin to flower around March, extending to May, and are called winter avocado in Hawaii.

Pollination and Fruit Set

The protogynous, diurnally synchronous dichogamy in avocado flowers is the normal behaviour and occurs if warm weather prevails during flowering. Synchronous dichogamy of flowers of the same cultivar restricts self-pollination and encourages cross-pollination between those of complementary groups. Exposure to sun, wind and temperature changes can cause variability in the time of floral anthesis, prolonging the open period and time between the two open periods and increasing the overlap of female and male opening on a single tree. However, with moderate day temperatures (25°C) and cool nights (20°C), there is sufficient overlap of flower types to permit self-pollination of up to 93%. The possibility of self-pollination is enhanced by relatively long pollen viability. There is also evidence that the stigma remains white and is still receptive on the second day (Davenport *et al.*, 1994). In spite of the fact that some studies have shown relatively high rates of self-pollination, most investigators have advocated interplanting of pollinator cultivars of the complementary class in orchards in order to increase fruit set (Bower, 1981c). The avocado can be insect-pollinated, although self- or wind pollination is possibly the norm (Davenport *et al.*, 1994). Insects can carry the sticky pollen on their bodies and are capable of depositing viable pollen from one flower cycle to another within the same cultivar.

Some cultivars are more effective in increasing fruit set than others. This may indicate the presence of various degrees of incompatibility among cultivars, with some complementary cultivars superior to others. Mexican cultivars are more efficient pollenizers in California and Israel, while Guatemalan cultivars are more effective in Florida because of greater cold tolerance of pollen improving the chances in the fertilization process (Gazit, 1976). Under warmer conditions, West Indian hybrids are good as pollen donors, probably due to their greater tolerance of higher temperatures. However, fruit set may be a poor measure of final matured fruit, as all cultivars lose flowers and developing fruitlets, regardless of pollination.

Following pollination, considerable fruit drop occurs, due to poor pollination and excessive vegetative growth (Wolstenholme and Whiley, 1992). There is probably sufficient carbohydrate available from mature leaves to support the growth of both developing fruitlets and young leaves (Finazzo *et al.*, 1994). This supply during early fruit development does not limit fruitlet growth or stimulate fruitlet abscission. However, the lack of an adequate spring shoot growth, though competitive, will mean that fruit set and early fruit growth will be dependent upon older leaves and stored carbohydrate, which may be depleted by flowering. The later vegetative growth flush near harvest (Plate 7) is crucial for final fruit growth, build-up of carbohydrates and root growth (Fig. 4.3). This is discussed further with respect to maturity of fruit at harvest and biennial bearing.

Fruit

The avocado fruit is a one-seeded berry (Fig. 4.2). The single large seed is composed of two cotyledons enclosing an embryo and is surrounded by a thick fleshy mesocarp. The skin varies in thickness to 0.65 cm, depending upon the race (Table 4.1) and has 20,000–30,000 stomata per fruit, less than on a leaf. The skin colour of the ripe fruit ranges from various shades of green to yellow-green and from reddish to maroon and light to dark purple. The buttery flesh (mesocarp) is greenish yellow to bright yellow to creamish when ripe. Oil content ranges from 7.8 to 40.7% on a fresh-weight basis (Kawano *et al.*, 1976). Size varies from small fruit of some Mexican types, about 227 g or less, to the large Guatemalan types, 1.4–2 kg or more (Table 4.1). In shape, the fruit is usually pyriform to oval and round.

In West Indian and Mexican cultivars, the fruit matures 150–240 days after anthesis while Guatemalan cultivars take more than 250 days (Fig. 4.5). All show a typical sigmoid growth curve (Fig. 4.6). There is little delay before growth occurs in fruit, seed and pericarp thickness.

CULTIVAR DEVELOPMENT

Cytogenetics and Genetics

Persea species generally have a diploid number of 24 chromosomes. Among *P. americana* materials examined, Garcia (1975) found one triploid (2n = 36) and a tetraploid (2n = 48). Karyotype in *Persea* is asymmetric, with chromosomes ranging in size from 2.3 μm to 6.1 μm.

Garcia and Tsunewaki (1977), using peroxidase isozyme analysis, found significant variations between Mexican strains within and between nine states in Mexico. The three races showed significant variation in their pattern of activities. The frequency of significant variations within strains of the three

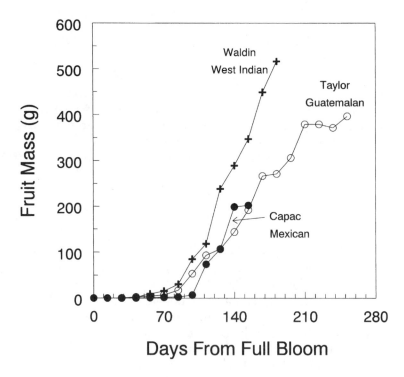

Fig. 4.5. Increase in fruit mass of three cultivars from different ecological races, showing the difference in time for fruit to reach maturity and final fruit size relative to the ecological race (redrawn from Valmayer, 1967).

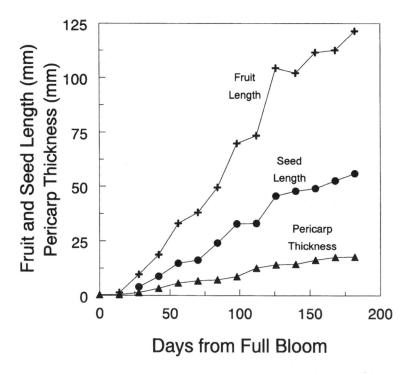

Fig. 4.6. Rate of total fruit and seed growth and the increase in pericarp thickness of the West Indian cultivar 'Waldin' (redrawn from Valmayer, 1967).

races is high in the Guatemalan (62.5%), intermediate in the West Indian (50.0%) and relatively low in the Mexican race (39.0%). Isozymes have also been used as genetic markers to detect different genotypes resulting from natural outcrossing and controlled pollinations (Bergh and Ellstrand, 1987). Restriction maps suggest a very complex evolution from a number of different sources (Furnier *et al.*, 1990).

Skin colour, flowering group and anise scent in the leaves are probably coded by several loci, having several alleles in each loci (Lavi *et al.*, 1993a). The juvenile period also has a major genetic component. Many other quantitative traits, such as tree size, flowering intensity, fruit weight, fruit density and harvest duration, show additive and non-additive genetic variance (Lavi *et al.*, 1991).

Breeding

Intensive avocado breeding has been conducted over many years in the USA, Israel, South Africa and other countries. However, until recently, cultivars in the avocado-growing countries of the world originated as chance seedlings (Bergh, 1976).

Breeding methods may be by self-pollination or by hybridizing two selected parental cultivars, the choice depending upon objectives of the breeding programme. The avocado is highly heterozygous, because of enforced cross-pollination between complementary cultivars, and widely segregating variability can be obtained among self-pollinated progenies (Lavi *et al.*, 1993b). Selfing can be achieved by the use of cages with pollinator bees enclosed within the tree or by using isolated trees of a desired parent cultivar (Bergh, 1976). Outcrossing is reported to be nil where trees of a cultivar are separated from other genetic lines by 100 m or more. Variations in daily weather conditions during the flowering period may disrupt the diurnal synchrony of the flowers allowing self-pollination.

Cross-pollination is necessary when two or more desirable traits are to be combined or when traits intermediate between two cultivars are desirable. The major cultivars in subtropical areas are hybrids of the Guatemalan and Mexican races, as the Mexican race is the source of cold-hardiness. In tropical regions, hybrids of the Guatemalan and West Indian races prevail.

Selection and Evaluation

Breeding objectives may be divided into traits that are universally desired and those which are regionally specific. Tree characteristics sought after in all growing areas are ease in propagation, vigour, precocity, spreading, short fruit-maturation period, heavy and regular bearing, and wide adaptability with resistance to disease and insect pests. Traits specific to regions are cold or heat tolerance and salinity tolerance. Other qualities, such as dwarfing or semidwarfing, genetic uniformity, freedom from sun-blotch disease, caused by a virus, and drought resistance, are also constantly evaluated (Bergh, 1976).

Universal fruit characteristics are resistance to diseases, pests and blemishes, long tree storage, uniform ripening, minimum fibres and long shelf-life (Currier, 1992). Traits that differ due to specific climatic conditions or market preferences are fruit size and shape, skin colour, oil content and chill tolerance. Market preference is usually based upon consumer familiarity with certain traits. Fruit size is a good example. Consumers in California generally prefer a size range of 200–300 g, while in most Latin American countries larger fruit are preferred.

Breeding for cold-hardiness is a major objective in the subtropics. 'Brooksville', a seedling of the Mexican race, has shown outstanding cold-hardiness, withstanding artificial chilling to −8.5°C. 'Gainesville', another Mexican type, withstands field temperature as low as −9.4°C. The cold-hardiness of these cultivars has to be combined with genes for dependable productivity and fruit market acceptance (Knight, 1976).

Avocado rootstock breeding is an important aspect of the total breeding programme. The avocado root rot, caused by *Phytophthora cinnamomi*, has led to testing thousands of seeds and bud wood from avocado trees that have resisted the disease over many years and other *Persea* species collected in their native habitats. Small-fruited species, such as *Persea caerulea* (Ruiz y Paron) Mez., *Persea donnell-smithii* Mez., *Persea pachypoda* Nan and Mart. and *Persea cinerascens* Blake, though having strong resistance, are graft-incompatible with the avocado. Hybridization between the small-fruited *Persea* species and avocado has not been successful. The cultivar 'Duke', a Mexican-race avocado, has provided two seedlings, now propagated vegetatively as 'Duke 6' and 'Duke 7', with moderate resistance and 'Duke 7' becoming the first commercially successful rootstock. Another resistant rootstock, 'Morton Grande' (G-775C), a hybrid from *P. americana* and *P. schiedeana,* has not performed well; 'Thomas', 'D9' and 'Barr Duke' (all Mexican) have greater resistance and are available (Ben Ya'acov and Michelson, 1995). The latter three were all found in infected fields. In the absence of root rot, the rootstock can significantly influence canopy growth and yield. 'Duke 7' and 'Tomas' produce significantly more fruit per cubic metre of canopy than other rootstocks, with 'D9' and 'Morton Grande' being the poorest performers (Arpaia *et al.*, 1992).

A rootstock with resistance to salinity is a focus of research in Israel, California, Australia and the north-western coastal areas of Peru. The three races differ in their ability to resist the uptake of salts (Table 4.1), in their ability to retain the toxic substances in the root system without translocating them to the scion part of the plant and in their tolerance within their tissues. The West Indian rootstocks have the highest level of tolerance to salinity, followed by Guatemalan and Mexican races, respectively (Ben-Ya'acov and Michelson, 1995). Israel has taken two approaches: (i) selection of types that produced a resistant seedling population; and (ii) selection of individual resistant plants for vegetative propagation. In the latter case, they have found that the true West Indian and high-salinity-resistant types are difficult to propagate by cuttings. In areas with salinity problems, there has been an increase in the use of West Indian rootstocks, although they are difficult to grow in some subtropical areas, such as California.

Cultivars

Some California and Florida cultivars have become international (Table 4.4), providing the basis for avocado development in many countries. 'Hass', a Guatemalan type, is the most widely distributed and grown, replacing 'Fuerte', a Mexican–Guatemalan hybrid. One or both have dominated the plantings in Australia, Israel, South Africa, Mexico, the Canary Islands and other areas (Table 4.4). In California, 'Hass' has gradually become the dominant cultivar, largely due to the relatively poor yields and inconsistent bearing of the type B cultivar 'Fuerte', whose flowering is disrupted by low temperatures. 'Zutano', 'Bacon', 'Gwen' and 'Pinkerton' are other commercial cultivars. In California, the fruit of most cultivars can be stored on the tree for months after reaching maturity, depending upon season, so not many

Table 4.4. A list of selected cultivars grown in major avocado-growing areas.

California	Florida	Australia	Israel	South Africa	Mexico	Spain
Fuerte	Pollock	Zutano	Fuerte	Fuerte	Fuerte	Hass
Hass	Simmonds	Sharwil	Hass	Hass	Hass	Bacon
Zutano	Nadir	Bacon	Nabal	Edranol	Bacon	Fuerte
Bacon	Booth 8	Fuerte	Ettinger	Ryan	Reed	Reed
Reed	Lula	Hass	Horshim		Criollo	Zutano
Pinkerton	Hardee		Wurtz		(Local)	Reed
Gwen	Ruehle				Zutano	Gwen

cultivars are needed to market fruit throughout the year. However, this tree storage can lead to more pronounced biennial bearing.

Florida cultivates a larger number of mostly West Indian and West Indian–Guatemalan hybrids to extend the market season (Campbell and Malo, 1976). Cultivars are grouped into early-, mid- and late-season. 'Pollock', 'Waldin', 'Simmonds' and 'Nadir' are some early-season types. Some midseason cultivars are 'Booth 7', 'Booth 8', 'Collinson', 'Hall' and 'Hickson', while late-season cultivars are 'Lula', 'Monroe', 'Choquette', 'Booth 3' and a few others (Table 4.4). A similar situation occurs in Hawaii, where a large number of cultivars are grown, including introductions from California, Florida, Australia and Mexico and selections of local origin. 'Sharwil', an Australian cultivar, has become a leading cultivar, with a recent release, 'Green Gold', increasing in acreage.

Mexico grows 'Fuerte' and 'Hass' as the main cultivars, along with other California and Florida ones. The 'Criollo' is mentioned as a selection of the Mexican race and is well adapted to median ecological areas of Mexico (Diaz-Avelar, 1979). Besides Mexico, Brazil has developed a substantial industry, mostly with introduced cultivars, such as 'Fuerte', 'Hass', 'Carlsbad', 'Corona', 'Edranol', 'Nabal' and 'Ryan', and cultivars of local origin, such as 'Solano', 'Ouro Verde', 'Quintal' and 'Fortuna'.

In Australia, 'Hass' and 'Fuerte' have been planted most frequently, especially along the coastal areas of northern New South Wales and southern Queensland. 'Sharwil', a local selection (Guatemalan–Mexican hybrid), has gained considerable popularity, although in some areas it had fluctuating yields. Under tropical conditions in the Northern Territory, the Mexican and Guatemalan cultivars yield poorly (Sedgley et al., 1985). South African production is largely from 'Hass' and other California cultivars. Israel's production was initially based on California cultivars, until an intensive selection programme over the years produced a large number of local cultivars, such as 'Ettinger' and 'Horshim'.

In Asia, the avocado has not attained the popularity it has in other areas, although it was introduced into Malaysia over 120 years ago. In Indonesia, avocado grows well at 200–1000 m on well-drained soils, and trees are generally propagated by seeds. In the Philippines, many cultivars have been introduced from the USA since 1903, and avocado is grown in nearly all parts of the country, although it has not attained the popularity of other fruit.

CULTURAL PRACTICES

Propagation and Nursery Management

Avocado is primarily propagated commercially by budding or grafting upon seedling rootstocks (Plate 8). However, the variability of seedling populations with respect to certain desirable characteristics, such as resistance to *Phytophthora* root rot and tolerance to salinity and calcareous soils, has posed problems (Ben-Ya'acov and Michelson, 1995). Seedlings of West Indian cultivars as rootstocks are more adapted to the warm tropics, while Mexican seedlings grow better in the cooler subtropics. Rootstocks used in Florida are largely West Indian or hybrids of West Indian and Guatemalan types. In Israel, where salinity tolerance is more important, West Indian cultivars highly tolerant to salinity and calcareous soils are used.

Seedling production has largely changed from grafting nursery-grown trees to grafting container-grown trees (Plate 8). Seeds are taken from fruit picked from trees free of sun-blotch virus and are treated in a water-bath at 49–50°C for 30 min, cooled and surface-dried in a partially shaded area. Seeds are planted (broad side down) in polyethylene bags, with a well-draining potting mix. Seeds germinate in about a month. Seedlings may be cleft or side-wedge grafted 2–4 weeks after germination. Scion wood from terminal growth from sun-blotch-free cultivars should be used. Propagation is usually done in shade houses, preferably with temperature control if the environment has a wide temperature range. Grafted plants must be hardened for approximately 2 weeks under full sunlight before field transplanting.

Air layers and cuttings have been successfully rooted; however, the variability in ease of rooting between races and even among cultivars of a race and the difficulty of large-scale production have discouraged commercial development. Generally, Mexican cultivars root most easily, followed by Guatemalan and West Indian, whether by air layering or by cuttings. Time required for rooting of air layers ranged from 146 to 518 days, depending upon cultivar and the time of the year.

Kadman and Ben-Ya'acov (1970b) tried leafy avocado cuttings under mist and found that some genotypes consistently root nearly 100% of the time under practically any conditions, while others did not root at all or rooted with difficulty. Generally, West Indian cultivars with strong resistance to salinity are difficult to root. Cuttings from mature trees are difficult to root and, in those that rooted well, they took 4–10 months to root. Cuttings from 1-year-old seedlings show a higher percentage of rooting in 4–12 weeks. A 50% light intensity in the intermittent-mist system during the summer is better for rooting than full sunlight.

The anatomy of the avocado stem provides a reason for the difficulty in rooting, with the fibre bundles and the sclereid ring being thicker in the West Indian types, intermediate for Guatemalan and hybrid types and least for the Mexican cultivars. Etiolated stems show few or no sclereid connections between fibre bundles, suggesting that the sclerenchyma ring may be acting as a barrier to root emergence (Gomez et al., 1973). Hence, shoots of most avocado cultivars produced in light do not root nearly as well as shoots produced in darkness. Seedlings produced upon scions of the desirable rootstock cultivar are grafted by tip grafting as close to the base as possible. Shoots of the scion are allowed to grow and then cut back to near the base. When new buds show signs of growth, the entire plant is placed in a dark room with the temperatures maintained at 21–23°C. When the new shoots reach about 8–10 cm in the dark, the plants are again placed in light, with a tar-paper collar placed around each etiolated stem and filled with vermiculite to continue exclusion of light from the base of the shoots. Only the tips of the shoots are exposed to light, in order to produce green leaves. This procedure is done under shade to prevent sunburn. Shoots are then allowed to grow until several leaves have matured. The collar is then removed and shoots are detached for rooting in propagation frames. Rooting hormones have shown no beneficial effects. Rooted shoots are transplanted into 10 cm peat pots and grown in an enclosed area for further root and top growth and gradually hardened. These rooted plants are then transplanted into larger polyethylene bags for more growth and hardening and are grafted when they attain appropriate size. The side-wedge technique is usually used, so some terminal foliage on the rootstock is retained until scion growth begins.

Field Preparation

Land preparation for avocado does not differ from that for other tree crops. Development of a drainage system is a prime consideration. Subsoiling or ripping down to at least 0.5 m or more with a subsoiler, preferably running diagonally across the slope to allow subsurface movement of water, aids drainage. If soil pH needs adjusting, this could be done during the final stages of land preparation. Cover crops, such as legumes or grain, can be preplanted a year before orchard planting to increase organic matter and minimize erosion and root rot.

Transplanting and Plant Spacing

The use of grafted container-grown trees has dramatically minimized transplanting mortality. Polyethylene bags can be removed with all the soil intact around the roots. Soil in the planting hole should be moist but not wet. In dry areas, application of water in the holes a few days before transplanting is advisable to moisten the soil. In the tropics, transplanting can be done during the dry season, if irrigation is readily available; otherwise, transplanting should be done at the beginning of the rainy season. In the cool subtropics, where low winter temperatures can be expected, young plants may need to be protected or transplanting delayed until the coldest period has passed.

The cultivar's natural growth habit (spreading or erect), vigour of the rootstock, and the environment and soil are major determinants of the mature tree size and this influences spacing and the continued productivity of the orchard. Close spacing is sometimes used, with later thinning to obtain maximum benefits per unit area. Initial close spacing can be justified only for precocious cultivars and is beneficial only if trees are judiciously thinned when this becomes necessary. For non-precocious cultivars, the yields obtained before tree thinning becomes necessary may not justify the additional cost. Traditional spacings in Florida are 7.5–10.5 m in rows and 7.5–12 m between rows, which precludes canopy overlap. Narrower spacings of 4.5–6 m in rows and 6–7.5 m between rows lead to higher yields while the trees are young. In South Africa, similar narrow spacing of 7.5 m × 7.5 m (177 trees ha^{-1}) is recommended for 'Fuerte' and 7 m × 7 m (204 trees ha^{-1}) for 'Hass', requiring later tree thinning or topping. At close spacing on the square, thinning is done by removing the trees in every second diagonal row (Bower, 1981b).

Placement of pollinizer trees in the orchard so that an adequate ratio can be maintained after thinning may need to be considered in some spacing schemes; the old guide is one pollinator for every nine trees. Large blocks of a single cultivar have shown good production, presumably by self-pollination, but interplanting pollinizer cultivars undoubtedly increases fruit set for some cultivars that may not have pollen (old Florida cultivar: Collinson) or are a poor pollen source. 'Ettinger' is an excellent pollen source for 'Hass' and there is a higher survival of fruit to maturity.

Irrigation

The avocado can tolerate neither water stress nor excess moisture, especially when drainage is inadequate. Water stress reduces yields, fruit size and tree vigour. The soil around the trees should be moist but not wet; the field condition, soil drainage and tree density, as well as canopy size, prevailing weather conditions and past irrigation records, provide the experience necessary to decide irrigation frequencies. Evapotranspiration data and a tensiometer provide more accurate means of determining time of irrigation. Evapotranspiration data are the average over a period of days. To give a seasonal rate, this is adjusted for efficiency of water use to determine irrigation application rate. Young non-bearing trees require light, more frequent irrigation. Older trees can withstand longer intervals between irrigations but never to a point of water stress. Only 50% of the tree's requirements should be given in the middle of the cool season and spring, in order to favour flowering rather than vegetative growth. When fruit set is completed, irrigation reverts to normal amounts during fruit development. High rates are necessary during flowering and may be necessary as the fruit approach maturity and if the weather is hot and dry (Fig. 4.3). Irrigation is necessary down to at least 60 cm. Total water applied per year is estimated at about 35–50 ha-cm for mature trees (Gustafson, 1976).

On level land, less efficient furrow irrigation or sprinklers can be used, with microirrigation systems (drip or microsprinklers) being more efficient in water use. Mature trees require eight or more drip emitters around the tree, providing greater control and higher efficiency of water use (up to 80%). Later plantings of avocado in California are on the sides of steep hills, which can be irrigated only by drip irrigation or minisprinklers.

Pruning

The growing tips should be pinched off young trees to develop a more compact tree. This is continued until the tree is too tall. Lower limbs are removed only if they interfere with irrigation and fertilization. Management should be keyed to avoiding vegetative and reproductive growth occurring at the same time. Since avocado is polyaxial, it must continue to increase in size to remain productive. Shoot growth in the warm season occurs in spurts of up to 1 m, which is needed in order to put on leaves required for fruit growth (Fig. 4.3). Tree trimming aims at a canopy that has fruit at all heights and reduces the competition for light. Pruning at the bud ring (several closely spaced buds without subtending leaves), formed at the conclusion of a shoot-growth flush, releases more buds and increases shoot complexity and hence bearing sites. Cutting below this ring depresses tree vigour and releases only one bud (Cutting *et al.*, 1994). Bud-ring pruning can extend the period before trimming is necessary. Pruning should be timed to the end of the autumn period. Tree growth can also be reduced by spraying with paclobutrazol, an inhibitor of gibberellin synthesis. Tree thinning is recommended in South Africa when 90% of the orchard floor is shaded. Topping (staghorning) is also practised on healthy trees to rejuvenate a crowded orchard, where the trees are cut back to 1–3 m from the ground. Tree thinning, topping or stumping is essential to avoid canopy crowding and the loss of bearing volume in the lower third or more of the tree. Severe topping to *c.* 3 m means the loss of production for 2–3 years, while less severe topping to *c.* 5 m can increase fruit yield and the amount of fruit set in the lower half of the tree (Crane *et al.*, 1992). Cultivars with upright vigorous growth habits may be severely affected by topping and orchards may have to be rejuvenated in sections over several years, by topping to 5 m.

Fertilization

Fertilizer practices differ in avocado-producing areas, due to differences in climate, soil, cultivars and management practices. Numerous attempts have been made to determine critical foliar nutrient levels in order to regulate the application of macro- and microelements. The following ranges of foliar levels for macronutrients are established: nitrogen (N) = 1.6–2.0%; phosphorus (P) = 0.07–0.20%; potassium (K) = 0.75–2.0%; calcium (Ca) = 1.0–3.0%; and magnesium (Mg) = 0.25–0.50% (Malo, 1976). The ranges for micronutrients are: iron (Fe) = 50–200 p.p.m.; zinc (Zn) = 50–150 p.p.m.; and manganese (Mn) = 30–70 p.p.m.

Critical levels have been difficult to establish, due to the highly variable yields of avocado. Nitrogen seems to be one of the controlling factors in avocado yields, as this is the only element that has shown a curvilinear relationship to yields. Maximum production of 'Fuerte' is found at a moderate level of N in the leaves, with reduced yields occurring at levels below and above the moderate level. In the 'Fuerte', a range from 1.6 to 2.0% N in the leaves in late summer is a desirable level in order to maintain high production. Nitrogen fertilization is not recommended during the cool season, with application delayed to the summer leaf and root flush (Fig. 4.3). This application should include P and K. A later application of P and K should occur near the peak of fruit set (Whiley *et al.*, 1988a).

For many tree crops, including avocado, a soil pH of 7.0 and above creates problems with Fe and Zn deficiencies, refered to as lime-induced chlorosis. Iron-chelate is used to correct Fe deficiencies. Soil application of Zn is more effective in acid soils than in alkaline soils and foliar application has not proved successful. In Florida, where deficiencies of Fe, Zn and Mn are common, good results are obtained by combining the chelates of these elements and applying them through the drip-irrigation system. Boron (B) deficiency occurs in some soil types and the 'Sharwil' appears to be more sensitive to B deficiency.

Pest Management

Diseases

A number of avocado diseases have been reported from producing areas around the world (Table 4.5), the most serious being root rot caused by *P. cinnamomi* Rands. This is suspected when trees show a gradual decline, with leaves becoming smaller, yellow-green in colour and shedding. In severe cases, twig dieback occurs. The destruction of the unsuberized feeder roots is associated with high soil moisture in poorly drained areas of the field, with *P. cinnamomi* thriving under wet soil conditions, especially when the temperature ranges from 21–30°C and with a soil pH of 6.5 (Zentmyer, 1976). Soil fumigants, fungicides and sanitation have been used, with a research

Table 4.5. Some important diseases of avocado.

Common name	Organism	Parts affected, symptoms	Region or country
Root rot	*Phytophthora cinnamomi*	Feeder roots, tree decline	Worldwide
Verticillium wilt	*Verticillium dahliae* (*Verticillium albo-atrum*)	Wilting of branches, death of trees	California, Australia, Florida, Israel, Spain
Armillaria root rot and crown root	*Armillaria mellea*	Large roots, gradual death of trees	California, Mexico
Sun blotch	Sun-blotch virus	Stunted, decumbent growth, distorted leaves, yellow or red streaks on fruit	Many areas
Anthracnose	*Colletotrichum gloeosporioides* (*Glomerella cingulata* var. *minor*)	Leaves, most serious on fruit	Worldwide
Stem-end rot	*Dothiorella aromatica* (*Dothiorella gregaria*) *Diplodia natalensis*	Purplish-brown spots on fruit surface, flesh discoloration, offensive odour	USA, South Africa, Israel, Australia, parts of South America, Caribbean
Cercospora spot	*Cercospora purpurea*	Leaves, young stems, fruit	Florida, other areas
Scab	*Sphaceloma perseae*	Foliage, fruit	Humid tropics and subtropics

emphasis on the development of resistant rootstocks, such as 'Morton Grandee', 'Thomas', 'Barr-Duke' and 'D9' (Gabor *et al.*, 1990). The use of resistant rootstocks is integrated with hygiene, sanitation and cultural methods (Coffey, 1987). In Australia, root rot has been minimized by building up heavy mulch with bagasse, grass or cereal straw. Trunk-injected phosphonate fungicide (Aliette-Fosetyl-A1) is timed to coincide with the shoot maturation (Fig. 4.3), the phosphonate being carried to the mature shoots and then translocated to the roots, peaking in the root in about 30 days (Whiley *et al.*, 1995).

Many of the diseases reported are not necessarily serious, although the potential of causing heavy losses exists. Avocado scab is considered to be an important disease of avocado fruit and foliage in Florida (McMillan, 1976). Among fruit diseases, anthracnose, stem-end rot and avocado scab can cause serious problems. Anthracnose requires control practices in the field during fruit development (Plate 9).

Sun blotch, caused by a viroid, is of concern. Trees are stunted, with cracked bark, necrotic streaks on branches and white or light green areas on the fruit. There are no known vectors and what makes it a potentially serious problem is the presence of many symptomless carrier trees. The disease is transmitted by use of seedling rootstocks from such trees. Scion wood from a healthy tree grafted upon a 'carrier' rootstock becomes infected. In the major avocado-growing countries, indexing techniques have been developed and used to identify healthy cultivars for scion wood, as well as those needed for rootstocks (Broadley, 1991).

Insects

There are many insect pests reported in avocado orchards, although they do not usually pose any serious problems (Table 4.6). Occasionally, a sudden increase of a specific insect can cause severe damage; this increase may be associated with a sudden change in weather conditions. Avocado red mite can cause significant leaf damage and reduction in photosynthesis and transpiration (Fig. 4.7), possibly leading to a reduction in yield. Insects such as scales, aphids, mealy bugs and various mite species are also commonly found in orchards, but natural enemies have been shown to provide satisfactory control.

Table 4.6. Some important insect pests of avocado.

Common name	Organism	Parts affected, symptoms	Region or country
Mediterranean fruit fly	*Ceratitis capitata*	Fruit	Hawaii
Oriental fruit fly	*Dacus dorsalis*	Fruit	Hawaii
Queensland fruit fly	*Dacus tryoni*	Fruit	Australia
Black coffee twig borer	*Xylosandrus compactus*	Branches, twigs	Hawaii
Chinese rose beetle	*Adoretus sinicus*	Leaves of young plants	Hawaii
Fuller rose beetle	*Pantomorus godmani*	Leaves of young plants	Hawaii
Spotting bug	*Amblypelta nitida*	Young fruits	Australia
Pine-tree thrips (greenhouse thrips)	*Heliothrips haemorrhoidalis*	Fruit, scarring	California, Florida, South Africa, Hawaii, Canary Islands (wide distribution)
Red-banded thrips	*Selenothrips rubrocinctus*	Leaves, fruit, scarring	Wide distribution

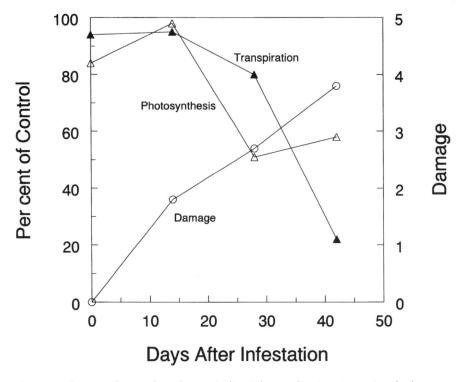

Fig. 4.7. Influence of avocado red mites (*Oligonichus yothersi* McGregor) on leaf transpiration, photosynthesis and damage (redrawn from Schaffer *et al.*, 1986).

In some producing areas, fruit flies may require some form of disinfestation procedure for fruit to be exported to some markets. Studies in Hawaii with 'Sharwil' avocado have shown that it is not normally a host to the Mediterranean fruit fly (*Ceratitis capitata*), melon fly (*Dacus cucurbitae*) and the oriental fruit fly (*Dacus dorsalis*) at the mature green harvest stage. Under dry conditions, eggs are deposited between the pedicel and the fruit, making it a host.

Weed Control

Young avocado trees are sensitive to herbicides, so weeding around the trees is usually done manually. Black polyethylene mulch around the plant is effective in preventing annual broadleaf weeds, but not effective against perennial grasses and nut-sedges (Nishimoto and Yee, 1980). In bearing orchards, the canopy provides enough shade to prevent weed growth. Interrows up to the periphery of the trees may be mowed or controlled with herbicides. Disc-harrowing of avocado orchards is discouraged, as the shallow feeder roots will be destroyed.

HARVESTING AND POSTHARVEST HANDLING

Harvesting and Handling

Some cultivars can be stored on the tree and harvested according to marketing schedules. The length of time fruit can be left on the tree depends on the cultivar and season. However, this 'on-tree storage' can lead to biennial bearing or crop failure in the following year. Generally, cultivars of the West Indian race have little or no tree-storage time, and hybrids, particularly Guatemalan × Mexican, have a longer tree-storage life. Some avocado will start changing skin colour and then fall from the tree when mature and a few lines can begin ripening on the tree.

A major problem is the stage of maturity for harvest, especially for cultivars that remain green upon ripening. Some green-coloured avocados develop a yellowish tint on the stem near the fruit. Maturity of cultivars that normally change skin colour from green to reddish or purplish, like the 'Hass', is easy to ascertain. Immature fruit, if harvested, take longer to soften and they shrivel upon storage, with the flesh becoming rubbery rather than buttery. Most countries with commercial avocado production have developed some sort of standards for determining maturity. The increase in oil content is correlated significantly with maturity. As the fruit advances towards maturity, oil content increases, the level depending upon the cultivar (Barmore, 1976). In California, oil content was used as a standard of maturity, with a minimum of 8% oil, based on fresh mass of fruit, exclusive of skin and seed. Oil content as a measure of maturity was impractical in Florida, due to wide variations in oil content among and within cultivars. Some cultivars, particularly those belonging to the West Indian race, never reach the 8% oil content, as in California. Florida uses minimum fruit weight and diameter as related to number of days from fruit set, thus establishing the earliest harvest date for each cultivar (Hatton and Campbell, 1959). Fruit originating from known bloom dates becomes progressively smaller in size from the earliest bloom date to the latest (Hatton and Reeder, 1972).

Moisture content is also used as a measure of maturity, as it does not involve oil determination and there is a negative correlation between oil and moisture content. As oil content increases during maturation, there is a similar decrease in moisture content. A 10 g sample of flesh from a total of ten fruit is collectively grated and spread on an open dish and dried in a microwave oven for a few minutes. When dry-matter percentage is greater than 21%, the fruit is considered to be mature; this can increase to 30% dry matter in more mature fruit. After moisture content has been determined, fruit comparable to the sample fruit can be picked and ripened at room temperature. If the sample fruit ripen within 8–10 days without shrivelling, they are considered to be mature and the grower can proceed to harvest comparable-sized fruit on the tree. Grade standards frequently set a minimum dry-matter percentage around 20% for 'Fuerte', 22% for 'Hass', 25% for 'Gwen' and 19% for 'Zutano'

(Rainey *et al.*, 1992). This method is rapid and less time-consuming and does not involve oil determination.

The stage of fruit maturity when harvested can significantly influence the occurrence of biennial bearing in some avocado (Whiley *et al.*, 1992). In the tropics, later harvesting of 'Fuerte' fruit having 30% dry matter results in pronounced biennial cropping. Results for a cool subtropical location in South Africa suggest no effect of later harvesting on subsequent yield (Kaiser and Wolstenholme, 1994). Other tree-storage limitations are fruit drop and development of an off-flavour or rancidity with overmaturity. A higher yield can also be obtained if harvesting is staggered, with 50% of fruit harvested with 21% dry matter and the remainder at 30%.

Mature fruit are harvested by cutting or snapping off the stem at the base of the fruit with about 12 mm of the stem attached. Aluminium picking poles are usually equipped with a cutter and a bag to catch the fruit. Large trees on suitable terrain can be partially mechanized by the use of a three- or four-wheeled, self-propelled, hydraulically powered platform from which pickers can use short poles. Pallets or bins should be placed under the shade of trees while waiting to be picked up for transport to the packing plant. This prevents overheating of fruit, as precooling is not generally practised.

Postharvest Treatments

Most packing houses are automated, with all debris removed automatically and fruit cleaned with roller brushes. Cleaned fruit pass through graders, where all diseased, injured and defective fruit are removed and fruit are separated into size lots for packing. Sizing is done by use of drop–roll (diameter) sizers or by weight sizers. Each producing country has its own grading standards and size classes. Standards for Europe allow for three classes of avocado, based on appearance, defects, tolerances, uniformity, packaging and marketing. The minimum mass is 125 g and a range of size from 125 to 1220 g in 14 ranges is used in all classes (Anon., 1995). Fruit with lower quality go to processing or are used for local sales.

Sizes of packing cartons differ in various countries. They are usually corrugated paperboard cartons for single- or double-layer packing, ventilated for good air circulation. Pads or styrofoam trays with cup impressions may be used to prevent bruising. Fruit may be waxed or wrapped and packed by hand. All fruit in a specific size container must weigh between a specified minimum and maximum range. Cartons are labelled with fruit number and size range. In California, packed cartons are cooled at 8–12°C, depending on cultivar, before loading into transport vehicles.

Storage temperature for delaying ripening varies with the cultivar. Storage temperatures of 12.5, 8 and 4°C for West Indian, Guatemalan and

Mexican cultivars, respectively, and a relative humidity of 80–90% are recommended (Lutz and Hardenburg, 1968). Other storage methods, such as controlled-atmosphere (CA) and hypobaric (low-pressure) storage, have been evaluated. With CA storage at 2% oxygen (O_2) and 10% carbon dioxide (CO_2) at 4.7°C and 98–100% relative humidity, 'Lula' avocados are still 100% marketable after 60 days (Spalding, 1976).

Fruit can be ripened at 25°C or by exposure to 10 p.p.m. ethylene at 15–17°C for 24 h and then transported to markets. The ethylene-treated fruit are marketed locally, as the current market strategy is to provide ripe fruit for immediate consumption.

UTILIZATION

The avocado is considered a nutrient-dense food (Rainey *et al.*, 1994). It has the highest fibre content of any fruit and is a source of food antioxidants (Bergh, 1992a). Avocado has a high oil content, although Florida cultivars are lower in oil content compared with cultivars from California. Major fatty acids are the monounsaturated oleic acid, followed by palmitic and linoleic acids (Bergh, 1992b). Palmitoleic acid is found to approximate percentages of linoleic acid and values are slightly higher or lower depending upon the cultivar analysed. The nutrient values of the different ecological races also vary, but, in general, it is a fair to good source of P, provitamin A, riboflavin and niacin (Table 4.7). The protein content of 1–2% is considered to be greater than in any other fresh fruit (Bergh, 1992a). Avocado also increases the diet's content of antioxidants, foliates, K and fibre (Rainey *et al.*, 1994).

Avocado is mainly used fresh in salads, its high fat content combining well with acid fruit and vegetables, such as pineapple, citrus and tomatoes, or with acid dressings. A major commercial avocado product is guacamole, used as a favourite dip with potato chips, tortilla chips and similar products. Avocado may be used to supply the fat content of frozen desserts, such as ice-cream and sherbets. Miller *et al.* (1965) provide numerous recipes for various types of salads, cocktails and desserts.

Deterioration of flavour and enzymatic discoloration are serious problems in the commercial processing of avocado (Ahmed and Barmore, 1980). Avocado is not conducive to heat processing as it results in an off-flavour. High polyphenoloxidase and total-phenol contents contribute to the processed product's browning potential, with differences in browning potentials occurring among cultivars. Enzymatic browning of avocado products can be minimized by processing a cultivar with low polyphenolase activity into an acidified product containing an antioxidant (e.g. ascorbic acid), packing under N or a vacuum and storing at low temperatures.

Table 4.7. Proximate analysis of avocado (Wenkam, 1990).

		Halumanu West Indian type	Nabal Guatemalan type
Proximate			
Water	g	82.8	69.9
Energy	kJ	431	874
Protein	g	1.5	1.03
Lipid	g	9.3	21.8
Carbohydrate	g	5.7	6.3
Fibre	g	1.6	2.0
Ash	g	0.8	0.9
Minerals			
Calcium	mg	8	77
Iron	mg	0.5	0.4
Magnesium	mg	–	–
Phosphorus	mg	34	42
Potassium	mg	–	–
Sodium	mg	–	–
Vitamins			
Ascorbic acid	mg	–	5.5
Thiamine	mg	0.03	0.09
Riboflavin	mg	0.09	0.14
Niacin	mg	1.23	–
Vitamin A	IU	–	802

FURTHER READING

Ben Ya'acov, A. and Michelson, E. (1995) Avocado rootstocks. *Horticultural Review* 17, 381–429.

Bower, J.P. and Cutting, J.C. (1988) Avocado fruit development and ripening physiology. *Horticultural Review.* 10, 229–271.

Whiley, A.W. and Schaffer, B. (1994) Avocado. In: Schaffer, B. and Anderson, P.C. (eds) *Handbook of Environmental Physiology of Fruit Crops,* Vol. II, *Subtropical and Tropical Crops.* CRC Press, Boca Raton, Florida, pp. 3–35.

Whiley, A.W., Saranah, J.B., Cull, B.W. and Pegg, K.G. (1988a) Manage avocado tree growth cycles for productivity gains. *Queensland Agricultural Journal* 114, 29–36.

5

BANANA

BOTANY

Introduction

The banana is important in the humid tropical lowlands, with year-round fruit production. The commercial banana is a giant, perennial, herbaceous monocotyledon, propagated vegetatively, belonging to the family *Musaceae*, genus *Musa*. The banana (English) has various names: bananier (French), pisang (Malay, Indonesian), kluai (Thailand), chuoi (Vietnam), xiang jiao (Chinese). *Ensete*, the other genus in the family, ranges from Asia to Africa, while *Musa* ranges from Africa through Asia to the Pacific (Reynolds, 1951). The genus *Musa* has four sections: *Eumusa*, *Callimusa*, *Rhodochlamys* and *Australimusa*, among which are wild seeded plants and edible clones, with overlapping geographical distributions. The genus *Musa* has about 25 species and some have numerous subspecies.

Genus *Musa*

Edible bananas are derived from either *Musa acuminata* (A) or *Musa balbisiana* (B) or a combination of both. Cultivars are diploid or triploid, with some new tetraploids developed by breeding. Considerable somatic variation has led to a great range of cultivars. Cultivars are described by their name and genomic make-up, e.g. 'Pisang Raja' AAB, the AAB indicating that it is a hybrid with two genomes of A and one genome of B. Most dessert bananas are AA or AAA, with the triploid AAA being the most important in the trade. The different groups and subgroups have somewhat distinct fruit characteristics (Table 5.1).

   M. acuminata has a number of morphological characters that separate it from *M. balbisiana*. For example, *M. acuminata* has an open petiolar canal, which in *M. balbisiana* is closed. *M. acuminata* has prominent bract scars,

Table 5.1. Fruit characteristics of some major cultivars within groups and subgroups.

Group	Subgroup	Characteristics
AA	Sucrier	Small fruit (8–12 cm long), thin golden skin, light orange firm flesh, very sweet, 5–9 hands per bunch, 12–18 fingers per hand
	Lakatan	Medium to large straight fruit (12–18 cm), golden yellow, flesh light orange, firm, dry, sweet and aromatic, 6–12 hands per bunch, 12–20 fingers per hand
AAA	Gros Michel	Medium to large fruit, thick skin, creamy white flesh, fine-textured, sweet and aromatic, 8–12 hands/bunch
	Cavendish	Medium to large fruit, yellow skin, white to creamy flesh, melting, sweet, aromatic, 14–20 hands per bunch, 16–20 fingers per hand. Susceptible to Panama disease race 4
AAB	Silk	Small to medium yellow fruit (10–15 cm), thin yellow-orange skin, white flesh, soft, slightly subacid, 5–9 hands per bunch, 12–16 fingers per hand, skin frequently has blemishes
	Pisang Raja	Large fruit (14–20 cm), thick skin, cooking banana, creamy orange flesh, coarse texture, 6–9 hands per bunch, 14–16 fingers per hand
	Plantain	Yellow skin, creamy orange firm flesh, 2 hands per bunch
ABB	Bluggoe	Medium to large cooking banana, thick coarse skin turns brownish red when ripe, creamy orange flesh, starchy, 7 hands per bunch
BBB	Saba	Stout, angular, medium to large cooking banana (10–15 cm), thick yellow skin, creamy white flesh, fine textured, 8–16 hands per bunch, 12–20 fingers per hand

bracts that are lanceolate and curl, and two regular rows of ovules, compared with four irregular rows in *M. balbisiana*. Using and scoring 15 morphological characters allows the relative contribution of the two species to be determined in hybrid cultivars. Triploids and tetraploids are larger and more robust than diploids.

Origin and Distribution

The primary centre of origin is thought to be Malesia (Malaysia, Indonesia, the Philippines, Borneo and Papua New Guinea). *M. acuminata* Colla. has seeded fruit. Dessert cultivars were developed from it via parthenocarpy and sterility, aided by human selection and vegetative propagation. *M. balbisiana* Colla. also has a wild seeded fruit, is more suited to drier areas and occurs from India to New Guinea and the Philippines, though absent from central Malesia (Espino *et al.*, 1992). It was similarly taken into cultivation, with the selection of natural diploid, triploid and tetraploid hybrids. Bananas are now cultivated throughout the tropics and in selected areas in the subtropics.

ECOLOGY

Soil

Deep friable loams with natural drainage and no soil compaction are preferred. High organic matter and fertility assure high yield. Soil pH between 4.5 and 7.5 is used, although 5.8–6.5 is recommended. Most exported bananas are produced on highly fertile alluvial loams. Soil textures ranging from sands to heavy clays are used.

Climate

Rainfall

Bananas require a regular water-supply that matches or slightly exceeds the free-water evaporation rate. Irrigation is essential for high yield if rainfall is less than evaporation. It also provides the advantage of fertilization via the irrigation water. Areas with very high rainfall may be too overcast for optimum photosynthesis, have more disease problems and require extensive drainage.

Temperature

A temperature range of 15–38°C occurs in most production areas, with the optimum temperature being *c.* 27°C. The optimum for dry-matter accumulation and fruit ripening is about 20°C and for the appearance of new leaves about 30°C. Growth ceases at 10°C and can lead to 'choke-throat' disorders, where inflorescence emergence is impeded and poor fruit development occurs. Temperatures lower than 15°C can be withstood for short periods, while temperatures less than 6°C cause severe damage (Turner, 1994). Frost causes rapid death. Temperature higher than 38°C causes growth cessation and leaf burn. Plants growing in the subtropics produce fewer leaves per year than those in the tropics and take longer to produce and develop fruit (Table 5.2).

Light

Full sunlight is required for best growth, although fruit sunburn can occur, especially if water-supply is low. Shaded or overcast conditions extend the growth cycle by up to 3 months and reduces bunch size (Fig. 5.1).

Photoperiod events

No evidence exists that photoperiod influences flowering. Increasing photoperiod (from 10 to 14 h) increases the rate of new leaf appearance, probably due to increased photosynthesis.

Table 5.2. Phenological differences between cultivars of Cavendish subgroup ratoon plantation in the humid tropics (Honduras: cv. 'Grand Nain', Stover and Simmonds, 1987) and subtropics (South Africa: cv. 'Williams', Robinson and Nel, 1985; Robinson and Human, 1988).

	Humid tropics	Subtropics
Mean number leaves per month (warm/cool season)	3.5/2.5	4/0.5
Total leaves per year	40	25
Planting to harvest (months)	9–11	15–20
Harvest to harvest (months)	6–8	11–13
Flowering to harvest (warm/cool season, days)	98/117	110/204

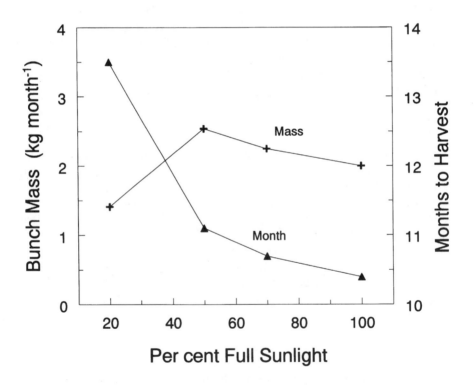

Fig. 5.1. Effect of shading on the bunch weight produced per month (kg month⁻¹) and months to harvest. Low light reduces the leaf area produced and a positive relationship exists between leaf area of the third mature leaf and final bunch weight: $y = 4.83x - 12.83$, $r^2 = 0.75$, significant at $P < 0.05$ (after Murray, 1961).

GENERAL CHARACTERISTICS

Plant

The 2–9 m perennial herb has an underground stem or rhizome. The corm has buds producing short rhizomes and shoots (suckers) near the parent (Fig. 5.2). Adventitious roots form a dense mat and spread extensively 4–5 m from the parent and down to 75 cm or more. The stem (pseudostem) consists of overlapping leaf sheaths with new leaves and finally the inflorescence, growing up through the centre (Fig. 5.2).

The large leaf lamina are 1.5–4 m long by 0.7–1 m wide with pronounced midribs and parallel veins (Fig. 5.2). Stomata occur on both surfaces, with three times more on the adaxial surface. The leaf takes 6–8 days to fully unroll from the tip and in the tropics the leaves last *c.* 50 days. Leaves emerging just before the inflorescence can live up to 150 days or more. Twenty-five to 50 leaves emerge, with 10–15 functional leaves present (total area 25 m^2) at inflorescence emergence. The number of leaves at flowering is positively correlated with bunch weight (Fig. 5.3), fewer than six to eight lead to significant reduction in bunch weight.

Flowers

One terminal inflorescence emerges through the pseudostem and bends downward after extrusion (Fig. 5.2). The flowering spike consists of groups of two rows of appressed flowers enclosed within large, ovate, reddish bracts at each node. The bracts reflex and are shed as the fruit starts to develop. The female flowers emerge first and the males are distal. Sometimes hermaphrodite flowers develop in the middle of the bunch, but they may abscise. There are 12–20 flowers per node (hand) and 5–15 hands with female flowers. The bracts open sequentially, about one per day. The peduncle continues to elongate up to 1.5 m, terminating in a male bud, which continues to produce male flowers enclosed within the bracts.

The 10 cm female flowers of cv. 'Cavendish' have an inferior ovary of three united carpels with a short perianth. The perianth consists of five fused and one free segment, forming a tube around the style and sterile androecium and three-lobed stigma. Male 'Cavendish' flowers are 6 cm long with five stamens, which rarely bear fertile pollen.

Pollination and Fruit Set

Natural pollination

The fruit develops parthenocarpically. The ovules shrivel early and are recognized as brown specks in the mature fruit along the axial placenta

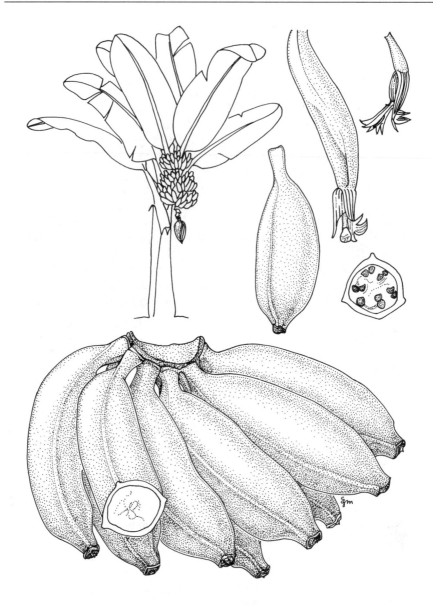

Fig. 5.2. Plant, leaves, bunch, flowers and fruit of banana. The transverse fruit section in the upper right section is a seeded species.

(Fig. 5.2). Very infrequently, seeds are found in mature fruit of edible cultivars, especially those with a B genome. The 'Cavendish' group has absolute female sterility with some viable pollen. 'Pisang Awak' (ABB) can sometimes be very seedy if pollen-bearing diploids are growing nearby.

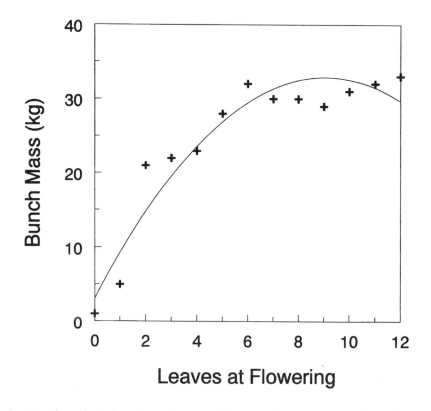

Fig. 5.3. The effect of number of functional leaves at flowering on bunch weight (redrawn from Turner, 1970).

Floral induction and fruit set

There are no external symptoms of the start of inflorescence development. The flowering stimulus is unknown. Externally, it is not temperature or photoperiodism, and internally it is not the number of leaves developed, as the number of leaves is more or less fixed depending on cultivars and environment.

Fruit

Morphology

The fruit, although it develops from an inferior ovary, is a berry. The exocarp is made up of the epidermis and the aerenchyma layer, with the flesh being the mesocarp (Fig. 5.2). The endocarp is composed of a thin lining next to the ovarian cavity. The axial placenta has numerous airspaces and ventral vascular bundles.

Each node has a double row of flowers forming a cluster of fruit that is commercially called a 'hand', with the individual fruit called a 'finger'. 'Cavendish' bananas can have 16 hands per bunch with up to about 30 fingers per hand, and the bunch can weigh up to 70 kg.

Growth and development

Pollen sterility is due to triploidy, while female sterility is due to at least three complementary dominant genes, plus modifier genes. These sterility genes are found in wild populations and have been selected for fruit edibility. Parthenocarpy is separate from sterility. For the first fortnight after anthesis the ovules increase in size (50% over initial), and later they shrivel and ovary growth slows. Parthenocarpic bananas that have seeds ('Pisang Awak' ABB) show a stimulation of fruit growth, due to the presence of developing seeds. Pollination can stimulate fruit growth, even without seed development (Israeli and Lahav, 1986).

There are periclinal and anticlinal divisions from 6 weeks before inflorescence emergence (anthesis) to 4 weeks after emergence. This division is followed by cell expansion for 4–12 weeks after emergence. Skin mass increases rapidly in the first 40 days after flowering, with the fruit pulp not beginning to develop until day 40 (Fig. 5.4). Starch accumulation parallels finger length and diameter increase (Lodh *et al.*, 1971). The fruit matures in the tropics 85–110 days after inflorescence emergence. Fruit development may take up to 210 days in the cooler subtropics or under overcast conditions (Table 5.2). Harvest maturity is a commercial stage (three-quarters round), with the fruit still having some angularity and being only 75% of its potential maximum size. Fruit allowed to fully develop to a round shape may show skin splitting. Depending upon the persistence and viability of the remaining leaves, fully developed fruit may also show sunburn.

Since fruit number and size decrease from the promixal to the distal (bottom) hands, fruit thinning is sometimes practised. Fruit on the inner whorl of a hand can be 15% smaller than those on the outer whorl. The male inflorescence can be removed soon after full development of the bunch without damaging the hands of the female fruit.

CULTIVAR DEVELOPMENT

Genetics and Cytogenics

The haploid number is n = 11, with 22, 33 and 44 chromosomes being found in the diploids, triploids and tetraploids. There are 200–300 clones, of which more than half are triploids. Triploids are more vigorous, easier to grow and higher-yielding than diploids (Gowen, 1995).

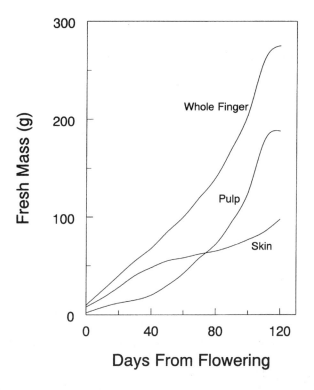

Fig. 5.4. Growth of finger, pulp and skin fresh mass of 'Gros Michel' in the Caribbean (Simmonds, 1982).

Problems in Breeding

The majority of breeding efforts have been directed at the AAA group, because it is the group with the highest export potential. However, female sterility and low numbers of viable pollen make conventional breeding very difficult. Some inferior cultivars can, under controlled conditions, produce seeds, however, it takes 3 years from seed to seed. Considerable breeding efforts and selection of crossed material may lead to the development and release of cultivars with suitable horticultural characteristics and increased disease resistance (Gowen, 1995). Apart from the efforts supported by the International Network for the Improvement of Bananas and Plantains (INIBAP) (Vuylsteke *et al.*, 1993), there have been a few efforts to develop better varieties for small and medium sized growers serving local markets with local varieties in Brazil, Africa and India.

Selection and Evaluation

Considerable efforts have been expended in characterizing various cultivars under comparable conditions (Rowe and Rosales, 1996). The use of somaclonal variation and mutation breeding is under way. The focus of these programmes is resistance to black Sigatoka, *Fusarium* wilt, bunchy-top virus and nematodes. Gene transfer from unrelated species also offers considerable opportunities for improved planting materials. Besides selection for resistance to the above diseases, other priorities include resistance to the weevil borer, dwarfism, tolerance to drought and cold, and improved bunch yield, harvest index, fruit quality and storability (Novak, 1992). Similar priorities exist for plantains, with susceptibility to black Sigatoka and low bunch yield being the major limitations, followed by some of the same priorities as for dessert bananas.

Major Cultivars

As mentioned above, there are at least 200–300 clones in various countries, many having different names in different localities (Table 5.3). There are numerous germ-plasm collections around the world, including those in Indonesia, Malaysia, Thailand, the Philippines, India, Honduras, Jamaica, Brazil, Cameroon and Nigeria. The large number of synonyms for many of the better cultivars makes for some confusion (Lebot *et al.*, 1993).

CULTURAL PRACTICES

Propagation and Nursery Management

Sexual

Seeds are only used in breeding programmes. Many of the important commercial cultivars are female-sterile.

Asexual

Suckers and a piece/section of corm with a growing point are used for planting material. A sucker is a lateral shoot with a rhizome and an apical growing point. It may be a young sucker just emerging from the soil or a large sucker with narrow leaves and a large rhizome. Corm material is a section of the mother rhizome with a lateral bud for regeneration. Before planting, the pieces are pared to remove old roots and disease, and are immersed in water at 52°C for approximately 20 min or treated with pesticide to control nematodes and borers. A rapid multiplication technique of plants has been set up in Brazil (Loyola Santos *et al.*, 1986).

Table 5.3. Major genomic groups and some cultivars. The seedless diploid (AA) and triploid (AAA) are regarded as desserts, while the seedless diploid (BB) and triploid (BBB) are cooking bananas. These and the many hybrids are all referred to as *Musa* spp. followed by the code and subgroup (SG) name (Cavendish, Gros Michel, plaintain, etc.).

AA	AAA	AAB	ABB	BBB	Other
Sucrier	Gros Michel (SG)	Silk	Pisang Awak (Indo, Mal)	Saba (Phil)	Atan (AAAB)
syn.	syn.	syn.	syn.	syn.	Kalamagol (AABB)
Pisang Mas (Mal, Indo)	Bluefields	Apple (Hawaii)	Ducase (Aust)	Cardaba (Phil)	Gold Finger (AAAB)
Kulai Khai (Thai)	Pisang Ambon (Mal)	Pisang Rastali (Mal)	Katali (Phil)	Kluai Hin (Thai)	
Amas (Aust)	Disu	Pisang Raya Serek (Indo)	Kluai Namwa (Thai)	Pisang Nipal (Mal)	
Susyakadali (India)	Kulai Hom Thong (Thai)	Latundan (Phil)	Pisang Klotok (Indo)		
Lakatan (Phil)	Cavendish (SG)	Woradong	Karpuravalli (India)		
syn.	Dwarf Cavendish	Cantong	Bluggoe		
Pisang Barangan (Indo)	syn.	Tundan	syn.		
Pisang Berangan (Mal)	Chinese (Hawaii)	Lady Finger	Pisang Kepok (Indo)		
Senorita	Canary Banana	syn.	Pisang Abu Keling		
	Dwarf Chinese	Pome Pacha Naadan	Kluai Hak Muk (Thai)		
	Basrai (India)	Pisang Raja (Mal, Indo)	Nalla Bontha (India)		
	Governor (West Ind) syn.	Radja (Phil)	Moko (Trinidad)		
	Enano (Latin Amer)	Larp	Da Jiao (China)		
	Giant Cavendish	Houdir	Fen Da Jiao (China)		
	syn.	Mysore (India)			
	Mons Mari (Old)	syn.			
	William (NSW)	Colombo			
	Grand Nain	Poovan (India)			
	Bongali Johaji (India)	Honderawala (India)			
	Beijiaw (China)	Plantain (SG)			
	Honchuchu	Horn			
	Pisang Ambon Putih (Malay)	syn.			
	syn.	Pisang Tanduk			
	Kluai Hom Dek Mai (Thai)	Tindok			
	Ambon (Phil)	Klui Nga Chang			
	Pisang Embun	French			
		syn.			
		Nendran (India)			

NSW, New South Wales.

Tissue culture (*in vitro* plantlets) allows for rapid multiplication of uniform, disease-free materials. Other advantages include very high field establishment rates, uniformity of harvest timing, precocity and high production, at least in the first crop cycle. These advantages have to be balanced against higher cost, extra care at multiplication and the transmission of viruses that have not been eliminated. Somatic variation is the major problem and care in multiplication is necessary to reduce the incidence to an acceptable level (< 3%) (Smith, 1988).

Field Preparation

Soil should be ripped, ploughed and disced before planting. This may involve cross-ripping to 1 m to break up any soil compaction. Lime, phosphorus (P) and potassium (K), as determined by soil analysis, should be added at this time. Sloping land should be prepared so as to avoid erosion. Drainage is vital to avoid waterlogging, which reduces yields. The water-table should be kept below 1.2 m in depth. Extra drainage can be installed during field preparation via trenches and drains.

Transplanting and Spacing

A planting hole slightly larger than the material is dug or a furrow made. The sucker or corms are usually covered with 100–200 mm of soil. Planting is scheduled in subtropical areas to meet certain harvest periods in the first crop cycle. This is rare in tropical localities, as annual production is more evenly distributed throughout the year than in the subtropics (Galan-Sauco, 1992). The concern is to avoid hot weather or a dry season; hence, the best time for planting is just before the wet season. In subtropical regions, the late summer through to winter is avoided, because of the low temperatures when the plants are very young.

Planting densities of 1000–3000 plants ha^{-1} are used, with rectangle, single- or double-row cropping. Double rows (2 m) combine higher density with a 3.5 m alley for access. The actual density depends upon cultivar and climate. Higher densities are used in hot, dry localities to generate the necessary shade and microclimate for maximum yields. The vigour of a plantation is related to the canopy characteristics, leaf-area index and yield. If only one growth cycle is to be used, a higher density may be planted (3000 plants ha^{-1}); for three or more growth cycles, a lower density is recommended (2000 plants ha^{-1}) for highest gross margin.

Irrigation Practices

Irrigation water can be provided by furrow or over-canopy sprinkler, with drip irrigation and microsprinklers increasing in use. Irrigation is used to supplement rainfall and its scheduling requires a calculation of amount and frequency. This schedule is based on pan evaporation, soil-water holding capacity, banana root depth, water depletion and crop water use. The important characteristics of banana are its: (i) high water-loss potential, associated with its large broad leaves; (ii) shallow root system – *c.* 90% of roots are in the top 300 mm; (iii) poor ability to absorb water from a drying soil; and (iv) rapid physiological response to water deficit. Soil-water potentials more negative than −20 to −40 kPa can adversely affect growth, with relative impact depending upon local climate – hot and dry (severe impact) to moist and humid (less impact).

Pruning and Bunch Propping

Other than removal of the male inflorescence, no other vegetative pruning is normally practised. Withered styles and perianths persisting at the end of the fruit are usually removed at the packing station after harvest, but are sometimes removed by hand 8–12 days after the bunch emerges to reduce fruit scarring and disease (cigar-end rot). Early removal of one or more hands from the distal end of the bunch to increase fruit size by reducing interfinger and hand competition, though common practice, is not borne out by research and therefore is not recommended.

Bunches falling from the plant or the whole plant falling over can lead to considerable bunch damage. Lodging is caused by poor anchorage, poor planting material or very large bunches. The problem is reduced if single or double poles are wedged against the throat of the plant under the curvature of the peduncle or twine guys are extended from this same point to lower positions on nearby plants.

Sucker Management and Leaf Removal

Sucker management is an essential step to remove unwanted suckers developing from the base of the parent rhizome and to select a suitable sucker to produce the ratoon crop. The strategy is to remove suckers that receive nutrients from the parent plant and which would extend its cycle and reduce its yield if they remained (Fig. 5.5). This operation is vital in the subtropics.

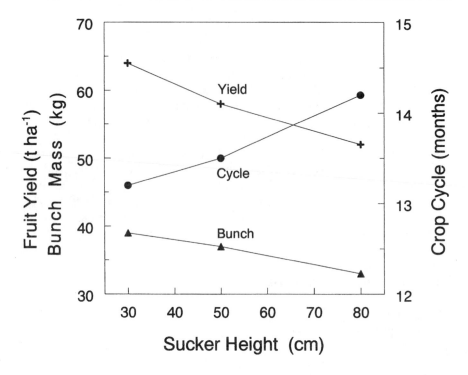

Fig. 5.5. Effect of the amount of unwanted sucker growth of 'Williams' banana before removal on remaining sucker (follower) bunch mass, annual yield in the first ratoon crop and the crop cycle duration between the first and second ratoon. Unwanted suckers were removed when they were 30 cm, 50 cm and 80 cm high, leaving one following sucker per plant mat. The leaf area of the 80 cm sucker was 39 times greater than that of the 30 cm sucker. (Redrawn from Robinson and Nel, 1990.)

Desuckering is done by hand by cutting and gouging every 4–6 weeks, and paraffin-oil injection so that suckers do not use too many of the resources available to the parent (Fig. 5.5). Selection of ratoon suckers is critical to maintaining yield, production and appropriate plant spacing. A sucker on the most open side is usually selected as the daughter plant and left as the follower, taking into consideration field spacing.

Removal of dead leaves is practised to reduce disease spread, to prevent senescent leaves from hanging over suckers and reducing light and to prevent fruit scarring. There should be at least six to eight healthy leaves remaining on the plant at flowering to ensure maximum bunch development (Fig. 5.3). Plants with severe leaf removal or damage (e.g. as a result of insect feeding) have reduced bunch weights. The green life of the harvested banana is also reduced by leaf loss.

Fertilization

Regular fertilizer practices are followed on large commercial plantations (15% of total world production) to maintain optimum productivity, while in the less intensively cultivated plantation less supplementary fertilization occurs (Martin-Prevel, 1990). Deficiency symptoms have been described for leaf blade and petioles. Mineral analysis of banana plants has been used and ranges of deficiency and adequacy suggested. For plant growth and fruit production, large amounts of nutrients are required. These nutrients come from the soil and decaying plant material, and the remainder comes from applied organic matter and fertilizer. The amount removed by fresh fruit of cv. 'Cavendish' (50 t ha^{-1} year^{-1}) includes 189 kg ha^{-1} nitrogen (N), 29 kg ha^{-1} P, 778 kg ha^{-1} K and 101 kg ha^{-1} calcium (Ca). As a proportion of the total nutrients taken up by the banana plant, this is equivalent to 49% of the N, 56% of the P, 54% of the K and 45% of the Ca. These total amounts and proportions are about halved for plantain (Table 5.4). Most of the data have been derived from experimentation with bananas produced for export, especially from varieties in the 'Cavendish' subgroup. Cultivar and subgroup differences have been recorded, with the optimum nutrient supply not being the same for all cultivars. Seasonal influences on fertilization are much greater on bananas growing in the subtropics, where temperature probably has the greatest influence. Intercropping needs should also be considered.

The large N and K requirements will be modified by local soil-nutrient concentrations. The rate of application depends on climate, soil type, variety, management practices and yield. Since vegetative and reproductive stages of development are found in one field at the same time, and as the initial stages of fruit inflorescence development are crucial for final yield, a constant supply of nutrients is essential for high yield. If the pseudostem is left standing after bunch harvest, up to 40% of the nutrients (especially N, P and K) can be removed by the following ratoon sucker (Fig. 5.6) and increase the bunch weight of this next generation. Delaying fertilization can have a significant impact on yield, reducing it by 40–50% or more; a 3-month delay makes it difficult for the plants to recover.

Because of the extensive root system, fertilizer should be applied away from the pseudostem, mostly on the side of the next ratoon. Solid fertilizer is applied three to four times per year, more frequently if there is high rainfall. Fertilizer applied through irrigation water (fertigation) is more efficient and gives better control of application time and rate to meet demands. Frequency of fertilizer application is increased when fertigation is used monthly, weekly or continuously, especially if drip irrigation is used. Foliar application is also used.

Numerous forms of fertilizers have been tested. If the different forms meet the management strategy and crop needs and do not lead to excessive runoff

Table 5.4. The quantity of nutrients in banana and plantain with yields of 50 t ha⁻¹ or 30 t ha⁻¹, respectively, from 2400 plant ha⁻¹ and bunch masses of 25 kg and 15 kg (Lahav and Turner, 1983), and critical nutrient concentrations in the lamina of the third youngest leaf of the vegetative plants (Stover and Simmonds, 1987).

Nutrient	Fresh fruit (kg ha⁻¹)		Remaining in plant (kg ha⁻¹)		Proportion removed in fruit (%)		Leaf lamina critical conc. in banana
	Banana	Plantain	Banana	Plantain	Banana	Plantain	
N	189	76	199	189	49	29	2.4%
P	29	11	23	16	56	41	0.15%
K	778	243	660	945	54	20	3–3.5%
Ca	101	9	126	149	45	6	0.45%
Mg	49	11	76	53	39	17	0.2–0.22%
S	23	9	50	19	32	32	n/a
Mn	0.5	0.2	12	7	4	3	60–70 p.p.m.
Fe	0.9	0.4	5	3	15	11	60–70 p.p.m.

N, nitrogen; P, phosphorus; K, potassium; Ca, calcium; Mg, magnesium; S, sulphur; Mn, manganese; Fe, iron; n/a, not available.

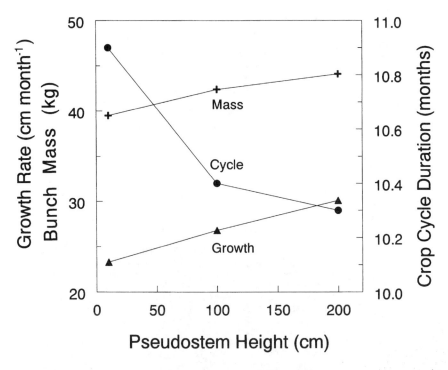

Fig. 5.6. The effect of cutting height of pseudostem at primary bunch harvest on the . growth rate of follower plant in the month after cutting, crop cycle in months to harvest of the follower and the weight of the follower bunch (after Daniells and O'Farrell, 1987).

or leaching, most are suitable. The most common forms of nitrogen are ammonium nitrate (NH_4NO_3), ammonium sulphate ($(NH_4)_2SO_4$) and urea, while, for potassium, potassium chloride (KCl), potassium sulphate (K_2SO_4), and potassium nitrate (KNO_3) are used. The rates vary widely; for example, using KNO_3, 100 kg ha^{-1} in South Africa is suitable with minimum leaching, while 250 kg ha^{-1} is needed in Israel and up to 600 kg ha^{-1} is used in other countries with high runoff leaching and yields. Under less intensive cultivation, organic matter plus 0.25 kg each of urea and KCl per mat (production unit) supplement is applied every 3 months. In some areas, no K is recommended, e.g. the Sula Valley of Honduras and the Jordan Valley of Israel.

Bunch Covers

Polyethylene bunch covers (30–40 μm thick) are almost universally used to improve yield and maintain fruit quality (Fig. 5.7). Some covers have a pesticide impregnated into them to reduce thrip or mite damage. The covers produce a microclimate around the bunch and prevent fingers inside the

Fig. 5.7. Bunch covers provide protection of developing fruit from pathogens and insect damage. The coloured plastic hanging from the bottom of the bunch shows date of emergence and is used to indicate harvesting.

bunch covers from being chafed by leaves and covered in dust. The higher temperatures and humidities generated inside the bunch cover are helpful in subtropical areas, though not required. For tropical areas, perforated covers for aeration and cooling are used. Covers are applied after the bracts have fallen and should hang 15 cm below the distal hand. Thinner covers are used in the tropics, where wind is less of a problem.

Disease and Pest Management

Diseases

Fruit diseases can cause severe problems if not controlled (Table 5.5). The most virulent disease of banana and plantain is black Sigatoka, which is found worldwide, except for subtropical areas and the Caribbean. *Fusarium* wilt (Panama disease) has caused severe disruption of banana and plantain

Table 5.5. Principal diseases of banana, causal organisms, distribution, varietal susceptibility and symptoms.

Disease	Causal organism	Distribution	Varietal susceptibility	Symptoms
Bacterial				
Moko	*Pseudomonas solanacearum:* only distinct strains of race 2 attack bananas and plantains	Central and South America Southern Caribbean Philippines Indonesia	Cavendish AAA susceptible Bluggoe ABB susceptible Horn AAB resistant Pelipeta ABB resistant	Root: progressive transient yellowing older leaves first, necrosis and collapse of base petiole, younger leaf panels flaccid and necrotic, premature bunch development, discoloured vascular bundles cream to brown or black in all parts of corm, roots, rhizome suckers Fruit: male bud blackens and shrivels, vascular discoloration extends up fruit, bunches blacken and rot and then it progresses down pseudostem
Blood disease	Possibly related to Moko	Indonesia	Bluggoe ABB susceptible	Similar symptoms to Moko leaf yellowing but more conspicuous, reddish tinge to discoloured vascular bundles, infects fruit via style
Rhizome rot	*Erwinia chrysanthemi*	Worldwide	Certain AAA and AA varieties susceptible incl. Cavendish	Rotting of rhizome, plant falls over. Soft, pale brown or yellow rot in outer cortical tissue. Entry through wounds, especially in wet conditions

Table 5.5. Continued.

Disease	Causal organism	Distribution	Varietal susceptibility	Symptoms
Fungal				
Fusarium wilt or Panama disease	*Fusarium oxysporum* fsp. *cuberse*	Race 1 worldwide Race 2 widely distributed Race 3 Central America Race 4 Canary Island, Taiwan, Australia, South Africa	Gros Michel AAA, Silk AAB susceptible Bluggoe ABB susceptible Only *Heliconia* spp., not banana Cavendish AAA susceptible	Vascular wilt disease, leads to collapse of crown and pseudostem. Can infect through the roots. Premature yellowing of older leaves, necrosis and collapse. Red-brown flecks or spots on inner surface of leaf sheath. Brown discoloration of vascular tissue. Fruit do not show symptoms
Sigatoka leaf spots Black Sigatoka Black leaf streak	*Mycosphaerella fijiensis* *Paracercospora fijiensis*	Worldwide except Caribbean and subtropical areas	Pisang Awak ABB tolerant Saba BBB tolerant Grand Nain AAA susceptible Plantain AAB susceptible	More destructive than yellow Sigatoka, rapid onset. Leaf shows small, translucent, pale yellow streaks, which turn brown and eventually become necrotic – light grey with surrounding yellow halo. Symptoms of reddish brown specks first appear on lower leaf surface, lesions appear in 8–10 days

Disease	Pathogen	Distribution	Susceptibility	Description
Yellow Sigatoka Banana leaf spot	*Mycosphaerella musicola* *Pseudocercospora musae*	Worldwide		Distinctive elliptical, sharply defined lesions. Slower to develop than black Sigatoka. First appearance of pale yellow specks on upper leaf surface, which lengthen to yellow streaks and turn brown with distinct yellow halo, more common on margins and leaf apex
Other leaf-spotting diseases				
Black spot (banana freckle)	*Phyllostictina musarum*	Asia, Pacific	Cavendish AAA susceptible	Small dark lesions, especially on older leaves. Brownish lesions on fruit. Important if fruit infected
Cordana	*Cordana musae*	Worldwide		Necrotic pale ovate leaf spot several cm in length. Brown lesion with chlorotic halo, not normally important
Deightoniella leaf spot	*Deightoniella torulosa*	Worldwide	Cavendish AAA susceptible	Small brown spots on stressed and older leaves. Also occurs on older fruit, where can be major problem

Table 5.5. Continued.

Disease	Causal organism	Distribution	Varietal susceptibility	Symptoms
Viral				
Banana bunchy-top virus (BBTV)	Vector banana aphid	South-east Asia Australia India Central Africa Pacific	Gros Michel AAA susceptible Cavendish AAA susceptible Saba BBB tolerant	Infected leaves become stunted and chlorotic at the margins. Bunching of leaves at the apex. No fruit production, suckers infected
Banana mosaic	Cucumber mosaic virus Many aphids as vector	Worldwide	Cavendish AAA susceptible Horn AAB susceptible	Sharply defined interveinal chlorosis of leaves can lead to rotting heart leaf and cylinder. Stunted plants
Banana streak virus (BSV)	Vector mealy bugs, infect planting material	Worldwide	Cavendish AAA susceptible Mysor AAB susceptible	Yellow streaks, continuous or broken, running across the leaf blade. More pronounced in warm weather

production where susceptible cultivars, such as 'Gros Michel', are grown. The two other major diseases are the bacterial disease 'Moko' and the viral bunchy-top disease. Expensive fungicides can be used to protect export crops against black Sigatoka, but may not provide a long-term solution for small growers and others serving local markets. Breeding for resistance is essential, using natural resistance in other banana varieties, or biotechnology to transfer resistance.

Preharvest diseases, including fruit freckle (*Phyllostictina musarum*) and speckle (*Deightoniella torulosa*), can cause fruit-skin spotting. 'Cigar-end rot' is associated with *Verticillium theobromae* and *Trachysphaera frutigena*, infecting the flower perianth and slowly developing along with fruit growth. The end of the fruit darkens and later is covered with spores. The disease is found in West Africa, Egypt and Australia. The pulp develops as a dry rot. Removal of pistil and perianth 8–12 days after bunch emergence and polyethylene bunch covers help to reduce incidence.

A complex of organisms are known to cause the most serious worldwide postharvest disease – crown rot. The wound caused when the bunch is dehanded is colonized by the organisms, the rot spreading into the pedicels during shipping and sometimes into the pulp. A number of organisms have been found, with the actual complex varying with region; most common isolates have been *Colletotrichum musae*, *Fusarium* spp., *V. theobromae*, *Botryodiplodia theobromae* and *Ceratocystis paradoxa*. Spores are carried by wind or rain splash; however, dehanding knives and delatexing tanks may be bigger sources. Good dehanding practices, with clean cuts using sharp knives, and sanitation of packing sheds and areas are essential.

Anthracnose on the fruit skin and neck rot are both caused by *C. musae* and can be a serious problem when the fruit becomes overripe. Sanitation and postharvest fungicide are used for control.

Insect pests and nematodes

Insect pests are generally of minor importance (Table 5.6). Occasionally, severe fruit scarring can occur. Borers are a problem where control practices, such as planting clean material and insecticide schedules, are not followed.

The continuous nature of banana production makes pests such as nematodes more important (Table 5.6). Nematodes can severely limit production, with the extent of the problem varying with cultivar, soil conditions, type of nematode and plant vigour. Nematocides are used for fruit destined for export and are applied every 4–6 months.

Weed Management

Weeds are a major problem during stand establishment before canopy closure occurs. Cultivation needs to be carefully carried out so as not to damage the

Table 5.6. Major pests of banana.

Pest	Latin name	Spread	Susceptibility	Symptoms
Banana borer/weevil	*Cosmopolites sordidus*	Worldwide	No clones resistant	Burrows into rhizome, network of tunnels, lays eggs at base of pseudostem, adults (12 mm long) feed on pseudostem, nocturnal
Bunch pests				
Thrips	*Chaetanaphothrips* spp. red rust thrip *Frankliniella* spp. Flower thrips	Central America, South America, Asia, Australia South-east Asia, Australia, Caribbean, Latin America		Rust blemish to fruit skin, especially between fingers of immature fruit Corky scab on developing fruit
Scab moth	*Nacoleia octasuma*	Australia, Pacific Islands		Brown scabs on developing fruit
Nematodes				
Burrowing nematode	*Radopholus similis*	Most areas except East Africa, Israel, Canary Islands, Egypt and Taiwan	AAA more tolerant than AAB	Elongated blade Lesions on root
Lesion nematode	*Pratylenchus goodleyi*	African highlands, Canary Islands	Cavendish susceptible	Similar to *R. similis*
	Pratylenchus coffeae	Tropics	AAB more commonly associated	
Root-knot nematodes	*Meloidogyne* spp.	Worldwide		Deformation and stunting of roots. Gall formation

surface feeder roots, using hand hoeing around the plants. Herbicides can be used once the banana canopy is sufficiently high to avoid contact with the leaves. Mulching and intercropping during early stand development can be used. Failure to weed can lead to severe yield decline.

Orchard Protection — Windbreaks

Banana plants tolerate wind up to 30 m s^{-1}. Higher speeds lead to tearing of leaf laminas. Windbreaks may be necessary if prevailing winds tear leaves into strips less than 5 cm wide (Turner, 1994). The need for a windbreak must be balanced against the effects of shading, the area needed, competition for water and nutrients and the limited protection on the leeward side. Plant propping is carried out to reduce the impact of winds.

HARVESTING AND POSTHARVEST HANDLING

Harvesting

Individual fingers increase in weight and begin to lose angularity in cross-section during maturation. Fruit for export are harvested while still green at 75% maturity with some angularity, 10–14 weeks after flower emergence in the tropics and up to 9 months in the cool subtropical areas. The middle finger of the outer whorl of the second hand is then 31–41 mm in diameter and is measured with calipers. More mature fruit have a shorter postharvest life and are more liable to be sunburnt and split (Plate 10).

Small growers tend to harvest bunches when more mature: three-quarters with little angularity to fully rounded stage. Other criteria used to judge maturity include drying of leaves, drying of stylar ends and days from bunch emergence. The days from bunch emergence can vary from 7 to 24 weeks, depending upon cultivar, season, crop management and environment. Other measures include pulp-to-peel ratio and skin firmness. Plantain maturity standards vary widely and lead to considerable variation in product quality.

The bunch is removed from the plant by cutting a notch in the pseudostem while supporting the bunch with a pole and slowly lowering it on to the shoulder pad of a harvester. The stem is then fully cut, leaving a 300 mm peduncle. The bunches are transported to the packing shed on padded trailers or on an overhead cable system. Dehanding can be performed in the field, with the hands transported to the packing shed on padded trailers (Plate 11). Care is essential in these steps to avoid any mechanical injury that would reduce fruit quality.

Postharvest Treatments

Bunch covers are removed in the field or after the bunches reach the covered packing shed. Bunches must be protected from exposure to direct sunlight to avoid sunburn injury during transportation and at the packing-shed holding area. Hands are removed with a sharp curved knife or curved chisel, leaving part of the crown attached. The hands, irrespective of whether dehanding takes place in the field or in the packing shed (Plate 11), are placed in clean water to remove dirt and latex exuding from the cut crown. Care is needed to avoid build-up of fungal spores in the water of the wash tank, by frequent changes and the use of chlorine.

The hands are removed from the tank and sometimes cut into clusters of adjacent fingers, with defective fingers being removed. The clusters are placed in trays that pass through a fungicide-treatment spray or dip on a conveyor. The conveyor then passes the trays to a packing station, where weight adjustment occurs and the fruit are packed into cardboard cartons, with a thin plastic liner to prevent chafing injury and water loss. Hands or clusters are tightly packed to avoid movement during shipping (Plate 12). Carton sizes vary from 12 to 20 kg, with Central American cartons holding 18 kg of fruit. Cartons may then be cooled to 13°C before shipping to market. In small local operations, bunches may be sold, with dehanding taking place upon sale to the consumer. Alternatively, hands may be shipped to market in a number of different containers. Export dessert bananas, because of the regular nature of the hands and curved nature of the fingers, can be easily packed. Many other cultivars are not as easily packed, as the fingers are not arranged in a regular manner, e.g. 'Pisang Mas' (AA).

Green bananas are shipped at 13–14°C to delay ripening. Lower temperature can lead to chilling injury, whose symptoms include a dull, grey skin colour, poor ripening, poor conversion of starch to sugar, poor flavour development and susceptibility to decay. Symptom development is dependent upon exposure temperature and time and susceptibility depends mostly on cultivar. Maturity, prior growing conditions and previous temperature exposure may also be factors. 'Dwarf Cavendish' show injury after 20 days at 11°C and 'Lacatan' after 12 days at 14°C . Plantain, such as 'Pisang Awak', are less susceptible to chilling injury than dessert types.

Marketing

Market quality standards vary widely, with export bananas having the most stringent standards. In large measure, this is related to consumer preferences and the condition of the bananas received at the market. In Western supermarkets, unblemished fruit are preferred, even required, while in markets in tropical areas postharvest handling is very abusive and fruit with blemishes are normal, with degree of fruit ripeness being a deciding factor.

Export quality standards are applied before and after shipping and include: blemishes and fruit shape, finger length and diameter, cluster size and arrangement, and carton weight. A minimum finger length on a hand is 203 mm for export from Central to South America, with the whole hand being culled if one finger does not meet this standard. When hands are cut into clusters, the cull fingers are carefully cut out. The diameter for 'Cavendish' fingers ranges from 31 to 41 mm. Fingers with obvious blemishes are culled. Such standards can lead to cull rates of up to one-third, with culled fruit being sold locally, processed or dumped.

Banana fruit are very susceptible to both abrasion and impact injury. The major marketing problem of export bananas is mechanical injury, which shows itself as black sunken areas on the skin after ripening. The thin plastic liner in export cartons minimizes chafing damage to fingers that rub against the side of the carton during handling. Latex allowed to dry on the skin oxidizes as brown and black stains and can lead to downgrading of fruit. Other concerns are diseases, particularly crown rot, which may affect the whole carton and promote uneven fruit ripening. Because of weak pedicels, the fingers of some cultivars fall from the hand during ripening, exposing the pulp.

Ripening

Dessert bananas are allowed to ripen and are chiefly eaten raw, when they have low starch, high sugar and developed flavour. Plantains have high starch content and are eaten when green or ripe after boiling, frying or roasting. The conversion of starch to sugar during ripening of plantain-type bananas is less complete than in dessert bananas.

The ripening process for the export dessert-type bananas is normally carried out by specialists under controlled conditions just before distribution and marketing to consumers. When bananas are allowed to ripen naturally, it is difficult to predict when fruit will be ready to eat. Under controlled ripening conditions, ethylene is supplied from compressed gas cylinders, ethylene generators or ethylene-generating chemicals, such as ethephon. Commercially, bananas are treated with about 100 p.p.m. ethylene for about 24 h under controlled temperature and humidity conditions and ventilation to prevent carbon dioxide (CO_2) build-up. Newer systems pressurize the room to allow uniform ethylene distribution and temperature. Temperature control allows fruit to be ripened on a specific schedule to colour stage 3 for distribution in from 4 days at 19°C to 10 days at 14.5°C. The humidity control has a significant impact on the final skin colour developed and flesh softening.

In local markets, fruit are ripened by covering with a tarpaulin or cloth after inserting a packet of calcium carbide (0.3 g L^{-1}) to generate acetylene, a 100 times less effective analogue of ethylene. A prolific ethylene-producing ripening fruit, such as avocado, can also be used as an ethylene source, as can burning of incense sticks or young leaves of a number of trees, such as gliricidia (*Gliricidia sepium* Stend) (Acedo and Bautista, 1991).

UTILIZATION

The flavour, texture, convenience, ease of eating and nutritional values have made dessert bananas very popular (Baldry *et al.*, 1981). Banana is a useful source of vitamins A, C and B$_6$ and has about twice the concentration of K compared with other ripe fruit (Table 5.7). Plantain as a staple provides considerable energy and protein, although the diet needs to be supplemented. In West and Central Africa, people derive about one-quarter of their energy requirements from plantains (Wainwright, 1992).

Bananas are used in special diets where ease of digestibility, low fat, no cholesterol, minerals (high K, low sodium (Na)) and vitamin content are required. The fruit does not cause digestive disturbance, it readily neutralizes free acid in the stomach and does not give rise to uric acid. These special diets are used for babies, the elderly and patients with stomach problems, gout and arthritis.

Table 5.7. Proximate fruit composition of banana and plantain in 100 g edible portion (Wenkam, 1990). Differences in mineral content can be due to cultivar, fertilization, water-supply and environment.

		Williams (AAA)	Brazilian (AAB)	Plantain (BB)
Proximate				
Water	g	71.3	79.2	64.1
Energy	kJ	418	301	523
Protein	g	1.08	1.75	1.28
Lipid	g	0.13	0.18	0.03
Carbohydrate	g	26.56	18.03	33.39
Fibre	g	0.11	0.25	0.43
Ash	g	0.9	0.82	0.87
Minerals				
Calcium	mg	5	2	4
Iron	mg	0.49	0.35	0.54
Magnesium	mg	40	26	35
Phosphorus	mg	18	13	21
Potassium	mg	494	302	393
Sodium	mg	1	1	3
Vitamins				
Ascorbic acid	mg	5.1	8	17.5
Thiamine	mg	0.044	0.026	0.038
Riboflavin	mg	0.045	0.041	0.064
Niacin	mg	0.690	0.610	0.430
Vitamin A	IU	88	82	273

Only a very minor proportion of the total world banana production is processed. The procedures used include canning as slices, drying as slices or flakes, freezing of juice, extraction, frying and fermentation. Purée is canned or frozen and used in baby foods, baking and drinks. Banana essence is a clear colourless liquid used in desserts, juices and drinks. Dried, ripe fruit can be made into a flour. The lack of acidity makes processing difficult, because of the need for pasteurization. A year-round supply of fresh fruit makes preservation less economic. A beer is brewed from plantains and consumed in Uganda and Tanzania.

Corms, shoots and male buds are also eaten as a starch source or vegetable. Immediately after harvest, the pseudostem still has considerable starch reserves. The male bud, with the outer fibrous bracts removed, is boiled in South-east Asia and eaten as a vegetable, after several changes of the water to remove astringency. Banana leaves are used as food wrappers for steaming.

FURTHER READING

Gowen, S. (ed.) (1995) *Bananas and Plantains*. Chapman and Hall, London, 612 pp.

Hassan, A. and Pantastico, F.B. (eds) (1990) *Banana – Fruit Development, Postharvest Physiology, Handling and Marketing in ASEAN*. ASEAN Food Handling Bureau, Kuala Lumpur, Malaysia, 147 pp.

Robinson, J.C. (1996) *Bananas and Plantains*. Crop Production Science in Horticulture, CAB International, Wallingford, UK, 238 pp.

Simmonds, N.W. (1982) *Bananas*, 2nd edn. Tropical Agricultural Series, Longmans, London, 512 pp.

Soto, M.S. (1992) *Bananos* (in Spanish). Lithography Imprenta LIL, San José, Costa Rica, 649 pp.

Turner, D.W. (1994) Bananas and plantains. In: Schaffer, B. and Andersen, P.C. (eds) *Handbook of Environmental Physiology of Fruit Crops*, Vol. II, *Subtropical and Tropical Crops*. CRC Press, Boca Raton, Florida, pp. 37–64.

Turner, D.W. (1997) Bananas and plantains. In: Mitra, S.K. (ed.) *Postharvest Physiology and Storage of Tropical and Subtropical Fruits*. CAB International, Wallingford, UK, pp. 47–83.

CARAMBOLA

BOTANY

Introduction

The family *Oxalidaceae* is primarily herbaceous, often with tubers and bulbs. There are some shrubs and two woody genera; only two of the woody species are of interest because of their fruit, *Averrhoa carambola* L. and *Averrhoa bilimbi* L. The best known twelfth-century Moslem doctor and philosopher in Christian Europe, Averroes, living in Cordoba, Spain, gave the genus its name. Common names for *A. carambola* are carambola, starfruit (English), belimbing (Malay, Indonesian), babingbing (Philippines), caramba, yang-tao (Chinese), carambolier (French), five fingers (English), ma fueng (Thailand), fuang (Laos) and khe (Vietnam). *Averrhoa bilimbi*, native to South-east Asia, produces very acid fruit and is used in curries, chutneys, pickles and preserves.

Area of Origin and Distribution

The centre of origin of carambola is not clear; Indochina, Malaysia and Indonesia are considered the most likely. Other suggestions include the Moluccan Archipelago, India and Sri Lanka. It probably does not exist in the wild state. Knight (1983) suggested a secondary centre of diversification in northern South America around Guyana, where it has been established for over 150 years.

Carambola was taken to the Philippines and India in prehistoric times. A Sanskrit name, Karmana, in India attests to its long presence there. It was reported in Rio de Janeiro in 1856 and in Florida, Trinidad and Tobago in 1887. The crop was probably introduced to Hawaii sometime in the later eighteenth century after 1789 and into Australia at the end of the nineteenth century. The first introduction into the Canary Islands occurred in 1881 (Galan Sauco, 1993).

ECOLOGY

Soil

Carambola thrives in almost any soil type, from sand to heavy clay loam and rocky calcareous soil, that is reasonably well drained. Low lying areas and areas where water ponds for more than 12 h are unsuitable. The pH range of 5.5–6.5 is preferred, but carambola can be grown in alkaline soil to pH 7.7 (Galan Sauco, 1993). It may not produce well under saline conditions.

Climate

The crop is best adapted to sea level to about 500 m in the tropics and warm subtropical areas. Young trees are damaged by frost, while older trees are more frost-tolerant.

Rainfall

A well-distributed rainfall between 1500 and 3000 mm is reported to be suitable, although it does grow well in dry areas and is tolerant of seasonal drought once established. Good-quality fruit have been obtained in the Canary Islands with rainfall and irrigation of 800 mm. Water stress limits root growth, leaf growth and development and initiates flowering (Fig. 6.1) and can significantly reduce yield (Fig. 6.2). Well-watered plants have delayed flowering for up to 3 months and extensive vegetative growth (Fig. 6.1).

Temperature

The ideal is considered to be between 21 and 32°C. Growth ceases and flower opening is restricted below 15°C, while young shoots are damaged by temperatures approaching 0°C and die below 0°C, with complete defoliation at −1°C. Adult trees can die if exposed for more than 24 h at −4°C. Fruit produced during the winter can have 6°C lower sugars than in the summer (Galan Sauco, 1993). Temperatures greater than 30–35°C during flowering reduce pollen germination and pollen-tube growth (Salakpetch, 1987).

Light and photoperiod

Maximum production is obtained in direct sunlight with more than 2000 hours sunlight per year being recommended (Watson *et al.*, 1988). Flowering occurs with photoperiods between 8 and 16 h, although 14–16 h daylight reduced flowering rate and 8–12 h increased the number of flowers per inflorescence (Salakpetch *et al.*, 1990).

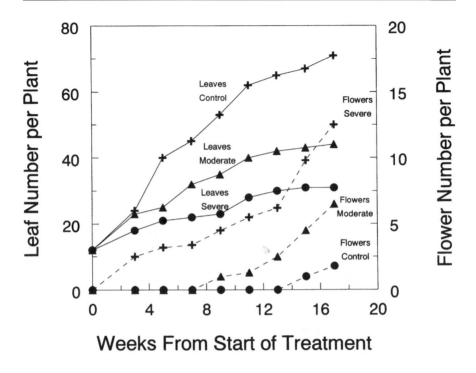

Fig. 6.1. Influence of different soil-moisture availability (88–100% control, 65–70% moderate, 42–48% severe) on leaf and flower number per plant of carambola B-17 (after Ismail *et al.*, 1996).

Wind

Hot and dry wind can harm or kill trees, while cold winds at 10°C and greater than 80 km h⁻¹ can defoliate trees, which require a few months to recover (Galan Sauco *et al.*, 1989). Fruit injury, due to rubbing on branches and other fruit, seriously reduces quality. Windbreaks, either temporary or permanent, are essential to produce high-quality unblemished fruit.

GENERAL CHARACTERISTICS

Tree

Carambola is a small, slow-growing, evergreen tropical tree, rarely more than 8–9 m in height. Young trees generally have a pyramid shape, changing to a symmetrical rounded top. Useful economic life is about 25 years (Galan Sauco, 1993). The trunk is smooth, greyish to dark, with a tendency to form low branches. The roots branch very close to the base of the trunk. These produce very thick lateral and deeply penetrating anchoring roots.

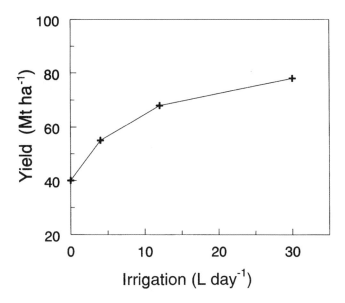

Fig. 6.2. Fruit yield of B-17 carambola trees, which were either not irrigated or irrigated with 4 litres, 12 litres or 30 litres tree^{-1} day^{-1}, only during the drought months, when the rainfall was less than evaporation or rain did not fall for 2 weeks (after Bookeri, 1996).

The leaves are arranged alternately (Fig. 6.3), petiolate and pinnate and may reach 20 cm. The young leaves are bronze-red and pale to dark green when mature. The pinnate leaflets vary in number from two to 11, usually three to six. Leaflet shape is ovoid, ovate–lanceolate or oblong elliptic (Fig. 6.3). The laminae are 2.5–7.5 cm long and 1–4 cm wide, pubescent on the upper surface and almost smooth on the underside.

Flowers

Grafted plants produce flowers in 9 months (Watson *et al.*, 1988), while seedlings may not flower until 4–6 years old. The flowers occur on a loose panicle, produced on basally branched slender twigs (1–8 cm long) towards the periphery of the tree. The panicles occur in leaf axils and occasionally on leafless branches and terminally on new shoot growth or from latent buds on older shoots. The terminal flowers on a peduncle open first. Flowers bloom during the daytime and wilt during the night of the same day (Shiesh *et al.*, 1985).

Perfect flowers have a calyx with five pink sepals (Fig. 6.3) surrounding the purple corolla and are 5–12 cm long, attached to a short (1 mm), round, dark red pedicel. Heterostyly (distyly) is a characteristic (Shiesh *et al.*, 1985). Some cultivars bear long-style (2 mm), short-stamen (3–4 mm) flowers, while others have short styles (0.5–1 mm) and long stamens (5–6 mm) (Table 6.1).

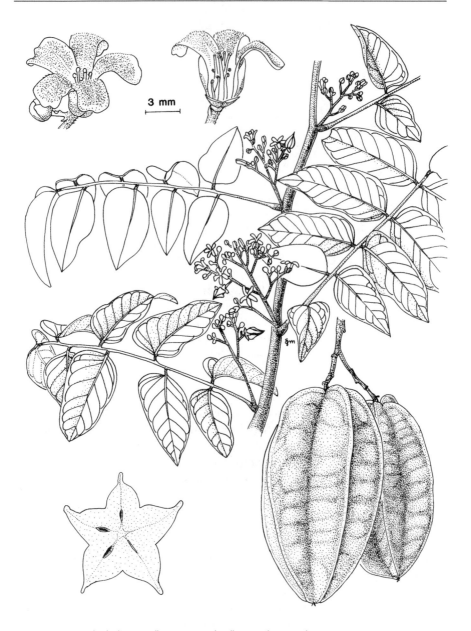

Fig. 6.3. Carambola leaves, flower panicle, flower, fruit and a transverse section through a fruit showing the star shape and seeds.

There are five sepals with ten stamens, of which the five adjacent to the petals are reduced to staminoids with no or aborted anthers. Four or five styles develop from the ovary and fuse with the stigmatic surface, having numerous papillae. The ovary is 15–25 mm long, with four or five loculi, each containing

Table 6.1. Flower-style length varies among carambola cultivars and is described as short or long (after Watson *et al.*, 1988). Cultivars starting with the letter B are from the Malaysian Agricultural Research and Development Institute (MARDI).

Short style	Long style
B-2	B-1
B-8	B-6
B-10	B-11
Fwang Tung	B-16
Lu Tho	Star King
Wheeler	Maha
Sri Kembangsaan	Arkin
Kara	Kary

two to four superimposed ovules (Galan Sauco, 1993).

Flowering can be induced by a period of either water stress or root restriction, or both, and is dependent upon cultivar. Continuous flowering occurs in the tropics, with different sections of the tree apparently flowering at random. The asynchrony of flowering can be reduced by management practices, such as shoot pruning.

Pollination and Fruit Set

From visible flower bud (1 mm) to open-flower stage takes about 2 weeks (Darshana, 1970). Flowers open between 8 a.m. and 10 a.m. and close between 2 p.m. and 6 p.m. the same day (Salakpetch *et al.*, 1990). Petals fall during the morning following anthesis. After opening, though no longer receptive, flowers may remain on the plant for up to 16 days. Insects are apparently needed for good pollination, although some wind pollination occurs. The brightly coloured flowers have nectar and are attractive to bees; one hive is recommended for each 0.4 ha of trees.

Heterostyly is a cause of pollination failure and self-incompatibility. Failure is almost complete in short × short-style crosses (*c.* 0.5% on the same tree and between flowers on different trees). Low fruit set occurs with long × long-style cross (*c.* 0.5%) and high fruit set with long × short-style crosses (3–20%). Microscopy indicates that, in non-compatible crosses, germination occurs and the pollen tube is usually inhibited in the style (Knight, 1982).

Fruit

The fruit is a large, indehiscent, fleshy berry between 5 and 12 cm long and 3–6 cm across. In cross-section, the fruit is a five (occasionally six)-pointed

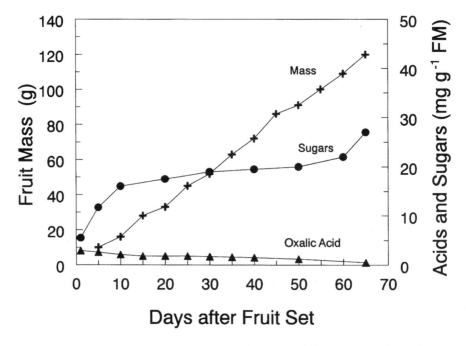

Fig. 6.4. Increase in 'Arkin' fruit mass and total sugars and decrease in oxalic acid content after fruit set (after Campbell and Koch, 1989). FM, fruit mass.

acute star (Fig. 6.3). It is yellow to orange when ripe, with 10–12 ovoid seeds *c.* 1 cm long. Within 7–10 days after pollination, fruit set is indicated by swelling of the ovary or shedding. Fruit set is regarded as having occurred when the petals have fallen, ovary expansion has begun and the colour changes from white to green. Initial fruit growth is via division until 12–15 mm long and then via expansion and elongation (Dave *et al.*, 1975). In Florida and Hawaii, fruit maturity occurs 60–65 days after fruit set (Fig. 6.4). Oxalic and malic acid levels decline during development and especially during fruit ripening, while total sugars increase. The tart or acid varieties do not show a similar marked decline in oxalic acids.

CULTIVAR DEVELOPMENT

Genetics, Cytogenics and Breeding

A. carambola L. is reported to have 2n = 22 or 24. There are limited data on gene frequencies, linkages or inheritance of desirable horticultural traits. Polymorphism has been found in isozyme alleles of a seedling population from controlled pollination (Schnell and Knight, 1989). Seeds of specific crosses can be obtained relatively simply. Anthers can be removed from short-styled flowers with forceps without damaging the flower (Knight, 1965). Long-styled flowers are more difficult to use as a female parent. Self-sterility further limits breeding.

Selection and Evaluation

Sweet and acid types of fruit were frequently used to separate seedlings before named varieties appeared. The most intensive selection programmes have been carried out in Malaysia, Taiwan, Thailand, Florida and Hawaii. The Malaysian Agricultural Research and Development Institute (MARDI) has a numbered series with different desirable characteristics, maintained at Selangor.

An extensive list of desirable characteristics for evaluation of carambola varieties includes early-bearing habit, abundant regular production and ease of training. Plasticity of growth, flowering and fruit set to withstand water stress, salinity and adverse soil and weather conditions would be helpful. The major focus of all selection has been for desirable fruit characteristics, including; weight, wing-to-body ratio, colour, sugar-to-acid ratio, low seed number, flesh texture, flesh browning and resistance to mechanical injury.

There is wide variation in fruit-quality characteristics, with colour varying from pale yellow to deep orange when ripe. The five-angled star shape, occasionally four- or six-angled, and rib depth can vary significantly, with shallow ribs being preferred for packing and avoidance of mechanical injury; the cultivars 'Fwang Tung' and 'B6' have bigger ribs than 'B-10' and 'Arkin'. Sugar content can vary from 5 to 13% sugars (Knight, 1989). A sugar level greater than 10% is preferred for consumers not to regard the fruit as bland.

The use of fruit is sometimes a factor in selection. Dessert fruit and minimally processed slices should preferably have a high sugar-to-acid ratio. Tart fruit are preferred in Taiwan for processing into juice. The acid level, mainly oxalic acid, shows the greatest variation between sweet and tart cultivars, ranging from 1 to 6 mg g^{-1} fresh mass. Oxalic levels are higher in the ribs than the body of the fruit and decline during ripening. The malic acid content is higher in sweet varieties, at 1.2 mg g^{-1} fresh mass, than in tart varieties, at 0.4 mg g^{-1}. The juice pH can vary from 2.4 to 5.0.

Major Cultivars

Seedling populations are very heterogeneous and cannot be relied upon for commercial production, with those from short-style types regarded as yielding greater diversity than long-style types (Watson *et al.*, 1988). Many named varieties are available with acceptable yield, fruit and disease tolerance (Table 6.2), although the climactic suitability of individual cultivars varies widely.

Table 6.2. Fruit characteristics of some carambola varieties grown in different countries (from Campbell, 1971; Sedgley, 1983; Watson et al., 1988; Galan Sauco, 1993; O'Hare, 1993). Flavour is ranked 1 = strongly dislike to 9 = strongly like.

Name	Other names, notes	Country	Weight (g)	Colour		Ribs		Flesh texture	Flavour	TSS(%)	Use
				Mature	Full ripe	Size	Angle				
Arkin	Star King, Sweetie, similar to B-2	US, Florida	90–200	Golden yellow	Yellow-orange	Thick	Large	Crisp	4.9	6.0–11.5	Fresh Processing
Golden Star		US, Florida	100–200		Golden yellow	Thick	Large	Crisp, very juicy			
Sri Kembangsaan	Collected in Malaysia selections, Kary and Kyra	US, Hawaii	150–200		Lemon-yellow	Thin	Sharp	Slightly crisp	6.5	7.0–13.0	
Kaput Lang Bak		Indonesia Singapore									
Cheng Tsey	Chum Choi	Taiwan	Up to 315	Light green	Orange	Thin	Sharp	Slightly crisp			
B-2	Maha 66, slow-growing tree, medium large fruit	Malaysia	100–200	Greenish yellow	Yellow	Deep furrow	Sharp	Crisp		7–8	
B-10	Ching Sing Keow, short styles, vigorous tree, large fruit, moderate resistance to fruit fly	Malaysia	100–200 to 315	Greenish yellow	Yellow to golden Reddish to orange	More compact than B-2	Slightly rounded	Crisp		9–12	Fresh Processing
B-17	Cristal Homey or Honey Carambola	Malaysia		Golden yellow	Golden yellow	Large			5.3	15–18	
Fwang Tung		Thailand	100–300					Crisp	5.5	9.2–12.6	

TSS, total soluble solids.

Taste similarly varies considerably with stage of maturity at harvest (Table 6.3) and this needs to be standardized before making comparisons. Some cultivars show different amounts of change in total soluble solids and titratable acidity during ripening (Table 6.3); for example, 'Arkin' has more titratable acidity and lower total soluble solids when fully ripe than 'Kyra'.

Table 6.3. Changes in total soluble solids and titratable acidity of three Hawaii cultivars and the Florida cultivar 'Arkin' of carambola during fruit ripening on the tree.

	Total soluble solids (%)				Titratable acidity (mequiv 100 mL^{-1})			
Stage of ripeness	Kajang	Arkin	Sri Kembangsaan	Kyra	Kajang	Arkin	Sri Kembangsaan	Kyra
Green	5.5	5.7	5.7	6.3	5.8	6.9	5.8	5.6
Green-yellow	5.9	5.5	6.2	5.9	5.3	6.5	6.3	6.0
Yellow-orange	6.4	6.4	7.1	6.3	5.2	5.4	5.2	5.5
Orange	7.6	7.3	9.6	8.1	4.1	5.0	4.3	3.1

CULTURAL PRACTICES

Propagation and Nursery Management

Mature seeds germinate readily within 7 days, if sown immediately after removal from the fruit. Seeds can be stored for about 2 weeks in the refrigerator, if kept moist. A well-drained medium should be used in trays, with the seedlings later transplanted to pots or bags when the first true leaf is mature.

Side-veneer grafting, chip budding, wedge grafting, modified Forkert bud grafting and splicing are used. Approach grafting is rarely used commercially, being slow and requiring the vegetative potted trees to be adjacent to each other. A range of seedling types are used as rootstock, with incompatibility not being reported, although seedlings vary in vigour. Defoliated, hardened (brown) scion wood is used, with leaves removed from the scion 3–4 days before collection. After 4 weeks, if the scion is still alive, the upper part of the tie is removed and the stock decapitated just above the graft. Watering is reduced to above one-third of normal until the scion makes significant growth. Air layering is difficult with slow root growth, and cuttings are not reliably rooted under mist.

Top working can be done with side-veneer grafting, or wedge or side-veneer grafting on new growth. Conventional propagation on regrowth from stump trees can be used. This allows existing trees to be top-worked when a better selection becomes available and can be expected to start yielding within 9–18 months.

Field Preparation

Deep ripping along the tree row is recommended, and should follow the contour if on hillsides. Drainage is essential and areas in which ponding occurs for more than 12 h should be avoided. Lime may be applied as required in acid soils, with sulphur (S) in alkaline soils, to achieve pH 5.5–6.5. Animal manure or organic matter should be dug into each plant hole, 4–6 months before planting. Fertilizer (nitrogen, phosphorus and potassium (NPK)) should be mixed with topsoil at transplanting.

Transplanting and Spacing

Holes should be dug just before planting and can be done at any time of the year. In the subtropics, spring and early summer are recommended. Eighteen-month-old seedlings, with the roots carefully pruned, if necessary, are used. The seedling ball of soil should be lowered into the hole and additional soil pressed gently into place and watered regularly during the first 2–4 weeks. Seedlings may need to be staked during the first year.

Spacing varies widely, depending upon the cultivar, soil type and pruning practices. Mature trees can reach 7.5 m in diameter. Minimum distances of 5–7 m between trees and 6–9 m between rows have been recommended. Spacings, however, can vary from 4.2 m × 3.6 m to 10 m × 10 m, the smaller spacing giving 900 trees ha^{-1}. Higher densities are used for younger trees, with surplus trees being removed later. Trellises allowing denser tree spacing are used in Taiwan and have been used in Florida and tested in Hawaii. Higher-density planting also shows promise. Intercropping with short season crops is frequently practised, leaving about 1 m around the base of the tree free.

Irrigation Practices

Irrigation is recommended, particularly following planting and during dry periods. Soil-moisture stress appears to be a key factor in inducing flowering in areas with low rainfall. Drought during and after flowering leads to poor flowering, early fruit abscission, reduced yield and small fruit. Applications up to 2000 L per mature tree per week during high-demand periods have been recommended, applied in two to three applications per week. Irrigation should be increased as flowering occurs and as the fruit develop and then reduced as the fruit begin to ripen. Water stress is avoided when trees are being cycled for flowering (Fig. 6.5), where pruning is the main management strategy. Mulching around the base of the tree is very beneficial.

Flood irrigation, sprinklers and microspot sprinklers are used, with the latter system being now more widely accepted. Flooding is to be avoided, as it reduces productivity of mature trees via severe leaf abscission.

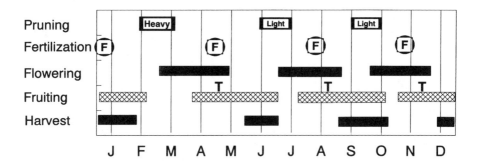

Fig. 6.5. Management practices utilized to induce three harvests per year in southern Taiwan. Double fertilization is carried out 1–2 months before the first pruning in every year during the cool dry season and then every 3 months. The time of pruning is varied to achieve year-round production. Other essential management practices include maintaining water-supply and fruit thinning (T) to improve fruit size and reduce the length of the harvest period.

Pruning

Three to five well-developed branches 50–80 cm from the soil, with wide angles, are allowed to develop. Annual pruning removes narrow-angled, low-hanging and overlapping branches. All side-shoots up to 1 m are removed. Pruning aims to retain recently maturing shoots and generate new shoots, which bear the highest fruit yield, although new shoots have smaller fruit (Wang, 1994). Flowers on main branches are less productive.

Tree height should be maintained at less than 2.5 m to allow fruit bagging, if practised, and harvesting. In Taiwan, trellis pruning of trees at 1.8 m is used to assist in maintaining tree height in an inverse conical shape, while, in Malaysia, trees are tied down to achieve a weeping-willow structure for ease of harvesting from the ground.

Heavy pruning is carried out at the end of the cool season in Taiwan to induce flower-bud initiation 7–15 days later. This heavy pruning reshapes the tree, reducing its height as well as removing slender and senescent branches. Lighter pruning at about 3-month intervals can then be used to develop further flowering and fruiting (Fig. 6.5), giving three harvests a year. The light pruning removes dry, slender and senescent branches to open up the canopy, produce new shoots and flower buds and prevent fruit drop. If water shoots develop between regular pruning, these are removed.

Flower and fruit thinning are frequently used to increase fruit size, reduce the occurrence of misshaped fruit and prevent interference with flower cycling. Thinning to three fruit per cluster 3–4 weeks after flowering has been recommended. If out-of-season flowering is required, all or most flowers and fruit from the previous cycle are removed (Fig. 6.5). When fruit thinning is practised frequently, the individual fruit are bagged when fruit are about 6 cm in diameter. This bagging avoids mechanical and insect sting damage. The bags can be made of white paper, newspaper or polyethylene. Commercial bags 20 cm × 10 cm with a wire tie at the open end are available and can be tied with one hand.

Fertilization

General recommendations can be adjusted after periodic soil and leaf analysis. Young bearing trees should receive 0.4–0.8 kg tree^{-1} year^{-1} of N : P : K (11:12 : 17 to 15:15:15), depending upon the soil analysis. Older trees (8 years) may require 6–25 kg tree^{-1} year^{-1}. The fertilizer can be applied at intervals of about 3 months during fruit production (Fig. 6.5). Fertilization should be limited during flowering. After the last fruit harvest of the year, application is made, followed by irrigation to stimulate new growth, and, when growth slows 1–2 months later, the trees are pruned heavily. Organic matter or manure, 10 kg tree^{-1} year^{-1} for young trees and 10–25 kg tree^{-1} year^{-1} for older trees, applied in a single application, is often recommended. The fertilizer is applied in a 1 m ring around the tree base or along the drip line.

Each tonne of fruit can remove 1.02 kg N, 0.12 kg P, 1.58 kg K, 0.1 kg magnesium (Mg), 0.1 kg S and 0.05 kg calcium (Ca). Leaf analysis from non-bearing trees has shown that they contain 1.4% N, 0.12% P, 0.12% K, 0.98% Ca, 0.64% Mg and 0.24% S. These figures are given as guidelines and may not directly relate to yield.

Pest Management

Diseases

Leaf spot (*Cercospora averrhoae* Petch.) can cause serious loss of leaves and can affect fruit. The small chlorotic spots (up to 5 mm) are at first brown and then turn greyish brown and lead to premature leaf loss. Cupric fungicides and difolatan provide control. Leaf spots caused by *Phomopsis* spp., *Phyllosticta* spp. and *Corynespora cassiicola* have been reported in different areas.

Fruit rot or blemishes due to *Botrytis* spp., *Ceratocystis* spp., *Colletotrichum* spp., *Aspergillus* spp., *Dothioriella* spp., *Alternaria* spp., *Phoma* spp. and *Phomopsis* spp. occur. Fruit blemish can lead to rejection on the packing line. Precooling and refrigeration reduce disease development. Anthracrose (*Colletotrichum* spp.) symptoms are thin, light brown patches, which enlarge and coalesce into salmon-coloured patches that blacken. There are no recommendations for postharvest disease control.

Insect Pests

Fruit flies, such as *Dacus dorsalis*, Oriental fruit fly, in Asia, are major commercial pests and a reason for fruit bagging. This and other fruit flies are also a quarantine restriction to export from tropical countries.

Several moths, including *Othreis* spp., pierce nearly mature to mature fruit and suck the juice. The area subsequently develops rot and the fruit falls prematurely. These are problems particularly in South-east Asia and Australia. Control measures are limited in their effectiveness. Various other beetles, fruit borers, thrips, mealy bugs and scales can attack the branches, fruit and flowers of carambola. Citrus red mite (*Panonychus citri*) is a severe problem, especially in the dry season, in southern Taiwan. Ants are sometimes found around the peduncle depression at the end of the fruit, collecting honeydew produced by tree hoppers (*Membracidae* spp.) feeding on the peduncle.

Non-pathogenic problems

A browning and rotting of the area between the ribs has been reported from Florida. This should not be confused with the surface browning frequently found in cooler months or following moisture loss. Chilling injury, leading to brown patches, has also been reported for carambola. This, however, may be related to dehydration and not low temperatures. Bird and bat damage can be a major problem. Cultivars with sharp rib edges are more susceptible to bruising and discoloration than cultivars with rounded edges. Bruising and rubbing are frequently caused by wind damage.

Weed Management

When trees are small, they are not good competitors and weed control is essential. Organic or plastic mulches are used to maintain a 1-m diameter free area around the base of the tree.

Orchard Protection – Windbreaks

Fruit are very susceptible to wind damage, characterized by rubbing and marking of fruit. The trees can also become defoliated, twigs die back and growth is stunted. Either wind barriers or a protected site or both are needed.

HARVESTING AND POSTHARVEST HANDLING

Harvesting

Grafted trees can start bearing within 9 months of planting, with satisfactory yields after another 2–3 years. Fruit thinning and bagging are used to achieve good yields of adequate-sized fruit, not misshapen and without wind or insect damage. Although trees can produce total yields of 20 kg, increasing up to 500 kg per tree when the tree matures in 6–7 years, 90% of the fruit can be rejected. Yields of 100–250 kg tree^{-1} year^{-1} are more common when there are two to three crops per year.

Maturity is assessed by experience, colour development and per cent sugar levels. Picking is based upon market needs; the minimum is the 'green mature' stage. Sugar increases and acidity decreases as skin colour develops on the tree (Fig. 6.4), with no further increase in sugars after harvest, although colour development continues. At the full-colour stage, the fruit ribs are more fragile and easily damaged; hence fruit are normally harvested at the colour break to 50% colour development, when they have a longer storage life.

Fruit are harvested by hand and should be carefully accumulated in the field and transported to the packing shed. It is necessary to avoid accumulating fruit in deep containers or buckets.

Postharvest Treatments

Fruit should be graded to remove small, misshapen and insect- and wind-damaged and diseased fruit, and then sorted as to size and degree of colour development. Single-layer carton trays or egg-box-type boxes are frequently used. Fruit are wrapped in tissue paper to avoid rubbing injury. Plastic wraps are sometimes used. Cooling as soon as possible after harvest is very effective in extending storage life.

Five weeks' storage is possible at 5°C and 2 weeks at 5–10°C (Yang and Wang, 1993), depending upon stage of maturity at the start of storage. These storage temperatures lead to minimum changes in sugars and a decline in organic acids. Moisture loss is a major problem during storage and can lead to skin browning, which has been mistakenly described as chilling injury. Waxing has been reported to be helpful in reducing desiccation. Plastic wraps, bags or liners in boxes are more effective.

Irradiation (250 Gy), cold treatment (12 days, 1°C) and heat treatments (fruit centre to *c.* 49°C) have potential for insect disinfestation to address quarantine concerns in importing countries. These treatments can occasionally lead to some skin discoloration, dehydration and pitting. Hot water can reduce shelf-life. Green–yellow-stage fruit are more easily damaged by irradiation and hot-water treatments.

Marketing

Careful handling can ensure that fruit reaches the consumer with a minimum of injury, rubbing or browning. Cartons holding 3.5 kg of fruit are used for export; within South-east Asia, larger cartons holding 14 kg are used. The carton should allow for display and still have sufficient padding of either paper or polyethylene to provide protection against crushing, bruising and desiccation.

UTILIZATION

Carambola can be consumed as jams, preserves, pickles, candy, juice and liquor (Matthews, 1989). Green fruit are sometimes consumed as a vegetable. Sliced fruit can be added to salads. The fruit (Table 6.4) is a good source of K and vitamin A and a moderate source of vitamin C (Wenkam, 1990).

Table 6.4. Proximate analysis of carambola (Wenkam, 1990).

Nutrient		Amount in 100 g edible portion
Main ingredients		
Water	g	90.23
Energy	kJ	155
Protein	g	0.85
Lipid	g	0.9
Carbohydrate	g	7.52
Fibre	g	1.47
Ash	g	0.5
Minerals		
Calcium	mg	1
Iron	mg	0.06
Magnesium	mg	9
Phosphorus	mg	11
Potassium	mg	145
Sodium	mg	2
Vitamins		
Ascorbic acid	mg	35
Thiamine	mg	0.04
Riboflavin	mg	0.044
Niacin	mg	0.71
Vitamin A	IU	21

FURTHER READING

Galan Sauco, V. (1993) *Carambola Cultivation*. FAO Plant Production and Protection Paper No. 108, FAO, Geneva, 74 pp.

Knight, R.J. (1989) Carambola cultivars and improvement programmes. *Proceedings of the Interamerican Society for Tropical Horticulture* 33, 72–78.

O'Hare, T.J. (1997) Carambola. In: Mitra, S.K. (ed.) *Postharvest Physiology and Storage of Tropical and Subtropical Fruits*. CAB International, Wallingford, UK, pp. 295–307.

7

GUAVA

BOTANY

Introduction

Guava belongs to the family *Myrtaceae*, which has more than 80 genera and 3000 species, distributed throughout the tropics and subtropics, mostly in the Americas, Asia and Australia. Leaves are usually opposite and simple; flowers are bisexual, with four or five calyx lobes, separated or united at the base, four or five petals and numerous stamens; the ovary is usually inferior, with one to three or more cells; the fruit is a berry or capsule, rarely a drupe or nut-like; and seeds are few to many. Species range from tall trees to shrubs and woody creepers. Many are cultivated as ornamentals and for products such as timber, oil, gum, tannin, resin, spices and fruit. Spices such as cloves, nutmeg, cinnamon and allspice are included in this family.

Important Genera and Species

There are four genera of interest because of the fruit: *Psidium*, *Eugenia*, *Syzygium* and *Feijoa*.

Psidium

This genus is composed of approximately 150 species of evergreen trees and shrubs in the American tropics. A good taxonomic classification of this genus is lacking. *Psidium guajava* L. is by far the most widely known and distributed. Common names include guava, guajaba, gujaba, goyavier, and jambu bat (Malaysian). *Psidium friedrichsthalianum* (O. Berg) Medenza, is known as the Costa Rican jelly guava, and is a shrub or small tree with square branchlets, glossy leaves above and pubescent below; the fruit is small (2.5–3.0 cm), sulphur-yellow and acid; it has some difficulty fruiting at sea level but does well at higher elevations. It is cultivated as a backyard plant in Costa Rica. *Psidium guineense* Swartz. (*Psidium araca* Raddi; *Psidium molle* Bertol), the Brazilian jelly guava, has yellow fruit, 2.5–3 cm diameter, when ripe and is highly acid, bears heavily at sea level and is quite tolerant to lower temperatures. *Psidium littorale* Raddi var. *longipes* (O. Berg) Fosb. and *P. littorale* Raddi var. *littorale* were formerly known as *Psidium cattleianum* Sab. and *P. cattleianum* var. *lucidum* (Degner) Fosb., respectively. The variety *longipes* is popularly known as the purple strawberry guava or Cattley guava, while var. *littorale* is known as the yellow strawberry guava or yellow Cattley, and they are vegetatively indistinguishable. Fruit of both varieties are about 1.5 to 2.5 cm in diameter, seedy, somewhat subacid but with a pleasant aroma.

Eugenia

This genus is a large and heterogeneous group of evergreen trees and shrubs of the American tropics and of the old world. It was revised to include most of the Asian species into the genus *Syzygium*, thus making *Eugenia* a smaller group. The *Eugenia* differ from *Syzygium* in having cotyledons usually united and the seed-coat smooth and free from the pericarp and the inflorescence is generally a raceme of pedicelled flowers. *Eugenia braziliensis* Lam (*Eugenia dombeyi* (K. Spreng.) Skeels) is commonly called Brazilian cherry and is cherry-size, dark red to black on trees up to 15 m (50 ft) tall. *Eugenia uniflora* L. (*Eugenia michelii* Lam), commonly called Surinam cherry or pitanga, is a native of Brazil, has an eight-carpellate (furrowed) fruit, is red to purple when fully ripe and is spicy and acid; it is frequently used as an ornamental.

Syzygium (*Jambosa* DC)

This genus contains about 400–500 species, mostly in the South-east Asian region. They are evergreen with opposite, simple leaves, pinnately veined. The fruit is a berry, usually one-seeded. *Syzygium aromaticum* (L.) Merrill & Perry (*Eugenia aromatica* (L) Baill.) is widely grown in South-east Asia for the sun-dried flower buds called cloves. *Syzygium jambos* (L) Alston (*Eugenia jambos* L.; *Eugenia malaccensis* Lour) is commonly called roseapple or malabar plum. Flowers and fruit are fragrant, dry and crisp. *Syzygium malaccense* (L.) Merrill & Perry (*Eugenia malaccensis* L.) is known as the wax or Malay apple, called mountain apple in Hawaii and is covered in a later chapter.

Feijoa

This genus is represented by two evergreen species in South America. *Feijoa sellowiana* O. Berg. is well known as Feijoa or pineapple guava. *Feijoa sellowiana* is found wild in southern Brazil, Paraguay, Uraguay and northern Australia. It grows well in parts of California under dry subtropical climates.

Origin and Distribution

The guava is native to the American tropics but the extent of dissemination in the pre-Columbian period is obscure. The English name guava probably came from the Haitian name, guajaba. The Spanish explorers took the guava to the Philippines and the Portuguese disseminated it from the Philippines to India. It spread easily and rapidly throughout the tropics because of the abundance of seeds with long viability and became naturalized to the extent that people in different countries considered the guava to be indigenous to their own region. It is now also grown in the subtropics. The hardiness of guava has made it a pest, especially in pasture lands.

ECOLOGY

Soil

Guava is adapted to a wide variety of soil types. Trees will thrive on shallow, infertile soils, although growth and production are poor. It responds well to soils with good drainage and high organic matter having a pH range from 5 to 7. Cultivation in soils with a pH less than 5 or higher than 7 has been observed, deficiency symptoms for zinc (Zn) and iron (Fe) can be expected at pH 7 or higher. Guava is fairly tolerant to salt.

Climate

Rainfall

The guava does best with abundant moisture – 1000–2000 mm is optimal – although it tolerates drought. The ideal rainfall pattern for guava is alternating dry and wet conditions. The concept of artificial cycling to induce flowering attempts to mimic this wet and dry cycle. Drought and very low humidity during flowering can drastically reduce fruit set. Low moisture conditions during fruit enlargement reduces fruit size and purée recovery, due to shrinkage of the inner pulp, which separates from the inner rind.

Temperature

Guava does best in warm areas with abundant moisture and it is grown from sea level to elevations exceeding 1500 m, if frost-free (Maggs, 1984). The optimum is reported to be between 23 and 28°C, with temperatures lower than 23°C and higher than 27°C during flowering reducing fruit set significantly (Fig. 7.1). Guava production in Hawaii thrives very well between 15.5 and 32°C. Young plants are reported to be killed at –2°C, if this

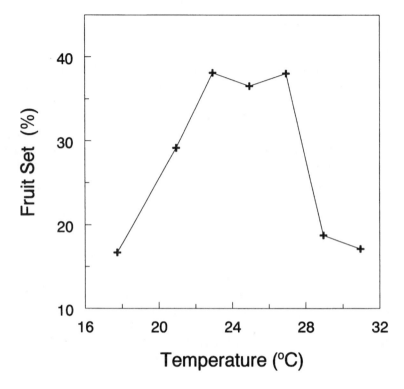

Fig. 7.1. Effect of temperature during flowering on guava fruit set (after Huang, 1961).

temperature is prolonged (Malo and Campbell, 1968). In areas where winter night temperatures are 5–7°C for a few hours a night, such as in Okinawa, growth ceases and leaves become purple. In subtropical regions with insufficient heat calories during the winter months, commercial production is difficult and the time from anthesis to fruit harvest can increase to 220 days (Fig. 7.2). The duration of harvest is shorter, with longer fruit-development time. Low winter temperatures during the dry season lead to natural defoliation, and flowering will commence as soon as warm weather and rainfall induce new growth flushes and fruit set.

Light

Light saturation for this typical C_3 plant is high, above 925 μmol m^{-2} s^{-1} photosynthetic photon flux.

Photoperiod

Guava has not shown any visible response to photoperiod. However, Choudhury (1963) reported some seedlings, grown under 15 h day length

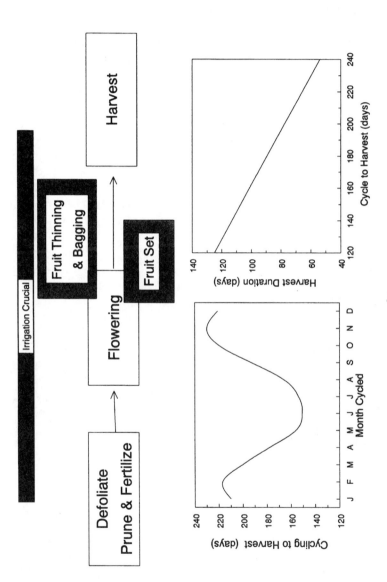

Fig. 7.2. The steps involved in cycling guava trees for fruit production. Relationship between month of cycling, induced by pruning and fertilizing, and days to fruit harvest (after Bittenbender and Kobayashi, 1990) and the relationship between days to harvest from cycling and the duration fruit are available for harvest (H.C. Bittenbender, unpublished) on the island of Kauai, Hawaii. Fruit thinning and bagging are only used for dessert-fruit production, not processing-fruit production.

from germination to 140 days and field-transplanted, which produced fruits within 376 days from sowing. These plants produced a second flowering at 507 days. Control seedlings under 10 h of day length did not produce flowers during this period; seedlings normally require 2–7 years before flowering. Seedlings normally take 2 years from seed sowing to flowering in Hawaii. Due to the absence of dark-interruption studies, these results could be due to greater solar radiation, rather than to day length *per se*. Other results have shown that greater sunlight duration leads to greater shoot growth.

Wind

The guava, though hardy, can benefit from windbreaks. Trees grafted upon seedling rootstocks have tap roots that provide substantial anchorage. However, trees produced from rooted cuttings are subject to uprooting by 65–80 km h^{-1} wind for the first 3 years, due probably to faster top growth than root growth. Trees exposed to prevailing winds of 16–32 km h^{-1} gradually develop branches away from the winds, with little branch growth against the wind. Windbreaks are crucial if high-quality dessert-type cultivars are being grown for the fresh market.

GENERAL CHARACTERISTICS

Tree

The guava is a shrub but, under high moisture conditions, grows to 6–9 m in height and spread, with trunk diameters of 30 cm or more. The trunk is short, freely branching from the base. Under cultivation, a single trunk tree is developed by proper pruning and training. The bark is smooth but peeling and greenish brown to brown in colour. Branches are pliable and hence are rarely broken by winds. The leaves, arranged in pairs, are oblong or oval, 10–18 cm in length, smooth on the upper surface, finely pubescent on the undersurface and prominently veined (Fig. 7.3). Young green twigs are square.

Flowers

Flowers occur singly or in clusters of two to three at the leaf axils of current and preceding growth (Fig. 7.3). The perfect bisexual flowers are white, 2.5–3.5 cm in diameter, with four or five petals, numerous stamens and one style. The tubular calyx encases the bud and splits into four or five segments at anthesis. The ovary is inferior with four or five carpels, each containing numerous ovules in axile placentation. Floral morphology favours self-pollination, but considerable cross-pollination occurs. Flowers open between 5 and 7 a.m., depending upon the cultivar and morning temperatures; the

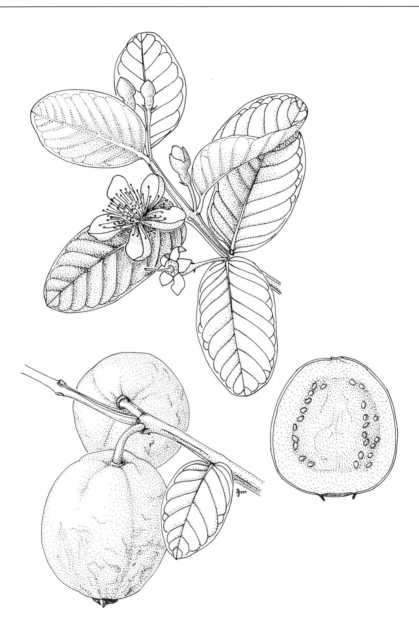

Fig. 7.3. Leaf, flower and fruit of guava.

calyx splits on the previous day. Usually the anthers dehisce at anthesis or shortly before. Bees are the principal pollinators. Two flowering peaks occur in Hawaii and India (Fig. 7.4). These natural peaks can be altered by changing weather conditions or by cultural manipulations (Fig. 7.2).

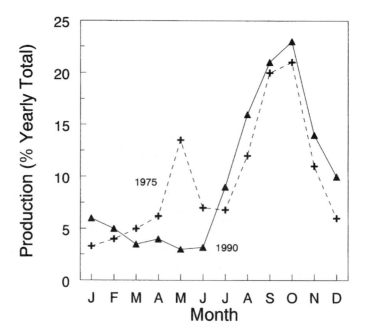

Fig. 7.4. Change in the monthly production of guava in 1975, when cycling was not used, and in 1990, when a major portion of the Hawaii production was cycled to give a single flowering peak.

Pollination and Fruit Set

Natural pollination

Few problems in fruit set occur with most guava clones. Guava generally have pollen with high rates of germination (Hirano and Nakasone, 1969b), except for triploid clones, such as 'Indonesian Seedless'. Fruit set in the triploid cultivars is good when grown together with diploid clones as a pollen source. Fruit set as high as 90% is obtained when 48-h-old flowers are pollinated. The period of stigma receptivity, in 'Beaumont', a Hawaii cultivar, is about 48 h. Post-fruit-set drop does occur as a result of factors other than pollination. Blossom end rot, caused by calcium (Ca) deficiency, can occasionally cause heavy drops.

Some degree of self- and cross-incompatibility among guava clones has been shown; some combinations are totally incompatible while reciprocal crosses produce some fruit (Table 7.1). Self-pollinated 'Beaumont' produces 100% fruit set and between 60 and 80% fruit set by cross-pollination with other guava cultivars (Ito and Nakasone, 1968). These incompatibilities are the result of inhibition of pollen-tube growth in the style.

Table 7.1. Cross-compatibility of some guava clones as percentage mature fruit from self- and cross-pollinations (Ito and Nakasone, 1968)

| Female parent | % Fruit set | | | | |
| | Male parent | | | | |
	7197	7199	Beaumont	Lucknow	Indonesian seedless
7197	67	33	33	29	0
7199	33	64	50	50	0
Beaumont	62	67	100	100	0
Lucknow	57	33	33	27	0
Indonesian seedless	57	14	0	0	0

Fruit set by chemicals

Growth regulators have been tried, primarily to produce seedless fruit. Indolebutyric acid, *p*-chlorophenoxyacetic acid and dichlorisobutyrate do not reduce fruit drop. Fruit set with 50 µg mL^{-1} gibberellic acid (GA) is greater and fruit contain fewer seeds and improved quality characteristics, such as total soluble solids, ascorbic acid and reduced fruit drop. When unopened flower buds are emasculated and styles cut off and the GA–lanolin mixture is applied to the cut surface, seedless fruit are obtained. Fruit obtained by this treatment are larger than fruit from untreated flowers (Shanmugvelu, 1962). The fruit, however, show six to eight prominent ridges and swelling at the calyx end.

Fruit

The fruit is botanically a many-seeded berry, varying in size from 2.5 to 10 cm in diameter (Fig. 7.3). The shape can be globose, ovoid, elongated or pear-shaped. Skin colour is yellow when ripe, but flesh colour may be pink, salmon, white or yellow (Plate 13). Skin texture may be smooth or rough. The inner wall of the carpels is fleshy and of varying thickness and seeds are embedded in the pulp (Fig. 7.3). Flavour and aroma vary widely among seedling populations. There are low-acid, sweet types, bland types that are low in both sugars and acidity and high-acid types. The undesirable musky aroma is more pronounced in the fully ripened low-acid, sweet types.

Fruit growth follows a simple sigmoid curve (Fig. 7.5) and pulp growth parallels total fruit growth. The days from anthesis to harvest can vary from about 120 to over 220 days (Fig. 7.2), depending upon temperature during fruit development. Cultivars also vary in the period to fruit maturity from

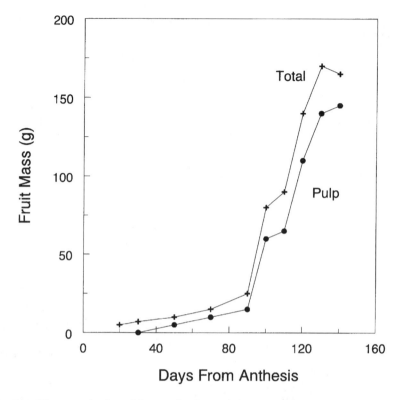

Fig. 7.5. The growth of total fruit and pulp mass from anthesis to maturity in Hawaii (after Paull and Goo, 1983).

anthesis, with up to 60 days' difference being reported. Where labour is available, young fruit destined for the fresh-fruit market are covered with a plastic or paper bag or wrapped with newspaper to protect them from insects (Plate 14). This procedure does not alter the growth rate or final fruit quality (Lam, 1987). If a plastic or paper bag is used, sometimes a polyethylene foam sock is first applied to provide additional protection from bruising. Prior to application of the bag, the fruit are sprayed with approved preharvest fungicide.

CULTIVAR DEVELOPMENT

Cytogenetics and Genetics

The genus _Psidium_, except for _P. guajava_, is represented by di-, tetra-, hexa- and octoploid species (2n = 22). _P. cujavillus, P. guineense_ and _P. friedrichsthalianum_ from El Salvador are tetraploids (2n = 24), while the latter species from Costa Rica is a hexaploid (2n = 66). Vegetatively, plants of the two separate accessions show no visible differences. *P. littorale* var. *littorale* (*P. cattleianum* var. *lucidum*) is hexaploid and *P. littorale* var. *longipes* (*P. cattleianum*) is heptaploid ($2n = 77$) (Hirano and Nakasone, 1969a).

Seedlings produced by the hexaploid var. *littorale* (*lucidum*) are extremely uniform in plant and fruit characteristics, as expected. Heptaploids are not expected to produce uniform progenies but var. *longipes* (*P. cattleianum*) produces uniform progenies. Controlled crosses between var. *longipes* (red-fruited) as the maternal parent and var. *littorale* (yellow-fruited) produced F_1 plants all with red fruit. Seedlings of the reciprocal cross produce yellow fruit. All seedlings look alike and variations in fruit characteristics, such as total soluble solids, total titratable acidity and size among seedlings and parents, are no greater than variations found among fruit of a single plant, indicating that seeds produced by both parents are apomictic rather than of sexual origin.

Plants of *P. guajava* are preponderantly diploid, but triploidy, tetraploidy and aneuploidy have been reported. Seedless cultivars have highly abnormal meiotic division, with univalents, bivalents and laggards, as well as bridging of chromosomes, indicating hybrid origin. Pollen is functional, although pollen-tube growth is inhibited in the style, indicating a gametophytic type of self-incompatibility (Table 7.1). Pollination stimulates fruit development.

Breeding

The flower morphology of guava favours self-pollination, with 35% outcrossing reported. This provides a heterozygous, open-pollinated seedling population, with adequate genetic variation for selection of desirable commercial types. Because of the many-seeded fruit, the guava lends itself to controlled hybridization if there are specific objectives, such as disease or insect resistance. Otherwise, open-pollinated progenies are adequate for cultivar development. A collection of germ-plasm is a necessary part of a breeding programme. An effort should be made to introduce the best cultivars from major breeding centres, that fit the needs of the local industry. Selection from wild populations can often be rewarding.

Selection and Evaluation

A set of selection criteria or objectives in a breeding programme would include all or some of the following. Fruit selection criteria include: (i) large (200–340 g), with few seeds and thick pulp; (ii) white for dessert or dark pink pulp colour, particularly for processing; (iii) flavour and aroma characteristic of fresh guava, with no woody off flavour or muskiness; (iv) more than 10% total soluble solids; (v) for processing, an acidity of 1.25–1.50% and, for dessert guava, 0.2–0.6%; (vi) vitamin C content of 300 g kg^{-1} or higher; (vii) minimum number of stone cells; (viii) good postharvest shelf-life; and (ix) resistance to fruit diseases and insects. Tree selection criteria include: (i) vigorous, spreading, low-growth type; (ii) resistant to tree diseases and pests; (iii) high yield; and (iv) dwarfing rootstock.

Cultivars

India has developed low-acid cultivars, such as 'Allahabad Safeda' and 'Apple Colour', with white flesh (Table 7.2). South Africa's 'Malherbe' and 'Fan Retief' are mild, sweet, dessert types, with light pink colour, for out-of-hand consumption. They are also suitable for canning as halved fruit. For processing, 'Beaumont', 'Kahua Kula' and 'Waiakea' are recommended for Hawaii.

'Beaumont' was the first processing cultivar introduced to the Hawaii industry and was the only recommended processing guava until the introduction of a selection 'Kahua Kula' in 1976 (Nakasone *et al.*, 1976). 'Beaumont' produces large fruit, ranging from 145 to 235 g, averaging about 170 g (Table 7.2) and a 78% purée yield. Fruit are mildly acid, with total titratable acidity ranging from less than 1.0% during the summer to about 1.25% in winter. Soluble solids range from about 7% in winter to 10% during summer. Fruit characteristics of 'Kahua Kula' are similar, with a slightly stronger pink colour and somewhat higher acidity. Both varieties are good yielders, exceeding 227 kg tree^{-1} year^{-1} after the fifth year with some pruning.

Florida has developed both dessert and processing cultivars of excellent quality. 'Ruby', 'Supreme' and the hybrid, 'Ruby' × 'Supreme', are excellent dessert types. Following crosses in 1945, one hybrid was released as 'Homestead' (6-29) in the early 1960s, the other was 10-30. 'Pink Acid' and 'Patillo' are two acid cultivars with dark pink colour, suitable for processing (Malo and Campbell, 1968).

CULTURAL PRACTICES

Propagation

Sexual

Seed germination is used to produce seedlings in breeding and selection programmes or to produce rootstocks for grafting of desirable cultivars. Seeds are sown in a well-draining media and more than 90% of fresh seeds germinate in 15–20 days.

Asexual

Container-grown seedlings may be budded or grafted when stem diameters are 12–20 mm, with greater diameter being especially suitable for budding.

Table 7.2. Fruit analyses of selected guava varieties (Nakasone et al., 1967).

Varieties	Origin	Fruit weight (g)	Flesh colour	Cavity diameter (cm)	Total diameter (cm)	Seeds (%)	Soluble solids (%)	Total acidity* (%)	Comments
Allahabad safeda	India	173	White	5.0	7.5	3.2	12.5	0.6	Sweet, lumpy surface
Apple	India	136	White	4.6	6.5	6.3	8.8	0.4	Sweet, seedy, musky
Beaumont	Hawaii	235	Pink	5.2	7.3	3.0	12.0	1.3	Processing, mild acid, large-fruited
Burma	Burma	210	White	4.9	6.4	4.5	10.7	–	Sweet, seedy
Hong Kong pink	Hong Kong	218	Pink	4.0	7.6	1.8	10.0	0.2	Sweet, thick-fleshed, few seeds
Hong Kong white	Hong Kong	181	White	4.7	6.7	4.2	12.5	0.4	Sweet, seedy
Indonesian white	Indonesia	105	White	4.1	6.3	1.4	10.5	0.3	Sweet, few seeds
Indonesian seedless	Indonesia	176	White	4.3	7.3	0.7	12.5	0.4	Sweet, thick-fleshed, no to few seeds
Lucknow 49	India	227	White	4.4	7.3	3.3	11.0	0.4	Sweet, thick-fleshed, medium seedy
Patillo (7197)	Florida	113	Pink	4.0	5.8	3.3	11.9	1.7	Processing, high acid
Pink acid (7198)	Florida	167	Dark pink	4.7	6.9	2.7	11.7	1.7	Processing, high acid, good colour
6362	Florida	176	Pink	4.5	6.7	2.0	10.5	0.4	Sweet, few seeds, good colour
6363	Florida	250	White	4.6	7.1	2.1	11.4	0.3	Sweet, large-fruited, few seeds
7199	Florida	153	Pink	5.1	6.9	1.9	11.3	0.4	Sweet, thin-fleshed, few seeds, slight muskiness

*Expressed as % citric acid.

Budding is preferred over other grafting techniques, inasmuch as bud growth is faster and each bud on a scion or bud wood is a potential plant. The patch-bud technique or the Forket modification method give good results. Success requires a vigorously growing seedling, where the bark peels readily, and well-prepared bud wood, with swollen axillary buds. A skilled propagator can achieve 90% or better success if done properly (Hamilton and Nakasone, 1967).

In grafting, the side-wedge method is used most frequently. The scion wood or bud wood should be prepared approximately 10–14 days before cutting by removing the leaves from the branch. This practice encourages axillary buds to enlarge and greatly accelerates growth when budded or grafted. Wood that is shedding or has already shed its bark and is smooth, greyish green in colour and without leaves gives good results in budding or grafting. Budding or grafting is useful if rootstocks have special attributes, such as disease or nematode resistance and dwarfing.

Rooting of greenwood cuttings, with two to four leaves retained and the basal end treated with root-inducing compounds and rooted in intermittent mist, has made it possible to produce large quantities of plants in a relatively short time for commercial orchard development (Paxton et al., 1980). Rooting can be achieved within 6 weeks to 2 months. There are clonal differences in rooting ability. In vitro propagation methods have also been developed to propagate guava rapidly.

Field Preparation

Field preparation follows conventional procedures. Soil pH is best maintained between 5 and 7 and lime should be incorporated during land preparation.

Transplanting and Plant Spacing

Well-grown rooted cuttings approximately 6–8 months old can be field-transplanted after hardening in the direct sun for several weeks. Transplanting can be done at any time of the year if irrigation facilities are installed. Planting holes should have approximately 225 g of a complete fertilizer, mixed into the soil at the bottom of the hole. Manure may also be used. Postplant application of 115 g of fertilizer is spread around the plant, mixed with the surface soil and irrigated.

Guava trees tend to have spreading growth habit and are relatively fast growers. In a commercial 15-year-old orchard in Hawaii requiring some pruning, trees are spaced 5.2 m × 7.6 m, with 252 trees ha^{-1}, height being controlled at about 3.7 m. In Australia, a recommended spacing is 4.0 m × 6.0 m, giving a tree density of 416 trees ha^{-1}. Closer spacing requires severe pruning and thinning; 600–1000 trees ha^{-1} are used in Taiwan in a north–

south orientation and are reported to give higher yield in the first 5 years. At the higher densities (73,000 plants ha^{-1}), trees grow in height with few laterals, as compared with the control, although the lowest density (27,000 plants ha^{-1}) produced the highest yield (Mohammed *et al.*, 1984). The spreading growth habits of guava do not appear to support such a high-density system with increasing tree age or the cultivar tested was unsuitable.

Irrigation

Guavas thrive in areas with long dry periods and over a wide range of rainfall. Adequate moisture is required during vegetative growth and for optimum flowering and fruit development. Almost complete postset drop is observed during drought. In the dry tropics, flowering is greatly influenced by water availability. Heavy flowering follows the onset of the rainy period.

Drip irrigation is being used increasingly in production to replenish daily water loss (25–50 mm per week). This method can deliver sufficient water if the entire orchard can be supplied on a daily basis. In large orchards, where irrigation is done by sections, the microjet or a low-sprinkler system is more desirable. These systems also make it easier to apply fertilizer for immediate effect.

Pruning

The objective of pruning is to open up the canopy, as more sunlight leads to more shoots and higher yield. Pruning begins at an early stage of plant growth to develop single trunk trees with well-spaced scaffold branches to form the framework. A single trunk tree with no interfering branches 0.5–1 m from the ground is desirable for hand-harvesting and shaker/catcher type of mechanical harvesting. Mechanical-harvesting tests have been conducted mostly on trees 5–10 years old, when trees may still be shaken with relative ease. Greater force is necessary on older trees, which can damage the trunk or roots. Scaffold branches above the main trunk can be shaken individually in these trees. Single trunk trees are easier to harvest manually than multiple-stemmed, bush-type growth.

Pruning of branches can be done immediately following harvest or the trees can be left for some months. Trees are pruned and fertilized to induce new axillary growth upon which flowers will be produced. Heading back branches induces new, long, whip-like shoot growth, which only has sparse flowering compared with cutting at a fork. Branches grown horizontally are far more productive than vertical ones. Pruning equipment can be large hand-loppers and handsaws or mechanically assisted small chain-saws or pneumatic or hydraulic loppers. These mechanical loppers are best used in teams of four or five cutters.

For dessert cultivars, an ideal tree shape is one with no branches 0.5–1.5 m from the ground and three to six horizontal branches. This flat shape allows for ease of orchard management, fruit thinning and fruit bagging and reduced labour for harvesting. This flat shape requires the end of the branches to be supported. Fruit are thinned and bagged to one or two fruit per new shoot and one or two fruit per node, 30–40 days after flower anthesis (Fig. 7.2).

Cycling

Guavas produce varying amount of fruit throughout the year in the tropics. In Hawaii, a small harvest peak occurs between April to May and a large harvest peak between September and November (Fig. 7.4). Under natural conditions of drought and/or low temperatures followed by irrigation or warm temperatures, there is prolific flowering, especially if the trees have shed their leaves. The fruiting peaks can deviate with prevailing weather conditions and cultural practices. Since flowers are produced on new growth, factors that stimulate new growth, such as irrigation, fertilization, pruning and defoliation, stimulate flowering (Fig. 7.6). The practice of pruning, fertilizing and irrigating at the end of harvest is essential in cycling that concentrates the harvest period (Fig. 7.4) and is enhanced by defoliation (Fig. 7.2). Pruning is repeated every 8–9 months, the period between one pruning and the end of harvest. Initial flowering after pruning occurs in about 5–6 weeks (Fig. 7.6).

Defoliation can be accomplished using urea plus ethephon and a detergent (Shigeura and Bullock, 1983). A 15–25% urea solution alone causes 90% of the leaves to abscise and is as effective as a 5% urea solution with ethephon and a wetting agent; the latter mixture is easier to prepare. A lower amount of ethephon is needed on warm, sunny days. Gibberellic acid acts synergistically with ethephon to induce abscission (Batten, 1984). Following defoliation, new leaves appear on developing shoots in about 3–4 weeks, with a peak of flowering occurring 9–12 weeks after defoliation (Fig. 7.6). This is followed by fruit set and a peak of abscission of young fruit. This fruit abscission may be related to the ability of the new leaves to develop all the fruit set, as the percentage fruit abscission is correlated with number of fruitlets set per tree.

Fertilization

It is a general practice in most areas to apply 110–225 g of a complete fertilizer three to four times a year for the first 2 years. When trees begin to produce commercial quantities of fruit, fertilizer is usually given after harvest, together with pruning and irrigation to encourage new axillary growth. Foliar spray of urea and phosphate solution at the preflowering stage increases mean

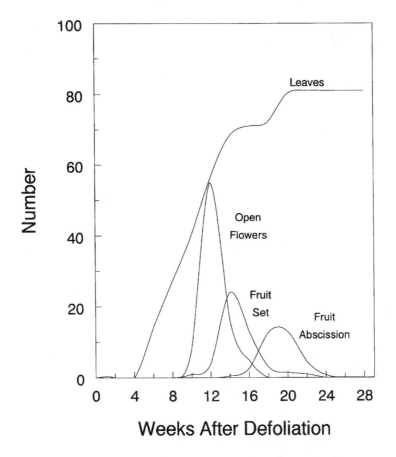

Fig. 7.6. The pattern of leaf production on new shoots after defoliation, flower production, fruit set and fruit abscission: 67% of the flowers abscised, followed by abscission of 69% of fruit, before harvest at Nambour, Queensland, Australia. Fruit matured 18 weeks after anthesis.

yields by as much as 45% and enhances fruit total soluble solids, sugars and ascorbic acid.

It takes approximately 30 days from floral initiation to anthesis and approximately 3.5 months from floral anthesis to fruit maturation (Fig. 7.2). Fertilizer should therefore be applied around 4.5–5 months before expected harvest periods, adjusted for prevailing temperatures and moisture; low temperatures prolong the period. In subtropical climates where winter temperatures are relatively low, only one major crop season occurs and fertilization is adjusted accordingly.

A tentative foliar-analysis guide to guava fertilization on the basis of dry matter is: nitrogen (N) = 1.7%, phosphorus (P) = 0.25%, potassium (K) = 1.5%, Ca = 1.25% and magnesium (Mg) = 0.25%; microelements Zn = 20 p.p.m., manganese (Mn) = 60 p.p.m., copper (Cu) = 8 p.p.m. and boron (B) = 20 p.p.m. (Shigeura and Bullock, 1983).

Yields

Fruit yields depend upon cultivar potential, plant density, weather conditions and all the factors involved in management. The average production in Hawaii is 26.9 Mt ha^{-1}. In experimental planting of 5-year-old 'Beaumont', annual production was estimated at 50.7 Mt ha^{-1} at a plant density of 198 trees ha^{-1}. Management practices for fruit cycling greatly increase production. To emphasize the importance of selecting high-yielding cultivars, in Australian trials, at a plant density of 805 trees ha^{-1}, a yield of 52.3 Mt ha^{-1} was obtained (Chapman *et al.*, 1986). Using the open-centre flat-pruning method in Taiwan, a yield of up to 120 kg tree^{-1} can be obtained, the average being 80–90 kg tree^{-1}; with 600–800 trees ha^{-1}, this gives an average of 59.5 Mt ha^{-1}.

Pest Management

Diseases

Some diseases and causal organisms are specific to certain countries and others are widespread where guavas are grown (Table 7.3). Anthracnose is widespread and is considered an important disease in most countries. Algal spots are very common but are not usually serious, except where fruit of the dessert types are sold in the fresh market, because of the unsightly appearance of the fruit. Other types of fruit rots are attributed to a number of organisms. *Guignardia* fruit rot becomes serious in Hawaii when fruit is left to overripen on the tree or on the ground. Wilting of guava trees is reported from South Africa and India and attributed to different organisms.

Mucor fruit rot first appears as a water-soaked area and later becomes covered with yellowish, fuzzy mycelia and fruiting bodies. Infection rate can be as high as 80–90% and, as a wound parasite, it is commonly associated with fruit-fly oviposition wounds (Ito *et al.*, 1979). Culture control is possible by removing fallen fruit from the field at 2–4-day intervals or by crushing under foot during harvest or lightly rolling the orchard floor. Low acid, sweet cultivars are more tolerant to this disease than acid types.

Blossom end rot of fruit appears to be widespread. In Hawaii, no organism has been isolated and fungicidal sprays have been ineffective. Calcium application to guavas largely alleviates this disease.

Insects

Many insects are common to all guava-growing areas (Table 7.4), most being present in small numbers and causing little damage. In Hawaii, Mitchell

Table 7.3. Some important diseases of guava.

Common name	Organism	Parts affected	Region
Anthracnose	*Colletotrichum gloeosporioides*	Fruit	Worldwide
Glomerella fruit rot	*Glomerella cingulata* (perfect stage)	Fruit	Puerto Rico
Blossom end rot	*Botrytis cinerea*	Fruit	South Africa
	Phomopsis psidii	Fruit	India, South Africa
	Physiological	Fruit	Hawaii, Australia
Fruit canker	*Pestalotia psidii*	Fruit	Australia
Fruit rot	*Macrophomina* sp.	Fruit	Caribbean
Guava fruit rot	*Rhizopus stolonifer*	Fruit	Hawaii
Guava wilt	*Gliocladium* sp.	Root	South Africa, Australia
Fusarium wilt	*Fusarium solani*	Root	India
Guignardia rot	*Guignardia* sp.	Fruit	Hawaii
Mucor rot	*Mucor hiemalis*	Fruit	Hawaii
Algal spots	*Cephaleuros virescens*	Leaf, fruit	Florida, Hawaii, Caribbean

Table 7.4. Some important insect and nematode pests of guava.

Common name	Organism	Parts affected	Region
Mediterranean fruit fly	*Ceratitis capitata*	Fruit	Hawaii, South Africa
Natal fruit fly	*Ceratiits rosa*	Fruit	South Africa
Oriental fruit fly	*Dacus dorsalis*	Fruit	Hawaii
Melon fly	*Dacus cucurbitace*	Fruit	Hawaii
Caribbean fruit fly	*Anastrepha striata*	Fruit	Caribbean, American tropics
No common name	*Monolepta australis*	Leaf	Australia
Red-banded thrips	*Selenothrips rubrocinctus*	Leaf, fruit	Universal
Bark-eating caterpillar	*Indarbela quadrinotata*	Bark	India
Root-knot nematode	*Meliodogyne incognita*	Root	Australia
	Meliodogyne aruenaria	Root	Caribbean
	Meliodogyne acrita	Root	Caribbean

(1973) compiled a list of approximately 45 species of insects and six species of mites attacking guavas. Included among them are species of aphids, thrips, scales, mealy bugs, beetles, moth larvae, false spider, eriophyid and spider mites. There are several species of parasitic wasps and predators that keep scale insects and mealy bugs under reasonable control. Fruit flies cause serious damage and fruit rot within a day or two upon ripening. Fruit bagging, along with thinning 30–40 days after anthesis can significantly reduce the problem and produce high-quality blemish-free dessert fruit.

Thrips can be most troublesome, causing silvering of leaves and scarification of the fruit. When young developing fruit are damaged severely, they often fail to develop and become mummified. Natural enemies can keep thrips under fair control, although outbreaks do occur, especially during the fruiting season. Skin of scarred fruit becomes russeted by disruptions of the epidermal layer and corky-tissue development in the subepidermal area. This corky tissue must be screened out during the processing; otherwise the purée from russeted fruit appears muddy pink in colour. Several sweet cultivars, such as 'Allahabad Safeda', 'Ruby' × 'Supreme' and 'Lucknow-49', have a higher degree of resistance to thrips than others. This resistance increases somewhat with increasing foliar levels of N and K (Foss, 1980).

Weed Control

Weed control is crucial during the first 2–3 years of orchard establishment. After that, the canopy of the trees provides adequate shade to minimize interference by weeds. Mulching with black polyethylene sheets or heavy mulching with organic materials, such as straw, dried grass, wood shavings or whatever material is available, immediately surrounding the plant drastically reduces weed growth. Herbicides are generally not recommended in young orchards, due to the possibility of causing severe damage by spray drift or direct contact. Herbicide such as glyphosate may be applied by rope wicks or rollers saturated with the herbicide solution and wiped on the weed. Grass alleys for harvest vehicles are maintained by mowing.

Orchard Protection

Windbreaks are recommended as rooted cuttings can be uprooted, especially during the first 3 years, with wind velocities of 65–80 km h^{-1}. No trees are lost to winds after the fourth year. If dessert-type fruit are being produced, windbreaks are essential, as the fruit skin is easily damaged by rubbing.

HARVESTING AND POSTHARVEST HANDLING

Harvesting

Harvesting is done manually, although tests with machine harvesting have been conducted in some countries. Dessert fruit are hand-harvested mature green and carefully handled to avoid injury, graded to size and packed carefully in cartons for shipment. Processing fruit should be picked at the firm yellow to half-ripe stage. Overripe fruit and those severely infected with fruit flies and diseases should be destroyed rather than left to fall and rot in the field, as these fruit become the source of continuous field infection. When only fully ripe fruit are harvested on a 3-day cycle, losses between 35 and 40% can occur, as fruit ripen so rapidly and abscise. Harvesting fruit showing some yellow skin to the half-ripe fruit stage allows the interval between harvest cycles to be lengthened to about 3 days with minimum losses.

Fruit are harvested into plastic buckets or packing bags worn by the harvester with bottom delivery. Fruit are transferred into larger bins with capacity for approximately 227 kg of fruit or into small wooden boxes. Harvested fruit should be shaded until delivered to the processing plant. Sorting of fruit according to maturity in the field allows fruit less than fully ripe to be held at ambient temperatures for ripening. When the supply of fruit exceeds processing capacity, ripe full-yellow fruit may be stored at 2.2°C for as long as 2 weeks without injury and for a week at 7.2°C.

Fruit detachment force is related to the stage of fruit development and fruit quality (Fig. 7.7) and is an important consideration if mechanical harvesting is to be used (Paull and Goo, 1983). Fruit requiring a detachment force of 75 N or more are mature green, the force declining to 10 N in overripe fruit with a decline in fruit softness and titratable acidity and a decline then increase in total soluble solids (Paull and Goo, 1983). Any mechanical harvester needs to be adjusted so that the force detaches only yellowing fruit and minimizes loss of green fruit. The difficulty of mechanical harvesting is compounded by the multiple harvests per tree, at 3–4-day intervals.

Postharvest Treatment

Guava is generally grown for processing but, in areas free from fruit flies or where fruit bagging is practised, low-acid cultivars can be grown for fresh consumption. Mature green fruit held at 20°C develops full-yellow skin colour in 6–8 days. Fruit for processing should be held at 15°C to allow gradual ripening and delay the deterioration of quarter- and half-ripe fruit. Ethylene

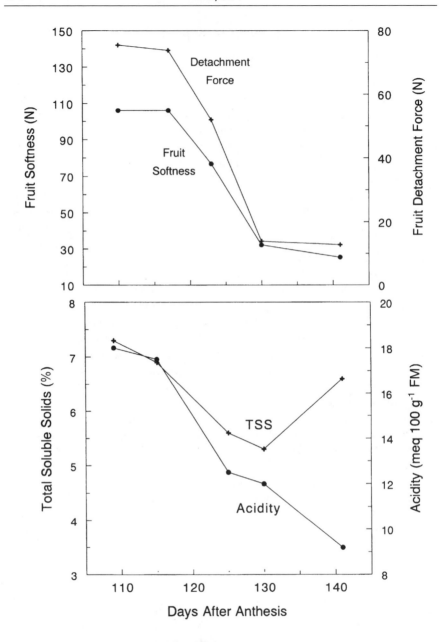

Fig. 7.7. Relationship between fruit-detachment force, fruit softness, total soluble solids and titratable acidity of 'Kahua Kula' guava and days from anthesis (after Paull and Goo, 1983). FM, fresh mass.

can be used on mature green fruit to accelerate ripening. Immature fruit do not ripen properly and develop a 'gummy' texture (Reyes and Paull, 1995). Marketability decreases as storage period increases beyond 10 days. Fruit packed in polyethylene bags can be stored at temperatures of 8–10°C for 14 days and be 100% marketable.

UTILIZATION

Guava is grown widely throughout the tropics and subtropics and is either consumed fresh or processed. India and Mexico are by far the largest world producers, the former producing dessert guavas that are consumed totally within the country. Mexico produces both the low-acid and the processing cultivars, consumed mostly within the country.

Guava is an excellent source of ascorbic acid, dietary fibre, vitamin A and Ca (Table 7.5). The ascorbic acid is mainly located in the skin and a slightly lower concentration is found in the flesh. Concentration ranges from 0.6 to 6 g kg^{-1} edible flesh. The flesh is high in pectin, making it useful for jams and jelly, although pectin necessitates dilution when guava purée is used in juice.

Conventional processing requires hot filling of containers or freezing as a means of preserving the products without the addition of preservatives. To

Table 7.5. Proximate composition of common guava, with seeds removed (Wenkam, 1990).

Amounts in 100 g edible portions		
Proximate		
Moisture	%	8.5
Energy	kJ	230
Protein	g	0.3
Fat	g	0.1
Carbohydrate	g	15
Fibre	g	2.4
Ash	g	0.5
Minerals		
Calcium	mg	15
Phosphorus	mg	16
Iron	mg	0.3
Potassium	mg	292
Sodium	mg	6
Vitamins		
Vitamin A	IU	109
Thiamine	mg	0.06
Riboflavin	mg	0.06
Niacin	mg	1.3
Ascorbic acid	mg	190

eliminate preservatives, aseptic methods with reduced heating that affect quality or the more costly freezing are used. The aseptic-processed product requires no refrigeration and the container is disposable. The purée shows little loss of ascorbic acid and flavour, but there is a significant, though not objectionable, change in colour. Loss of the pink purée colour in storage is more rapid at 38°C than at ambient temperature (Chan and Cavaletto, 1982). Lowering the storage temperature of aseptically processed guava juice to 10°C retards, but does not prevent, colour change and the slight ascorbic acid loss. The purée is used in juice, cakes, puddings, sauces, ice-cream, jam and jelly. Guavas are also dehydrated and powdered.

FURTHER READING

Ali, Z.M. and Lazan, H. (1997) Guava. In: Mitra, S.K. (ed.) *Postharvest Physiology and Storage of Tropical and Subtropical Fruits*. CAB International, Wallingford, UK, pp. 145–165.

Bittenbender, H.C. and Kobayashi, K.D. (1990) Predicting the harvest of cycled 'Beaumont' guava. *Acta Horticulturae* 269, 197–204.

Menzel, C.M. and Paxton, B.F. (1986) The pattern of growth, flowering, and fruiting of guava varieties in subtropical Queensland. *Australian Journal of Experimental Agriculture* 26, 123–128.

Reyes, M.U. and Paull, R.E. (1995) Effect of storage temperature and ethylene treatment on guava (*Psidium guajava* L.) fruit ripening. *Postharvest Biology and Technology* 6, 357–365.

8

LITCHI, LONGAN AND RAMBUTAN

BOTANY

Introduction to Family

The common family name soapberry refers to the tree (*Sapindus saponaria* L.), a native of tropical America which produces a fruit containing a soap-like substance (37% saponin). This dicotyledonous family is composed of around 150 genera and 2000 species of trees, shrubs and a few herbs and vines, usually monoecious, distributed widely in the warm tropics. Among the numerous genera, four related genera and five species are of interest to the fruit horticulturist.

Blighia sapida Koenig

The Jamaican akee is native to Guinea and is commonly grown in the Caribbean islands, particularly in Jamaica. The brightly coloured yellow to red fruit has thick walls with three parts, each containing a white nut-flavoured pulp with a shiny black seed attached at the tip. Pulp from immature or overripe fruit and the pink raphe attaching the aril to the seed are poisonous. When the pods open naturally, the pulp can be fried or boiled. When fried, it resembles scrambled eggs.

Dimocarpus longan (Lour.) Steud.

Longan, lungan, langngan, dragon eye (English), lengkeng (Malaysia, Indonesia), longanier, oeil de dragon (French), lamyai pa (Thailand) or nhan (Vietnam) is reported to have originated in north-eastern India, Burma or southern China in Yunan province. Commonly called dragon eye, the trees attain heights of 9–12 m under suitable environments and are more vigorous than the litchi (Menzel *et al.*, 1988). The longan is generally grown in similar areas to that of litchi in China and Thailand but climatic requirements are less exacting.

Litchi chinensis Sonn., ssp. *chinensis*

The litchi is a subtropical species indigenous to south China. It is principally cultivated in Guangdong, Fujian, Guangxi and Xichuan provinces. Groff (1921) discusses in detail the origin of the names litchi and longan. The Chinese characters for litchi convey the idea that the fruit of the litchi must be removed from the tree by means of knives with the twigs attached. Many cultivar names are pronunciations of various southern Chinese dialects and spellings, and this has led to some confusion. The most recent Chinese descriptions uses the spelling litchi and the national dialect name (Mandarin, the common language). This will be followed in this book, with a version of the other pronunciations in parentheses. Other spellings and common names are lichee, litchee, leechee, lychee, lin-chi (Thailand), laici (Malaysia), lici, litsi, (Indonesia), letsias (Philippines), li-chi or lizhi (China), vai, cayvai or tu hu (Vietnam).

There are two other subspecies; one is found in the Philippines (*L. c.* ssp. *philippinensis*) and the other in Indonesia (*L. c.* ssp. *javanensis*). The Philippine subspecies has a long oval-shaped fruit with thorn-like protuberances, which split when ripe, exposing the inedible aril, partially covering the seed. The Indonesian subspecies is similar to the Chinese species and crops more regularly in hot equatorial areas, although it has not been exploited commercially.

Nephelium lappacium L.

The rambutan (Indonesia, Malaysia, the Philippines, English), litchi chevelu (French), ngoh, phruan (Thailand) is a bushy, open-crowned west Malaysian tree growing to 18–20 m high, in strictly tropical conditions. Rambutan is well distributed in South-east Asia, grown mostly in humid, high-rainfall areas.

Nephelium ramboutan-ake Bl.

The pulasan, ngo-khonsan (Thailand), kapalasan (Indonesia) is native to the Malaysian peninsula and is distributed widely in south-east Asia. Trees grow to 15–20 m in height and the leaves are smaller and narrower than those of rambutan. The fruit is 3.8 cm wide and 5–6 cm long with a dark red or occasionally yellow skin when ripe, covered with short thick spines or tubercles. The aril is translucent, sweet and not attached to the skin. The large light brown seeds easily detach from the flesh. Although the fruit is considered superior to rambutan, it is not as commonly seen in the market. Trees flower during June and August and fruit mature during October–December in Malaysia. Pulasan grows well below 350 m, with annual rainfall of 3000 mm, and in many ways is similar to rambutan in its climatic requirements, except that it is less tolerant of high light levels during establishment than rambutan.

Origin and Distribution

The cultivation of litchi in China antedates the beginning of the Christian era, according to Chinese writings going back to 1766 BC (Storey, 1973). Early distribution into other tropical and subtropical regions occurred during the sixteenth and seventeenth centuries. The litchi was introduced into India in 1798 and has developed into a significant industry. Litchi and longan reached Europe during the early part of the nineteenth century and are mentioned as having reached Trinidad before 1880, with a few trees being grown in Florida as early as 1883 (Groff, 1921). The first litchi tree is said to have been brought to Hawaii about the year 1873 by Mr Ching Chock, a Chinese merchant, and was planted on the property of Mr Chun Afong and became known as the 'Afong' tree, identified as Gui Wei (Kwai Mi), later reidentified as being similar to Da zao. The Office of Foreign Seed and Plant Introduction in the US Department of Agriculture began introducing litchi plants in 1907. The late Professor G.W. Groff, who resided in south China almost continuously from 1907 to 1941 as Dean of the College of Agriculture, Lingnan University, Guangzhou, was responsible for the introduction of a large number of cultivars into the USA. His contributions number 17 out of the 25 cultivars and clones now established in Hawaii. Today, the litchi is found in most countries of Latin America, the Caribbean, India, Mauritius, Africa, Australia, Israel, the Canary Islands, the Madeira Islands and many other countries.

ECOLOGY

The litchi and longan are evergreen subtropical plants, indigenous to areas where cool, dry winters, and warm, wet and humid summers prevail. Such conditions are found mostly in subtropical regions and at higher elevations in the tropics. In Guangzhou, summers are known for their rains and high humidity (> 80%) during the night and day. The winter is cool and dry, frost being rare. Litchi is described as a water-loving, low-elevation plant that is planted largely along the banks and dykes in the Pearl River delta in Guangdong. However, there are upland cultivars (mountain litchi) grown on hillsides or in orchard systems, with very little tillage after the trees mature.

Soil

Observations indicate that the litchi, longan and rambutan thrive on a wide variety of soil types, as long as drainage is good enough to prevent waterlogging. Satisfactory litchi growth occurs in India and Florida in neutral to slightly alkaline soils, with free lime present. In south China, the litchi soils are alluvial silty loams to clays in the Pearl River delta or laterite soils on the hills.

Longan and rambutan thrive best on deep clay loam soils and all three species prefer a slightly acid (pH 5.0–6.5) soil.

Climate

Rainfall and moisture

High rainfall and humidity induce good growth in litchi, longan and rambutan, with litchi being able to resist floods. Roots of litchi trees covered with water for 8 days showed no deleterious effects, even when fruiting. In China and Thailand, litchi are also grown on banks with high water-tables beneath them.

Ample annual rainfall is considered to be around 1250 to 2000 mm for litchi and 1500–3000 mm for longan and rambutan. When rainfall is lower, irrigation is essential. A dry autumn and winter are important to prevent litchi vegetative growth (Fig. 8.1), which is essential for good flowering (Nakata and Watanabe, 1966). In south China, litchi anthesis occurs during the rainy season (130–375 mm month⁻¹) and when relative humidity is higher than 80%. Too much rain during anthesis, however, can reduce flower opening and insect activities needed for pollination. A major problem of litchi culture in Hawaii, with the exception of the Kona district of Hawaii, is that rainfall comes in late autumn and winter, making it difficult to control vegetative flushing. Irrigation after a period of soil water stress is reported to aid in longan flowering (Fig. 8.1), even though the tree flowers profusely in areas with high water tables.

Rambutan flowering is correlated with the end of the dry season, leading to a second smaller crop per year on those branches that did not bear fruit in the previous fruiting. A dry period of at least a month is essential to initiate rambutan flowering (Fig. 8.1). The intensity of flowering appears to be closely associated with duration of water stress. Management practices in Thailand are directed towards producing a large crop in the May–June period. The cultivars 'Jitlee' and 'R137' are more suited to areas with some water stress than 'R4' or 'R7'. There are numerous stomata on the fruit spinterms, and low humidities during fruit development can lead to fruit desiccation. All three crops require adequate moisture, from rainfall or irrigation, during fruit set and growth.

Temperature

Young litchi and longan trees are severely damaged or even killed at −2 to −3°C, with dormant and mature trees having more tolerance to low temperatures. Periodic cold is required for flowering, although in Guangzhou air temperatures rarely drop below 0°C. The total duration of relatively low temperatures, rather than the frequency or time of occurrence during a critical period, seems to be the determining factor (Fig. 8.1). When trees are deeply rooted, low air temperature is more important than soil moisture (Menzel *et al.*, 1989). There is considerable cultivar difference in response to temperature

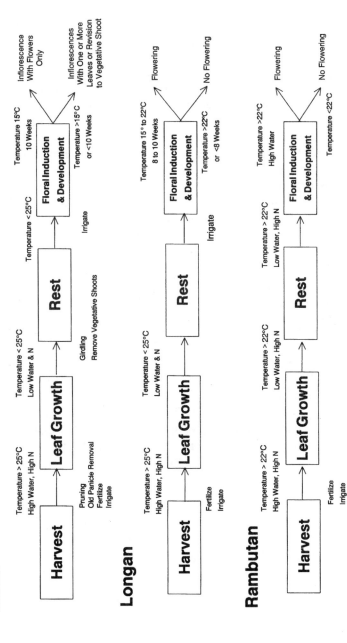

Fig. 8.1. Fruiting cycle of litchi (after Menzel and Simpson, 1994), longan and rambutan, as affected by temperature, nitrogen (N) fertilization and water availability. Litchi-leaf low nitrogen levels need to be in the range of 1.75–1.8%. There can be more than one flush of leaf growth during the year.

(Menzel *et al.*, 1989). In a test involving seven cultivars, all cultivars showed panicle emergence after 6 weeks at 15/10°C (day/night) or at 20/15°C; however, there are differences in the degrees of flowering between the cultivars at 20/15°C (Table 8.1). There is no flowering at 25/20°C or 30/25°C. Temperature also influences the time from emergence to anthesis at different temperatures; 14–16 weeks at 20/15°C versus 6–8 weeks at 15/10°C. The greatest number of inflorescences per branch occurs if low temperatures are maintained until anthesis. If temperatures are increased from a varying period at 15/10°C to 30/25°C after emergence and before anthesis, there is a reduced number of female flowers and some floral buds revert to vegetative growth (Fig. 8.2). Pollination is optimum at about 19–22°C and occurs in 5–7 days. At lower temperatures, pollen-tube growth is strongly inhibited. Hot, dry conditions may reduce yield, as poor fruit set occurs at 33°C. In hot equatorial areas, both litchi and longan grow vegetatively and seldom flower.

Longan is less demanding than litchi for flowering, requiring 15–22°C for 2–3 months for flower induction (Fig. 8.1). The exact chilling and cultivar differences in this requirement are not known. Rambutan has no cold requirement for flowering and is suited to tropical areas with a temperature range of 22–30°C. Temperatures less than 22°C reduce leaf flushing and can extend fruit development to nearly 6 months.

Light intensity and photoperiod

Litchi is a day-neutral plant and floral initiation is directly associated with the starch content of leaves and stems. Shading, such as when litchi trees begin to crowd each other in an orchard, leads to a decline in production, due to smaller panicles and a decrease in the number of panicles on a tree (Lin *et al.*,

Table 8.1. Effect of temperature and cultivar on percentage of litchi terminal branches flowering. Trees at 25/20°C (day/night) and 30/25°C did not flower (Menzel and Simpson, 1988). Panicles can be of two types; one is all flowers, while the other has a mixture of leaves and flowers and is referred to as a leafy panicle.

| | | | Branches flowering % | | | | | |
| Cultivar | Vegetative shoot (%) | | Leafy panicles | | Leafless panicles | | Flowering panicles | |
	15/10°C	20/15°C	15/10°C	20/15°C	15/10°C	20/15°C	15/10°C	20/15°C
Da zao	0	50	93	50	7	0	100	50
Bengal	0	49	71	32	29	9	100	41
Shui dong	0	58	39	3	61	28	100	31
Gui wei Pink	0	20	55	69	46	8	100	77
Gui wei Red	0	80	7	3	93	5	100	8
Salathiel	0	17	0	20	100	58	100	78
Huai zhi	0	4	0	14	100	81	100	95

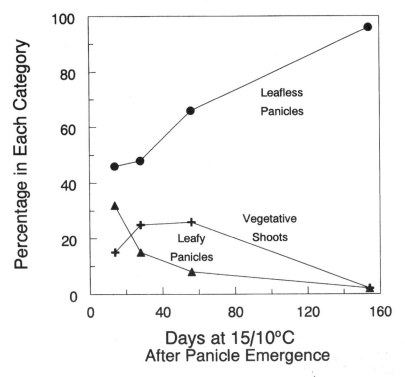

Fig. 8.2. Effect of postpanicle emergence temperature response to 15/10°C (day/night) before transfer to 30/25°C on 'Gui wei' litchi panicle development (after Menzel and Simpson, 1991).

1991). Shading to 50% of full sunlight causes a 40% reduction in the panicles formed and a 65% reduction in panicle length. There is little information on the response of longan and rambutan to varying light levels; both are thought to be day-neutral.

GENERAL CHARACTERISTICS

Tree

The litchi is a spreading evergreen tree, growing to 10–12 m under favourable conditions. It is generally considered to be a slow grower with a long life. A Chinese writer, Wen Hsun Chen, said that, at the time of his writing (before 1924), the cultivar 'Chenzi' ('Chen Family Purple', 'Brewster') was already 300 years old and still thriving (Groff, 1921). Depending upon cultivar, trees may be broad with low-hanging branches or have upright branches and a compact, rounded head. Branches are brittle and even large limbs are easily broken by winds.

Leaves are arranged alternately, pinnately compound with two to five
pairs of leaflets arranged in opposite positions or slightly obliquely along the
rachis (Fig. 8.3). The leaflets are elliptical to lanceolate, 2.5–6.4 cm wide, 7.6–
17 cm long and deep green. Young emerging flushes range from pale green to

3 mm

Fig. 8.3. Litchi leaves, panicle, flower and fruit.

pinkish to a copperish red (Plate 15). Growth flushes can occur several times a year. Vegetative flushing is encouraged immediately after harvest in summer by fertilizing and irrigating (Fig. 8.1). There is a close relationship between the number of leaves associated with a panicle and the number of fruit on that panicle (Fig. 8.4). Hence, a minimum number of two leaves are regarded as being required per fruit.

The compound longan leaves are around 25 cm long with six to 12 leaflets, alternate or nearly opposite, oblong, 5–15 cm long with blunt or pointed ends (Fig. 8.5). Rambutan leaves are 40 cm long, divided into two to four pairs of leaflets, usually alternate (Fig. 8.6). Leaflets are 7.6–23 cm by 3.8–9.0 cm, pale when young and becoming medium to dark green.

Flowers

The small litchi flowers are 3.0–6.0 mm in length when fully expanded (Fig. 8.3), and borne in profusion on leafless or leafy, well-branched terminal panicles (Plate 16). In the northern hemisphere, appearance of panicles can

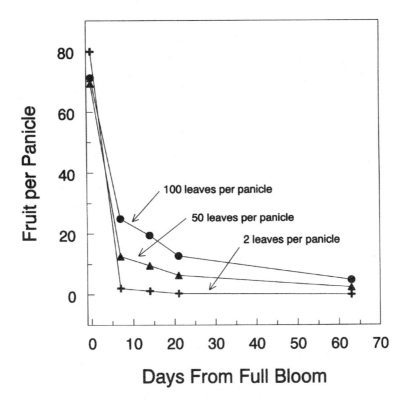

Fig. 8.4. Effect of leaf number on the litchi panicle branch to number of fruit set per panicle for cultivar 'H-1224' (after Yuan and Huang, 1988).

3 mm

Fig. 8.5. Longan leaves, panicle, flower and fruit.

begin as early as November to as late as April, depending upon cultivar and
environmental conditions. Flowers are yellow-green or brownish yellow,
apetalous with small valvate sepals (Fig. 8.3), a fleshy disc and up to eight
stamens. Among the numerous, apetalous flowers on the panicle, three sex

2 mm

Fig. 8.6. Rambutan leaves, panicle, flower and fruit.

types are described for litchi. These sex types are designated types I, II and III, based on the chronological order of development. Type I is morphologically and functionally a male (Fig. 8.3), possessing six to eight stamens producing an abundance of pollen and lacking ovule differentiation. The type II flower is morphologically a hermaphrodite, functioning mostly as female with a well-developed two carpel pistil and a two-lobed stigma (Fig. 8.3). The five to eight stamens do not dehisce normally and have little viable pollen. The type III flower functions as a male, although it is morphologically a hermaphrodite with a rudimentary pistil lacking style and stigma. The six to eight stamens produce numerous viable pollen grains. Approximately two-thirds of the flowers on a panicle are composed of type III and only about 20% of the flowers are potentially fruit-producing (type II). Observations indicate that the ratio of functional females (type II) to functionally staminate flowers (types I and III) varies with cultivar (Table 8.2), with 10–30% being female (Limangkura, 1966). Similar sex types occur in longan (Lian and Chen, 1965), with small, greenish yellow to brownish, self-incompatible flowers having five sepals and five petals (Fig. 8.5). Insects are the major pollinators (Alexander *et al.*, 1982).

Rambutan trees are normally classified into the three types: male trees producing only staminate flowers (40–60% of a seedling population), trees with hermaphrodite flowers that are functionally female and trees with hermaphroditic flowers some of which are functionally female and others functionally male (Fig. 8.6). The last type is more desirable and is most commonly found in some rambutan cultivar selections, with male flowers in the range 0.5–0.9% (Lam and Kosiyachinda, 1987).

Litchi floral anthesis occurs in cycles of sex types (Mustard, 1954), with a period of about 10 days during which only functional male flowers are open. This is followed by a transitional period of 2–3 days when both staminate and female (type II) flowers open and then 2 days of only female flowers. This is followed by 2–3 days of transitional types and finally male flowers for a period of 7–10 days. The number of days to complete each of the cycles appears to differ somewhat among inflorescences, between cultivars and with environmental conditions. Inflorescences produced around the same time generally produce the same sex types. The senior author has observed trees where the initial flowers are all functionally male in one season, while in the following season initial flowers at one site are all functionally female and at the other site all functionally male. Thus, the inflorescence may begin floral anthesis at any sexual phase of the cycle, going through the above sequence. The expression of sex type may be related to temperature, with lower

Table 8.2. Distribution of the three sex types of flowers per panicle of two litchi cultivars (Limangkura, 1966).

Cultivar	Flowers per panicle	Functional flower type (%)			Potential fruiting flowers (%)
		I	II	III	
Groff	964	14	20	66	20
Hei ye	1042	22	19	59	19

temperatures leading to a higher percentage of type II flowers. Longan show a similar progression of flower types: functionally male, hermaphrodite and finally male. As with litchi, there is an overlap of types on one tree, since not all panicles open at the same time.

Flowers of rambutan are small, less than 6 mm wide, greenish, apetalous, with four to six sepals, and are produced in panicles (Fig. 8.6) that are axillary near the branch apex (Allen, 1967). There can be 3000 male flowers on a panicle, with 500 or so open on a specific day. Trees classified as 'female' produce two types of hermaphrodite flowers, of about 500 flowers per panicle, with 100 open per day. The male-functional flowers have well-developed stamens and pistils that fail to function normally, as the bifid stigmas split but remain erect, preventing the exposure of the sticky stigmatic section required for pollination. The female-functional flowers have well-developed pistils with non-functional stamens, reduced to five to seven staminodes. The bifid stigmas split open and curve downward to expose the sticky stigmatic surface and remain receptive for a day. Nectar production for both flower types begins at anthesis.

Pollination and Fruit Set

Natural pollination

Controlled pollination studies have shown the presence of a relatively high degree of self-sterility in litchi and longan. Limangkura (1966) self-pollinated litchi flowers and produced 4.2% set, while cross-pollinated flowers yielded 6.9% set. Litchi inflorescences caged to prevent insect visitation had 0.026–0.105% fruit set compared with 0.17–11.25% fruit set on inflorescences open to insect visitation (Pandey and Yadava, 1970). Bees (*Apis* and *Mellifera* spp.) comprised 98–99% of the total insect visitors to the flowers, mostly during the morning hours, when nectar secretion occurs. The stigmas of litchi flowers are receptive to pollination for 72 h from the time they divide into the two lobes, and the anthers remain functional up to 1–3 days after anthesis.

Only 1–2% of flowers form fruit on a rambutan panicle. Since no pollen is apparent on the hermaphrodite flower anthers, pollen transfer or apomictic fruit development may occur. In mixed plantings with no male trees, pollination problems do not occur if flowering periods overlap (Plate 17). The cultivar 'Seechompoo', having low functional-male hermaphrodite flowers, is notorious for poor fruit set in a pure stand.

Floral induction, fruit set and chemicals

Growth regulators and other horticultural techniques have been tried by many researchers to overcome the problem of irregular litchi flowering, to enhance fruit set and to reduce seed size (Joubert, 1986). Sodium naphthalene acetate (SNA) at 200 and 400 p.p.m. promotes blossoming in 'Chenzi' only when climatic conditions favour vegetative growth. The inhibition of vegetative growth is a requisite for floral initiation and SNA mimics low temperatures, which inhibit growth. There is no increase in floral initiation and yields by SNA during years with dry autumn months and on trees that yielded heavily during the previous season. Foliar-applied paclobutrazol and ethephon increased earlier rambutan bud formation and increased percentage flowering.

A combination of ethephon and kinetin is more effective in reducing litchi shoot length and inducing high proportions of floral buds than either chemical alone (Chen and Ku, 1988), while unsprayed shoots remain vegetative. Killing winter flushes by the use of chemicals to promote litchi flowering is also practised. A 25% urea solution, ethephon (1500 mg l⁻¹), maleic hydrazide (0.125%) and dinitro-*o*-cresol (DNOC) at 0.1–0.5% force young winter-flush leaf abscission with subsequent emergence of axillary flowering. A 0.5% urea spray applied to late-summer growth flushes causes the soft leaves to turn green, mature quickly and produce flowers in the spring. The higher the concentration of ethephon applied at the time when trees approached flowering, the lower the percentage of flowers produced.

Girdling (cincturing) branches or the trunk in late summer increases flowering and fruiting (40–800%) in litchi trees that would have flowered poorly in the spring, when a tree has not produced a growth flush in the 6 months before the floral initiation period, is in poor condition or has a late autumn vegetative flush (Menzel and Simpson, 1987). Trees to be girdled should be fertilized immediately after harvest and irrigated to induce flushing and then girdled in late summer when the leaves are fully matured (Fig. 8.1). Girdling of limbs and trunks of trees that start flushing in late summer has no significant effect on trees that would have flowered profusely on their own. Early recovery from girdling is associated with vegetative flushing during winter (Menzel and Paxton, 1986b).

Retention of litchi fruit after setting is a universal problem. Fruit development is punctuated by four phases of abscission (Fig. 8.7), the first occurring within 15 days of anthesis, in which more than 92% of pollinated flowers drop due to endosperm degeneration and subsequent embryo abortion. The second period of abscission is possibly due to degeneration of the embryo sacs. Phase IV does not occur in fruit that have aborted seeds. There are many causes of fruit drop, including unfavourable climatic conditions and nutrient deficiencies. Newly set fruit sprayed twice at 15-day intervals with 35–100 p.p.m. of 2,4,5-T and naphthalene acetic acid (NAA) progressively reduces fruit drop, improves fruit size and even hastens ripening (Prasad and Jauhari, 1963). Lower concentrations also reduce litchi fruit splitting, another serious problem in many areas. Growth regulators (GA 100 p.p.m., NAA 20 p.p.m., 2,4,5-T 10 p.p.m., chlormequate 250 p.p.m.) sprayed on litchi cv. 'Rose Scented' at the pea stage had reduced fruit drop (Khan *et al.*, 1976). Spraying 'Early Seedless' and 'Calcuttia' with 20 p.p.m. IAA enhances fruit set, 50

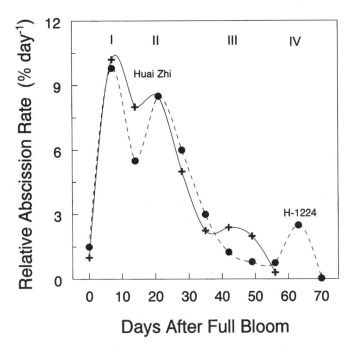

Fig. 8.7. Litchi fruit abscission rate, as a mean of 2 years, occurs in at least four stages (I–IV). Normal-seeded 'Huai zhi' (+) harvested 75–80 days after full bloom and aborted-seeded 'H-1224' (•) harvested at 70–75 days after full bloom (after Yuan and Huang, 1988).

p.p.m. GA-3 increases fruit retention and 100 p.p.m. GA-3 improves fruit weight, with a combination of IAA and GA-3 being suggested (Singh and Lal, 1980). Rambutan panicles sprayed with SNA or NAA had enhanced fruit set.

Reduction of seed size or the production of seedless fruit has also been an objective in the use of growth regulators. Kadman and Gazit (1970), using 2,4,5-trichlorophenoxypropionic acid (2,4,5-TP), prevented fruit drop to a higher degree and this also resulted in over 75% of the litchi fruit having small seeds. However, if the inflorescence is first treated with 2,4,5-TP and then sprayed with a combination of 2,4,5-TP and GA, fruit are 50–100% larger than those treated only once, and 90–100% of the fruit are seedless.

Fruit

Litchi fruit takes from 80 to 112 days to reach maturity from anthesis, depending upon cultivar and environment (Fig. 8.8). Longan growth is similar to litchi, with fruit reaching maturity in 100 days (Fig. 8.9). Litchi aril growth begins about 25 days after anthesis, while longan arils start about 50 days after anthesis, and then both aril growths parallel total fruit growth. The fruit

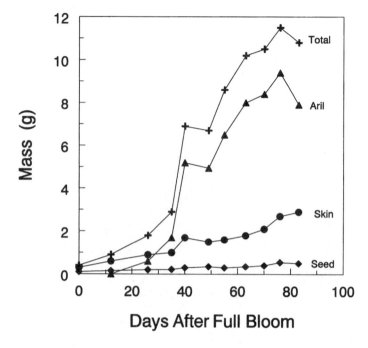

Fig. 8.8. The growth pattern of litchi cultivar 'Groff' fruit (after Paull *et al.,* 1984).

shows a dramatic increase in total sugars, from about 3 to 16–20% during the rapid growth period of the litchi and longan aril. The tart flavour of immature litchi fruit is due to high organic acids. The increase in sugar/acid ratio as fruit matures is due to the decrease in organic acids and an increase in sugars (Paull *et al.,* 1984), contributing significantly to the litchi flavour. Peel anthocyanins increase and chlorophyll decreases during this same late growth phase. Cultivar differences do not allow peel colour to be an accurate measure of maturity. Late-season cultivars show a longer initial growth phase than the early and midseason cultivars.

Rambutan fruits occur on woody stalks in clusters of 10–18, are large, ovoid or globose, about 4.5 cm long and 2.5–3.7 cm broad (Plate 17). The outer skin on the pericarp is 2–4 mm thick and covered with soft, long spines (spinterns), which turn red or yellow when ripe. Early rambutan fruit growth (Fig. 8.10) is dominated by skin growth; the aril is attached to the single large seed and does not begin to grow until 60 days after anthesis. The major phase of growth occurs 50–80 days after anthesis. Like litchi, there is a rapid decline in titratable acidity, from over 1% to less than 0.5%, and an increase in total soluble solids, from *c.* 13% to 20%, in the last 35 days of fruit growth. Total starch declines during the same period and fruit begin to change colour.

Usually only one of the two locules in the litchi, longan, and rambutan ovary develops to form a normal fruit. The other aborts and remains at the

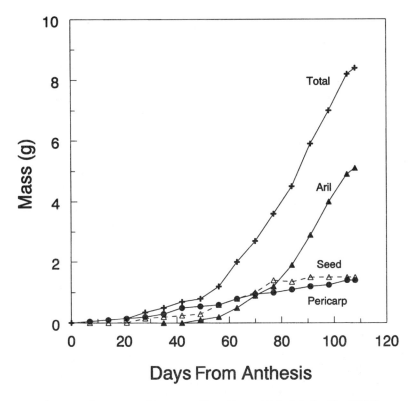

Fig. 8.9. The growth pattern of longan cultivar 'Dong bi' fruit (after Ke, 1990).

base of the normal fruit. Occasionally, both carpels develop equally to form two normal fruit. The aril tissue in litchi and longan develops from an obturator formed by an outgrowth of the funiculus. Aril development is not uniform and thus overlaps at the chalazal end of the seed. The aril is therefore free from the seed in both litchi and longan, while rambutan has the aril developing attached to the seed-coat. In some rambutan cultivars, the hard testa comes away with the aril and makes these cultivars less desirable.

The size and shape of the litchi fruit are characteristic for different cultivars 'Chenzi' has fruit averaging 23 g, while other cultivar fruit can weigh more than 30 g. The ratio of edible aril to seed is more important than large fruit size. Most cultivars have medium to large seeds, occasionally producing fruit with small, shrivelled seeds. The Hawaiian cultivar 'Groff' has perhaps the smallest-sized fruit (approximately 12.5 g fruit^{-1}), but produces small, shrivelled seeds in about 95% of the fruit, greatly increasing the edible portion to near 80%. In cultivars with shrivelled seeds, little embryo development occurs and the seed growth that does occur stops 30–40 days from anthesis.

Longan fruit are brownish, globose to ovoid, 1.5–2.0 cm in diameter; the

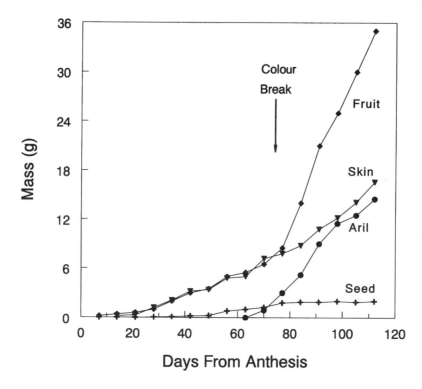

Fig. 8.10. The growth pattern of rambutan cultivar 'Seechompoo' fruit and its parts from anthesis (after Wanichkul, 1980).

skin is thin without protuberances; the aril is white, translucent and sweet, with an aromatic spiciness. There is little published information on fruit growth and development. The major portion of seed development occurs between 50 and 70 days from anthesis. Sugar accumulation in the aril begins after the main period of seed development, 70–100 days postanthesis (Ke, 1990).

CULTIVAR DEVELOPMENT

Cytogenetics and Genetics

There are suggestions that the litchi may have been derived from more than one wild progenitor. Haploid chromosome numbers of 14, 15, 16 and, rarely, 17 have been reported, with diploid numbers of 28, 30 or 32. The diploid number is 30 and 22 for longan and rambutan, respectively. It is possible to generate intergeneric hybrids, using litchi as the female parent with longan. The hybrid plants are similar to the maternal parent with smaller leaves.

Breeding and Selection

Litchi genetic diversity is indicated by the large number of cultivars in China and India, and these provide the basis for the development of new cultivars. Singh and Singh (1954) reported that the 33 or so cultivars grown in India are all selected from seedling populations derived from introduced Chinese cultivars.

Like most single-seeded fruit, the litchi does not lend itself to breeding by a controlled pollination system between two selected parents. The highly heterozygous nature of the species means that planting open-pollinated seeds derived from within a cultivar collection is the usual procedure used in new cultivar development. 'Kaimana' is a selected seedling of open-pollinated seeds of 'Hei Ye'.

Selection and Evaluation

Galan Sauco (1989) has presented a detailed list of characteristics for a litchi cultivar:

Fruit: Large, with small or shrivelled seed and a high proportion of edible aril, bright red skin colour, long shelf-life and ability to retain skin colour under storage conditions, firm flesh with acceptable sugar/acid ratio and resistance to diseases.

Tree: Vigorous, precocious, regular and high-yielding, resistant to water stress, wind, soil salinity, diseases and insects and adapted to regular flowering under warmer temperatures in the tropics.

Sometimes a selection is elevated to cultivar status by satisfying only a few of the above characteristics (Table 8.3). Such a selection is 'Groff', by virtue of consistent bearing over a 7-year period, producing flowers usually several weeks later than other Chinese cultivars grown in Hawaii and producing aborted seeds in about 95% of the fruit. Average fruit size is 12.5 g, smaller than other Chinese cultivars grown in Hawaii, but there is approximately 20% more edible aril for the same weight of fruit (Storey *et al.*, 1953).

Breeding for adaptation of cultivars to the higher temperature conditions in the warm tropics is an important criterion. Menzel and Paxton (1986a), using glasshouses at high temperatures (30°C day/25°C night) to evaluate litchi seedlings for low vigour, found this to be a useful technique for initial screening, prior to field evaluation of adapted genotypes that would flower and fruit under warm conditions. The rate of flushing in the glasshouse at high temperatures agreed with the relative vigour and consistency of flowering of the parent cultivars in the field.

Similar criteria have been developed for longan and rambutan (Table 8.3). Longan fruit criteria include retention of total soluble solids after harvest, crisp and smooth flesh, good taste, uniform size and high aril content. In

Table 8.3. Fruit characteristics of selected litchi, longan and rambutan cultivars.

Cultivar	Synonyms	Season	Vigour	Fruit quality	Fruit mass (g)	Edible %	TSS	Flesh colour	Skin colour	Seed size	Cropping regularity	
											Tropics	Sub-tropics
Litchi												
Da zao	Tai So	Early	Vigorous	Fair	24	72–77	15–18	Milky white	Bright red	Large	Regular	Irregular
Hei ye	Hak Ip, Blackleaf, Groff	Mid	Moderate	Good	19	78	16–20	Milky white	Dull red	Large	Irregular	Irregular
Huai zhi	Wai chee	Late	Dwarf	Good	21	72	15–18	Milky white	Dull red	Large	Irregular	Regular
Gui wei	Kwai May (Mi), Bosworth	Mid	Vigorous	Very good	17	70–83	18–21	White	Light red	Mostly small	Irregular	Irregular
No mi ci	No Mai Chee	Mid to late	Vigorous	Excellent	25	82–86	18–21	Milky white	Bright red	Small	Irregular	Irregular
San yue hong	Sum Yee Hong	Early	Medium	Fair	40	62–68	15–20	Milky white	Bright red	Large		
Chen zi	Chen Family Purple, Brewster	Early	Vigorous	Good	20	54–58	15–18	Waxy white	Deep red	Large or medium	Irregular	Irregular
Longan												
Kohala		Mid		Spicy	15						Regular	
Sao-an-liao		Mid			15	62	13		Dark brown	Small		
Duan ru		Mid			13		14					
Yang tao ye	Carambola leaf	Mid			16		11					

Fen ke	Mid		Crispy	11	66	20	White	Dark yellow	Medium		Irregular
Daw			Good							Regular	Irregular
Chompoo			Excellent							Regular	Irregular
Biew Kiew			Excellent								Regular
Fuyan / Lucky eye	Late	Vigorous	Good	12	67	15	Transparent crispy	Dark brown	Small		Regular
Wuyan / Blackround	Mid	Vigorous	Good	17	71	14	White	Light brown	Large		Regular
Shi xia / Shek Yip, Shi Yuan	Mid	Vigorous	Excellent	10	69	19	White crispy	Dark brown	Small		Regular
Rambutan											
Jit Lee / Deli		Medium		30–55	35	20–22		Orange/red, green tips	–		
Binjai		Medium/small		32–41	41	18–21		Orange/red, green tips	–		
Peng Ting Ching / R3		Medium		22–33	46	22.5		Red	–		
Tau Po Cheng / R9		Medium/large		40–51	36	20–23		Pink/crimson red	–		
Gajah Mati / R137		Medium/large		40–45	49	22–23		Red	–		
Roengrean / Rongrien		Medium/large		40–50	41	18–21		Dark red, green tips	–		
Chompoo / Seechompoo		Dwarf		28–35	40	18–20		Orange	–		

TSS, total soluble solids.

addition, for rambutan, adherence of the aril to the testa is an important criterion.

Cultivars

According to Groff (1921), Chinese nurserymen recognized the difficulties in perpetuating the desirable characteristics of highly regarded cultivars under conditions other than those in which the fruit originated. The environment profoundly influences cultivar characteristics and this may explain the large number of cultivars, with 74 cultivars mentioned in Wu Ying Kuei's list of very ancient origin. Among the many known cultivars in China, the cultivars 'San yue hong' ('Sum Yee Hong'), 'Shui dong' ('Souey Tung'), 'Fe zi xiao' ('Fay Zee Siu'), 'Hei ye' ('Haak Yip'), 'Gui wei' ('Kwai May'), 'No mi ci' ('No Mai Chee'), and 'Huai zhi' ('Wai Chee') (in the Mandarin, with the Cantonese in parentheses), are most widely grown (Table 8.4). Sixty-seven cultivars and selections are described for Guangdong, with 26 of these being major commercial cultivars (Anon., 1985). The 'Brewster' ('Chen Zi', 'Chen Family Purple') trees are vigorous, large and one of the faster-growing cultivars. 'Hei Ye' has been observed to be a slow grower, low in comparison with cultivars such as 'Chen Zi' and 'Gui Wei' of the same age, but it becomes spreading.

Cultivars in Thailand are 'Da zao', 'Hei ye', 'Huai zhi' and 'Hong huey' (syn. 'Mau mong') (Table 8.4). Other cultivars are 'Kim cheng', 'Hong thai', 'Ohia (syn. 'Ouw', 'Hei ye', 'Baidum'), 'Gui wei', 'Groff', 'No mi ci', 'Chen zi' and 'Mauritius'. In Australia, the cultivars most often planted are 'Da zao', 'Bengal', 'Hei ye', 'Huai zhi' and 'No mi ci'. In the USA, Florida cultivates primarily 'Brewster', 'Mauritius' and 'Sweet Cliff'. The 'Mauritius' in Hawaii is different from the Florida 'Mauritius'. In Hawaii, most of the plantings consist of 'Da zao', 'Groff', 'Hei ye' and 'Brewster'. In subtropical South Africa, particularly in the Transvaal lowveld, the Soutpansberg and Natal coast areas, 'Mauritius' is the principal commercial cultivar and is a synonym of 'Gui Wei'. The genetic diversity has led to many synonyms and confusion, along with mislabelling of different cultivars, but isozyme analysis is beginning to separate the different lines (Aradhya et al., 1995).

A longan selection 'Kohala' has been released in Hawaii and it has been propagated asexually. This cultivar has shown prolific vegetative growth at Kona, Hawaii, and produces heavy crops at the research centre and in commercial orchards at Homestead, Florida. There are numerous longan and rambutan cultivars (Salma, 1986), with a wide range of characteristics (Table 8.3). Some Chinese longan cultivars have been described (Chian et al., 1996).

Table 8.4. Major litchi and longan cultivars grown in different countries (after Menzel and Simpson, 1990), name in national Chinese pronunciation unless otherwise indicated (*).

Country	Main growing area	Major litchi cultivars	Major longan cultivars
China	Guangdong	Huai zhi, Hei ye, Sun yue hong, Gui wei, No mi ci	
	Guangxi	Huaizhi, Heiye, Dazao, Sanyuehong, Gui wei, No mi ci	
	Fujian	Shui dong, Hei ye, Da zao, Chen zi, Fu yan, Wu yuan, Shi xia	
Taiwan	Taichung	Heiye, Nomici, Yuhebao, Huaizhi, Sakengzong, Sanyuehong	Chienliou*, Yangtaoye, Saoandiao, Duanru, Fenke
Thailand	Chiangmai, Lamphum, Fang	Da zao, Huai zhi, Hei ye	Daw*, Chompoo* (Seechompoo). Haew*, Biew kiew*, Dang*, Baidom*
India	Bihar State	Shahi*, Rose Scented*, China*	
Madagascar	Wet coastal belt	Da zao	
South Africa	Transvaal – Lowveld	Da zao	
Reunion		Da zao	
Mauritius		Da zao	
Australia	Eastern coastal strip	Da zao, Bengal*, Huaizhi, Gui wei, Salathiel*	
USA	Hawaii	Da zao, Kaimana*, Hei ye	Kohala*, Ilao*, Wai*, Sweeney*
	Florida	Brewster* (Chen zi)	Kohala*, Chompoo*, Homestead No. 1* and No. 2*

CULTURAL PRACTICES

Propagation

Sexual

Seed propagation is used to produce litchi seedlings from selected cultivars for breeding and selection studies and for production of rootstocks for grafting (Menzel, 1985). Litchi seeds remain viable in the fruit for about a month after harvest; viability is lost within 4–5 days once removed from the fruit. Seeds packed in moist sphagnum moss may be shipped, although seeds frequently germinate in the packing within a week. Germination and seedling development are improved by the use of the large seeds and mycorrhizal soil for germination. Seeds germinate within 4–10 days in soil, sand, vermiculite, perlite, peat, wood shavings or mixtures of these materials, with adequate moisture and aeration. Rambutan seeds rapidly lose viability after removal from the fruit and germinate in 7–10 days.

Asexual

Litchi and longan are usually propagated by air layering, the method used by Chinese propagators for centuries. Moist peat or sphagnum moss is wrapped around the girdled area of a branch and held in place with plastic sheets cut to about 25–30 cm^2 with the ends fastened with string or some taping material, such as plastic electrician's tape. Sufficient roots are formed in 8–10 weeks and the success rate is nearly 100%. There is no need to apply growth regulators to enhance rooting, although some recommend a lanolin paste with 250 p.p.m. IBA, 100 p.p.m. 2,4,5-T and IAA or IBA at 500 p.p.m.

Cuttings with two to three leaves from girdled and ungirdled branches of 'Brewster', 'Gui wei' and 'Hei ye', treated with the sodium salt of NAA, IBA and their mixtures and misted or not in a polyethylene enclosure, give poor results. Others have obtained 93% rooting and 100% survival by using green, woody cuttings dipped in 5000 p.p.m. IBA and rooted under intermittent mist. There has been little success reported for rooting of rambutan cuttings; numerous budding techniques have been successfully applied and are the most common methods of propagation.

The variable and often low rate of success in litchi grafting is attributed to stock–scion incompatibility, poor cambial contact, grafting at the wrong physiological stage and poor management after grafting (Menzel, 1985). Graft incompatibility is suggested when only 20% of girdled 'Brewster' scion and 10% of 'Gui wei' are successful on rootstocks to the same cultivars. Histological studies of the graft unions between 'Hei ye' and the above two cultivars shows a slow rate of proliferation of callus tissues. Currently, there is generally a lack of knowledge on the graft compatibility between litchi cultivars. Proper culture of rootstocks and preparation of scion wood (girdling for starch accumulation), proper stage of maturity of scion wood, appropriate postgraft environment, particularly high humidity, and other factors contribute to more consistent success in grafting. Poor grafting results are obtained with rambutan, although approach grafting is successful.

Field Preparation

Deep ripping may be necessary if the soil is compacted. Liming to pH 5.0–5.5 should be carried out and manure incorporated before planting. Litchi trees tolerate poor drainage but do not grow well in standing water. Surface drainage should be installed if such conditions exist. If nematodes are expected to be a problem, fumigation should be carried out. Longan and rambutan require well-drained alluvial soils for good development.

Transplanting and Spacing

Sun-hardened trees (6–12 months old) can be planted out at any time in the year if there is sufficient moisture available and there is no chance of frost. Light mulching (100 mm) after fertilization is beneficial for young litchi trees. Seedlings should start to produce fruit in 6–10 years and asexually propagated trees in 3–5 years.

Spacings of 5–12 m, depending upon cultivar size, are used. 'Gui wei' is a very vigorous tree, 'Hei ye' being moderate, while 'No mi ci' is a smaller cultivar. In Florida, litchi are planted at 7 m by 4 m, with regular mechanical pruning to avoid overcrowding (Campbell and Knight, 1987), and in subtropical areas 6 m by 6 m spacing is possible (Galan Sauco, 1990). Rambutan are planted at 5–10 m between trees and 8–10 m between rows. Often one male rambutan tree is planted per 8–10 functional-female hermaphrodite trees. A similar between-tree spacing is used for longan. Rambutan are frequently interplanted with other tree crops, such as durian.

Windbreaks are necessary for young and mature trees of the three crops. This is essential because trees propagated by air layering do not have a tap root and are easily uprooted by winds.

Irrigation

Regular irrigation is essential during the growth of young trees. During dry periods, irrigation may be needed at 2- or 3-week intervals. Moisture control is applied when the trees reach the flowering age. In China, litchi are grown along canals, on dykes around paddy-fields and around village ponds, and no irrigation is practised. Moisture should be withheld during autumn and winter to discourage vegetative growth and promote flowering (Fig. 8.1). High soil-moisture tension for 6 months beginning midsummer, or even for 3 months beginning in late summer, promotes floral initiation and fruit set by inhibiting emergence of vegetative flushes (Nakata and Suehisa, 1969). Once the inflorescence has appeared and throughout fruit development, adequate moisture is necessary. Irrigation should be terminated a few weeks before harvesting. Moisture stress under non-irrigated situations can cause developing fruit to ripen prematurely, even before the aril has fully developed. Fruit growth will resume if water is provided, leading to fruit splitting. Copious irrigation immediately after harvest is desirable to encourage new shoot growth (Fig. 8.1). In areas with dry summers, irrigation is necessary, followed by limb girdling, in late summer to discourage further vegetative flushing and to accumulate carbohydrates in the trees during the autumn months (Nakata, 1953). This practice is necessary in areas where climatic conditions do not enforce a natural period of tree dormancy in the winter.

Irrigation is required for rambutan grown for export (Table 8.5), as water stress during flower and fruit development leads to reduction in yield and fruit size. Irrigation is also essential during the vegetative flushing stage after harvest (Fig. 8.1), and is withheld during floral initiation. Preflowering water stress that does not induce leaf wilting has earlier flowering and improved harvest synchrony, without affecting yield. High rainfall during blooming can lead to poor fruit set. Rambutan have a shallow root system, with 80% of the roots in the top 15 cm, which does not extend beyond the tree canopy sometimes necessitating irrigating up to three times a week. Severe leaf loss can occur within 4–10 days of withholding irrigation. The amount of irrigation should, at a minimum, replace that lost by evaporation.

Pruning

Little research has been done on the need for and effectiveness of pruning longan and rambutan trees, although there are some data on litchi. During the first 2–3 years, all side-shoots up to 0.8–1.0 m should be removed from litchi, as these will form low-hanging branches, which interfere with cultivation. In bearing litchi trees, even branches 1.5 m high will bend to the ground when laden with fruit. In Australia, the main terminals of 3- to 4-year-old trees are headed back in the spring (prior to flushing) in vigorous cultivars to create a more compact tree with many terminals, thus increasing potential bearing surface (Menzel *et al.*, 1988). Approximately 16 cm from the tip is pruned and, from each pruned branch, an average of three new terminals are produced. A pruning system is practised in Taiwan to maintain small tree size (< 2 m) to obtain a wineglass shape, in order to encourage regular flowering and for ease of management and harvesting (Yen, 1995). An inverted-cone approach is also used, wide at the bottom and narrow at the top (3–4 m high), with mechanical trimming to maintain shape. Tree thinning is also practised after harvest to open the canopy for light penetration, which is beneficial for inflorescence formation (Lin *et al.*, 1991).

Table 8.5. Seasonal changes in rambutan water use (Diczbalis *et al.*, 1996).

Storage	Average water use (mm week $^{-1}$)	Crop factor (water use/evaporation)
Preflowering	32	0.65
Flowering–fruit set	56	1.07
Fruit development	68	1.21

Light pruning sometimes encourages several weak flushes and results in poor flowering. Severe heading-back of a tree is harmful to future inflorescence formation. Little or no pruning is done in India and yields are said to be heavier than with the pruning practised in Taiwan. Canopy thinning after harvest is done in India. Panicle thinning or pruning is sometimes practised. Girdling is practised in October to inhibit new flushes; if too severe, the next year's growth will be reduced (Yen, 1995).

Early training of rambutan to achieve a strong framework of branches is essential to encourage secondary lateral branches. Annual pruning is designed to remove water sprouts, pest- and disease-infected shoots and dead branches, along with crossing branches. In some areas, the panicle is pruned after harvest to induce vigorous caopy regrowth.

Fertilization

Fertilization practices in commercial litchi (Table 8.6), longan and rambutan orchards differ, due to differences in climate, soil, availability of different kinds of organic and inorganic fertilizers and other factors. The Chinese practice for litchi on alluvial or red basaltic soils provides 300–400 g of urea, 100 g superphosphate and 25–50 kg animal manure to large trees just after harvest, but not later than July. At blossoming, 100 g urea tree^{-1} or mixed nitrogen, phosphorus and potassium (NPK) is applied. To prevent fruit drop, trees are sprayed with 0.1–0.2% urea, sometimes with 0.2% magnesium sulphate.

A light fertilization should be applied with care immediately after field transplanting, due to the sensitivity of litchi roots to fertilizer burns. Heavy applications should be delayed a year or after one or two growth flushes. One-year-old trees are given about 30 g urea plant^{-1} month^{-1}. Also, 30 g of a high-analysis mixed fertilizer, such as 15–4–11, is given every 3 months with the urea. Fertilizer in subtropical areas is withheld during cold months to prevent flushing.

Table 8.6. General guide for litchi fertilization.

Year	Fertilizer formation	Time of application	Amount (g tree^{-1})
First year	14–14–14	At planting time	125
	14–14–14	At 4 months	125
	14–14–14	At 8 months	125
	14–14–14	At 12 months	125
Second year	14–14–14	Once every 4 months	250
Third year	14–14–14	Once every 3 months	455
Fourth year	14–14–14	Once every 3 months	455
Fifth year	10–20–20	Appearance of flowers (Feb., Mar.)	1000–1500
	14–14–14	After harvest (May–June)	1000–1500

In Australia, general recommendations for 5-year-old trees are: 150 g urea, 300 g single superphosphate and 150–200 g potassium sulphate per application. These amounts are increased by 20–30% each year until year 15, when each application rate increases to 1200 g urea, 1200 g superphosphate and 600–800 g potassium sulphate. No fertilizer is applied to bearing trees in the spring to prevent vegetative flushing in the autumn in order to increase the prospect of good flowering. Three applications of a complete fertilizer, such as triple-14, at 114 g tree^{-1} the first year and increasing to 227 g per application in the second year, are used in Hawaii. In the third year, two or three applications at 454 g per application are applied. For bearing trees, a general rule of thumb of 0.90 kg of a complete fertilizer for every 2.54 cm of trunk diameter is given upon completion of harvest to encourage summer flush. An application in late winter or early spring before floral initiation has begun can result in a vegetative flush rather than a flower flush. The main fertilization is applied immediately after harvest in the summer (Fig. 8.1), followed by irrigation and the annual limb girdling.

Rambutan is reported to have similar nutritional requirements to litchi. A hectare of rambutan (70–80 trees ha^{-1}) yielding 6720 kg of fruit removes 134 kg N, 1.8 kg P, 10.2 kg K, 4.84 kg calcium (Ca) and 2.47 kg magnesium (Mg); 50% of the P, Ca and Mg is in the fruit pericarp and 20–40% in the aril. Application of fertilizer needs to correspond with differing needs at various stages of the growth cycle (Fig. 8.1). Critical periods are before flowering and fruit set, several weeks after fruit set and after harvest. Rambutan leaf N should be *c.* 2%, while in litchi the level should be about 1%. Leaf samples for analysis are taken twice a year, once at the end of harvest and the second at flowering, with only P and K applied during fruit set to avoid vegetative growth from N application. Minor elements are applied once a year at the end of the cold period.

Erratic Bearing

Litchi trees 7–8 years old can produce up to 45 kg. Potential yields for 10-year-old trees are 50–70 kg, 15-year-old trees 100 kg and 20- to 24-year-old trees 150–180 kg (Campbell and Knight, 1987). Occasionally, yields can be as high as 500 kg in very old trees. Apparently, potential yields can continue to increase with older trees, as has been demonstrated in China.

The litchi can be a heavy bearer when conditions are favourable for fruiting; however, considerable year-to-year varation in yields occurs wherever grown (Fig. 8.11). Erratic flowering can be due to unfavourable weather conditions, and, even if flower production is consistent, fruit set problems contribute to variable yields (Chapman, 1984). Cultivars are very particular about their climatic requirement (Table 8.1); trees propagated from good bearing trees and planted in another location are known to have produced fruit for a few years and later ceased to produce for years. The senior

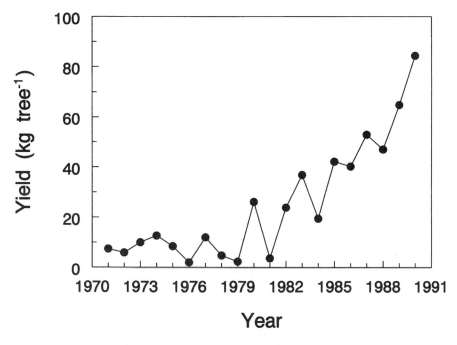

Fig. 8.11. Change in litchi production at Beilu, Guangxi, from 1971. Improved management practices were used after 1981, which led to increased production and reduced, but did not eliminate, year-to-year variation in yield per tree (Mo, 1992).

author is aware of trees that have not flowered for 30 years.

Fruit bearing is dependent upon a number of crucial phases, which need to be optimized throughout the year (Fig. 8.1). The influence of cultivar, pruning and tree management needs to be integrated. All three species need to be stimulated to leaf growth after harvest; this is followed by a period of rest and then floral induction and development. These flushes are induced by fertilization and rainfall or irrigation (Yen, 1995). There can be more than one flush of leaves during the 5 months after litchi and longan harvest, with leaves being forced to mature and growth restricted after this phase. While mature leaves are essential for flowering, the presence of young expanding foliage in the cool season is inhibitory to inflorescence formation; if removed, flowering capacity increases (Lin *et al.*, 1991). In autumn, young expanding leaves can be removed mechanically or with a low rate of ethephon (1500 p.p.m.); a higher rate will abscise more mature leaves. Tree girdling in late summer can be used to limit new shoot growth. The conditions for floral induction, besides tree condition, are temperature and water stress for litchi, temperature stress for longan and water stress for rambutan. The terminal nature of the inflorescence inhibits further main-axis development, while lateral shoots on branches that have fruited continue further shoot growth, with only 22% producing flowers in the following season.

In rambutan, shoots that have not fruited have vegetative shoots arising from the terminal buds and 57% flower and fruit in the following season. Poor synchronization of flushing and flowering within an orchard, especially of rambutan, increases management problems. Cultivars vary in their requirements for induction, leading to early- and late-season bearers.

Pest Management

Diseases

A number of disease organisms infecting the litchi fruit are listed in the literature but none are considered serious and all are of a postharvest nature. *Aspergillus* spp., *Botryodiplodia theobroma*, *Colletotrichum gloeosporioides* (anthracnose), *Cylindrocarpon tonkinense*, *Pestalotia* sp., *Penicillium* and *Alternaria* can cause some postharvest problems. Anthracnose is a problem in Florida and the cultivar 'Mauritius' is especially susceptible, the fungus affecting young fruit and causing fruit drop. Dipping fruit in 0.65% benomyl at 52°C for 2 min and enclosing fruit in plastic film have given good control of fruit diseases. Many of these fungal infections are secondary in nature, as the organisms enter the fruit only when the fruit skin is damaged. Puncture sites from fruit-fly oviposition are excellent points of entry.

The major diseases of rambutan are powdery mildew on young growth, stem canker, pink disease and sooty mould. Postharvest diseases are similar to those in litchi and include various fruit rots and anthracnose.

A serious disease of longan in China and Hong Kong is called the witches' broom, caused by phytoplasma, causing serious deformity of the inflorescence. It is transmitted by vegetative propagation and by seeds from diseased trees.

Nematodes have not been reported as a serious threat to litchi cultivation. Tree decline in Natal and Transvaal is associated with two species of nematodes. Roots become 'stubby' and brown in colour and secondary feeder-root development is inhibited.

Insects

Insects are of greater concern (Table 8.7) in litchi production than diseases. The erinose mite (*Aceria litchii*) is probably the most universal, having been reported from many widely scattered areas. It attacks the young leaflets, producing a thick growth of light brown to reddish brown felt-like hair on the undersurface. The worm-like mite is not visible to the naked eye and is very specific to litchi. Occurrence of this mite coincides with periods of new

Table 8.7. Insect pests of litchi.

Common name	Organism	Parts affected	Region
Erinose mite	*Aceria litchii* (*Eriophyes litchii*)	Foliage	Universal
Leaf miner (cacao moth)	*Acrocerocops cromerella*	Foliage	India
Rose beetle	*Adoratus sinicus*	Foliage of young plants	Hawaii
Trunk borer	*Anoplophora macularia*	Trunk	Taiwan
Bark borer	*Salagena* sp.	Bark/wood	South Africa
Shot hole borer	*Xyleborus fornicatus*	Branches	India
	Acrocerocops cromerella	Fruit	China
Litchi moth	*Argyroploce peltastica*	Fruit	South Africa
Macadamia nut borer	*Cryptophlebia ombrodelta*	Fruit	Australia
Litchi stink bug	*Tessartoma papillosa*	Fruit	China
Mediterranean fruit fly	*Ceratitis capitata*	Fruit	South Africa
Natal fly	*Ceratitis rosa*	Fruit	South Africa
Oriental fruit fly	*Dacus dorsalis*	Fruit	Hawaii

growth, regardless of season or locality, and in severe cases new shoots on the entire tree can be affected. It is effectively controlled by sprays of wettable sulphur at 2.3 kg per 378 l of water. It does not infest other relatives in *Sapindaceae*.

In Taiwan, a trunk borer (*Anoplophora macularia*) can cause death of trees within a few months by boring into the trunk and damaging the conducting tissues. The litchi stink bug can cause serious damage to fruit, but the bug is effectively controlled by a parasite, *Anastatus* sp. In Australia, the macadamia nutborer (*Cryptophlebia ombrodelta*) is considered a serious fruit-damaging insect. A number of insects infest rambutan but none are regarded as serious, most are controlled with minimal spraying programmes. The insects reported include loopers, caterpillars, a stink bug, a weevil, mites, thrips and mealy bugs. Scale insects and mealy bugs can be a problem, especially if they infest the fruit and have to be removed for export.

Mediterranean fruit fly (*Ceratitis capitata*) does not attack these three fruit except where the fruit skin has been broken by other means and the pulp is exposed. The Oriental fruit fly (*Dacus dorsalis*) does infest these three fruit, causing punctures, which are often the focus of entry of fungal organisms, which in turn cause fermentation and decomposition. Other fruit flies can be serious pests in other countries.

Other pests are birds and fruit bats. In Hawaii, the white eye (*Zosterops palpebrosus*) and the bulbul (*Pycnopus cafer*) can cause severe damage as fruit approach maturity. Fruit-bats (*Pteropus* sp.) or flying foxes have caused severe losses in South Africa and Australia.

Weed Control

Weed control is most important from time of field transplanting up to 3 or 4 years old. As trees grow and expand horizontally, there is a decreasing amount of weed growth underneath the canopy, due to shading. Use of polyethylene mulch about 1 m² around the plant at transplant time reduces weed growth near the plant. Organic mulches are highly recommended around the base of rambutan trees, while litchi are reported to be adversely affected by organic mulches around the base.

HARVESTING AND POSTHARVEST HANDLING

Harvesting

Fruit are ready for harvest when they attain full red colour. Fruit harvested before full maturity are acid and do not ripen further or improve in flavour. These are non-climacteric fruit and do not respond to ethylene. Litchi and longan are harvested by cutting or breaking off the entire panicle with the cluster of fruit. In many countries, the practice is to remove the panicle together with a variable portion of the last year's wood in order to enhance the production of a higher number of terminals, which may flower the next spring (Galan Sauco, 1989). Panicles without fruit are also cut, as the presence of old panicles is reported to delay the next flush. In small orchards, bamboo or aluminium tree-pruning poles and tall ladders are used. In large orchards, mechanical 'cherry-picker' platforms greatly facilitate harvesting. Harvesting fruit clusters on older, large trees is difficult and time-consuming.

Harvesting is not done in the afternoon in China, as fruit colour is said to become dull more rapidly, or on rainy days, as wet fruit is said to rapidly break down. Panicles of fruit are packed in straw baskets or cartons containing 15–20 kg of fruit. Green leaves are also packed with the fruit to reduce damage in transit and assist in providing moisture and preserving fruit colour. Harvested fruit are never exposed to sun to prevent rapid drying.

Postharvest Treatments

The major problem in marketing these three fresh fruit is the rapid skin browning, which makes them unattractive, although the aril remains edible (Paull and Chen, 1987). In many Asian countries, fruit are sold still attached to the panicle in bunches, while in Western countries individual fruit in punnets are marketed (Plate 18). Much of the postharvest research has been directed towards the prevention of pericarp browning in storage and marketing of fresh fruit. The field skin colour of all three fruit can be retained by a sulphur dioxide–acid treatment applied before insect disinfestation. All longan exported from Thailand are so treated, as are many of the litchi from some other countries. The treatment is less successful with rambutan but does produce a more appealing pericarp colour than the black desiccated fruit. The sulphur dioxide treatment is not approved in the USA.

Retention of litchi red pericarp colour for longer periods has been demonstrated by wrapping fruit in polyvinyl chloride (PVC) film or bagging in polyethylene bags and placing in storage temperatures of 0–10°C (Paull and Chen, 1987). Commercial quantities (5.4 kg) of litchi and longan packaged in large polyethylene bags and twisted closed or individual fruit in punnets placed in cardboard cartons and sealed are best stored at 2°C for up to 32 days. At the retail level, all fruit not on display should be kept under refrigeration, preferably as close as possible to 0–2°C. If surface transportation is used instead of air freight, properly packaged fruit should be shipped at 0–2°C. Skin colour and flavour of fruit is retained for at least 1 month. Rambutan should not be stored below 10°C, due to chilling-induced pericarp darkening; mass loss from the pericarp is a major problem with this fruit and overwrapping is essential (Landrigan *et al.*, 1986).

Grade standards are similar for all three fruit and are generally set between the shipper and buyer. Small, poorly coloured, immature or damaged fruit are culled. Rambutan spinterns are easily damaged and care is essential to avoid mechanical injury (O'Hare, 1995). Similar mechanical injury to litchi tubercles leads to a site for skin browning. Soluble solids should be greater than 15% and acid levels low.

Irradiation studies have shown that litchi, longan and rambutan for export can be irradiated at 250 Gy. Hot–cold disinfestation protocols and cold treatment for fruit-fly disinfestation can be used for litchi and longan (Paull *et al.*, 1995).

UTILIZATION

Important producing areas are China, India, Thailand, Taiwan, South Africa, Mauritius and Australia. Proximate composition of litchi and longan indicates that they are a poor source of calcium, iron, thiamine and riboflavin and a fair source of phosphorus, and do not have provitamin A. They are a good source of niacin and ascorbic acid (Table 8.8). All three fruit are usually eaten fresh, and can be preserved in a variety of ways. Dried litchi and longan, popularly known as a 'nut', and canning are the most common. Peeled litchi fruit may be canned in a 40% syrup, containing 0.2% citric acid. Hand-peeling of the skin is the usual method, although a hot-lye dip and mechanical peeling are also used. The canned litchi, if fresh fruit are not available, are added to rare and dainty dishes and restaurants serve delicious litchi dishes with meat or with syrup dressings.

Fresh litchi packed in polyethylene bags can be quick-frozen and stored for

Table 8.8. Proximate composition of 100 g edible portion of litchi ('Gui Wei'), longan and rambutan (Wills *et al.*, 1986; Morton, 1987; Wenkam, 1990). The edible portion of litchi and longan is *c.* 72% and rambutan *c.* 43%.

		Litchi	Longan	Rambutan
Proximate				
Water	g	77.6	83.5	79.5
Energy	kJ	335	–	297
Protein	g	0.94	0.84	1.05
Lipid	g	0.29	0.44	0.45
Carbohydrate	g	20.77	14.9	15.7
Fibre	g	0.16	0.32	2.8
Ash	g	0.37	0.43	0.3
Minerals				
Calcium	mg	4	9	8
Iron	mg	0.37	0.4	0.1
Magnesium	mg	16	–	10
Phosphorus	mg	35	36	12.9
Potassium	mg	255	170	140
Sodium	mg	7	3	2
Vitamins				
Ascorbic acid	mg	40.2	42	70
Thiamine	mg	0.035	–	0.015
Riboflavin	mg	0.084	0.05	0.065
Niacin	mg	1.91	–	0.8
Vitamin A	IU	0	–	0

more than a year and still be edible, with the skin colour and flavour reasonably well preserved. Fruit can be sun-dried in approximately 20 days, two-thirds of the moisture being removed. Oven-drying can be used to shorten the drying time, but care must be taken to use relatively low temperatures of around 40°C, as higher temperatures tend to caramelize the sugars and produce a burnt flavour and bitterness.

FURTHER READING

Galan Sauco, V. (1989) *Litchi Cultivation* (in Spanish) (Menini, U.G., FAO Coordinator). FAO Plant Production and Protection Paper No. 83, FAO, Rome, Italy, 136 pp.

Menzel, C.M. (1983) The control of floral initiation in lychee: a review. *Scientia Horticulturae* 21, 201–215.

Menzel, C.M., Watson, B.J. and Simpson, D.R. (1988a) The lychee in Australia. *Queensland Agriculture Journal* 114, 19–27.

Menzel, C.M., Watson, B.J. and Simpson, D.R. (1988b) Longans – a place in Queensland's horticulture? *Queensland Agriculture Journal* 115, 251–265.

Tindall, H.D. (1994) *Rambutan Cultivation.* FAO Plant Production and Protection Paper No. 121, FAO, Rome, Italy, 162 pp.

Underhill, S.J.R., Coates, L.M. and Salco, Y. (1997) Litchi. In: Mitra, S.K. (ed.) *Postharvest Physiology and Storage of Tropical and Subtropical Fruits.* CAB International, Wallingford, UK, pp. 191–208.

MANGO

BOTANY

Introduction

Mango (*Mangifera indica* L.) belongs to the family *Anacardiaceae*, also known as the cashew family, with about 75 genera and 700 species, mostly tropical, with some subtropical and temperate species. Common nut-bearing species include *Anacardium occidentale* L., cashew nut, and *Pistacia vera* L., pistachio nut.

Origin and Distribution

The genus *Mangifera* consists of 69 species of Asian origin; not all bear edible fruit. The mango, *M. indica* L., also known as manga (Tamil), mangga (the Philippines, Malaysia, Indonesia) and manguier (French), is the best known and most widely cultivated. The fruit is large, fleshy and sometimes fibrous (Fig. 9.1). Other species in the genus bearing edible fruit include *Mangifera altissima* Blanco., *Mangifera caesia* Jack, *Mangifera foetida* Lour., *Mangifera lagenifera* Griff., *Mangifera odorata* Griff., *Mangifera zeylanica* (Bl.) Hooker and *Mangifera sylvatica* Roxb.

The mango originated in the Indo-Burma region and has been cultivated in India for over 4000 years. This fruit is intimately associated with the Hindu religion and there are numerous ancient Sanskrit poems praising the blossoms and fruit (Singh, 1960). Indian traders and Buddhist priests probably introduced the mango into Malaysia and other east Asian countries during the 4th or 5th century BC and to the Philippines between AD 1400 and 1450. The Portuguese, the first Europeans to establish trade routes with India, transported the mango to East Africa and Brazil. Spanish traders took the mango from the Philippines to the west coast of Mexico before the English arrived in the Hawaiian Islands in 1778. The mango was introduced into

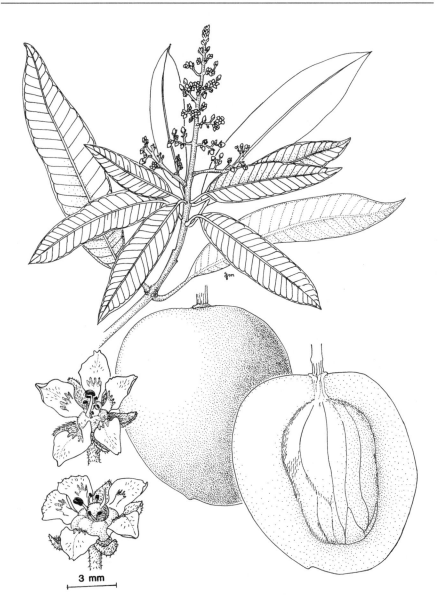

Fig. 9.1. Leaf, panicle, flower and fruit of mango.

Hawaii from the west coast of Mexico between 1800 and 1820, with credit being given to Don Francisco de Paula Marin, a Spanish horticulturist. Apparently, the Brazilian introduction was spread to Barbados and to other islands in the Caribbean area. Mango is now found in all tropical areas, as well as many subtropical regions of the world, attesting to its wide range of adaptability.

ECOLOGY

Soil

The tree is not exacting with regard to soil, although flat alluvial soils with a pH 5.5–7 and a soil depth of at least 1 m are preferred. Freely drained Oxisol, deep volcanic soils, in Java and the Philippines are favoured. Exchangeable aluminium should be less than 30 p.p.m. and available phosphorus (P) 720 p.p.m. There should be no hard layer to limit root penetration and the water-table should be no higher than 2.5 m. In fertile soil, minimal nutritional problems can be expected. The tree is sensitive to saline conditions. Calcareous soil with a high pH and salinity problems limit mango development. Suitable rootstocks have been selected and used in areas where these soil types exist.

Climate

Rainfall

Mangoes are very drought-tolerant and can withstand occasional flooding. Good rainfall distribution is crucial for flowering and fruit set, rather than total rainfall. A dry or, even more effective, cool period preceding flowering is necessary for reliable mango production, as it promotes flower induction (Fig. 9.2). In tropical high-rainfall areas, yields are low and there is excessive vegetative growth. Flowers are very susceptible to anthracnose under moist conditions, and low rainfall is preferred during flowering. Irrigation must be applied regularly to prevent water stress during early fruit development when cell division is occurring and to produce vegetative flushing after harvest.

Temperature

Mango can be grown to 1200 m in the tropics, although the best production is at less than 800 m. Air temperature in the range of 24–30°C is needed and the tree can endure up to 48°C during fruit development if sufficient irrigation is available. Pollen viability declines if it develops at higher than 35°C or below 15°C. Cold temperature limits growth, with no ability to acclimatize being shown; the minimum is 1–2°C. Frost can severely damage or kill young trees, with older trees being able to endure –4°C for a few hours with limited damage (Crane and Campbell, 1991).

Temperature plays a key role in mango flowering and the response varies with cultivar. For each degree of latitude north or south of the tropics, flowering is delayed by 4 days. In the subtropics, cold conditions, with temperatures below 15°C, may advance flowering. For each 125 m increase in elevation, flowering is also delayed 4 days. Panicle growth occurs at 12.5°C, when no vegetative shoots are produced (Schaffer *et al.*, 1994); however, low

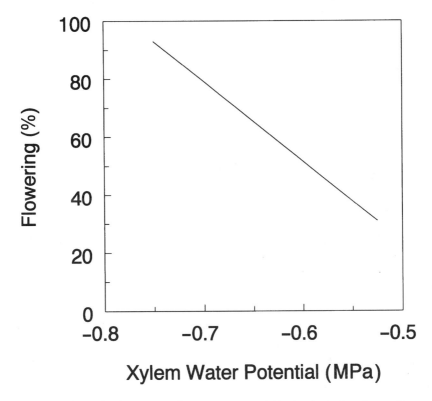

Fig. 9.2. Relationship between xylem water potential in the 6 weeks prior to 'Nam Dok Mai' panicle growth and the percentage of terminals flowering: $y = -121.83 - 286.7x$, $r^2 = 0.52$ (Pongsonboon, 1991, cited by Schaffer *et al.*, 1994).

temperature can lead to flower deformation and loss of pollen viability (Fig. 9.3). Ovule abortion can occur at low temperatures, around 14°C , leading to parthenocarpic fruit development.

Light

Shading can prevent or delay flower-bud formation and there are a higher number of perfect flowers on the side of the tree receiving direct sun. There are a significant number of shade leaves existing in a mango-tree canopy; pruning can increase light penetration, but how this influences yield is unknown.

Photoperiod

No photoperiod tested (10, 12, 14 h) with cv. 'Tommy Atkins' induces flowering at non-inductive temperatures (30°C day/25°C night). Flowering does occur at all photoperiods with inductive temperature (18°C/10°C). Failure to flower in the 'off season' of biennial cultivars in the subtropics and

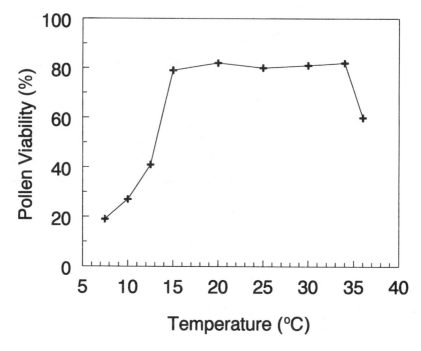

Fig. 9.3. Effect of temperature during pollen development on mature pollen viability (Issarakraisila *et al.*, 1993).

flowering occurring at the equator several times during a year indicate a day-neutral habit.

Wind

Yields are reduced if strong winds occur during flowering and early fruiting. The critical threshold for wind exposure is unknown.

GENERAL CHARACTERISTICS

Tree

The mango is an evergreen, symmetrical tree ranging in height from 8 to 25 m, bearing leathery, simple leaves. Growth occurs in vegetative flushes (Fig. 9.4) of the terminal, with 10–12 new leaves per flush. The number, frequency and length of these shoots per year depends upon cultivar, temperature, tree age, current fruit load and previous cropping history. When shoot elongation stops, a state of dormancy occurs until the terminal is matured and ready for the next vegetative growth flush or flowering (Plate 19). It is common to see different parts of a tree flushing at different times, especially in the hot tropics.

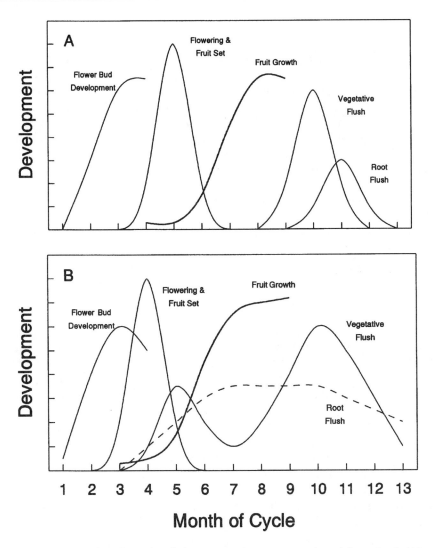

Fig. 9.4. Mango phenological cycle having synchronous growth and flowering in (A) the subtropics (after Cull, 1991) and (B) the tropics, with asynchronous growth and flowering and a long juvenile phase (after Galan Sauco, 1996).

Leaves of young flushes are usually copper-red or purplish in colour, gradually turning to dark green. The leaves persist on the tree for up to 4–5 years before being shed. In many tropical areas, young trees up to 5–6 years of age may produce several vegetative flushes on the same branches during the year. A biennial or irregular nature has been shown on several cultivars in Hawaii, vegetative flushes occurring during the June–July period are more conducive to flowering the following spring than flushes occurring earlier or later (Nakasone *et al.*, 1955).

Flowers

Night temperatures around 10–12°C and dry conditions promote flower initiation. The terminal buds of the stem produce a large, branched panicle (Fig. 9.1), with 300–3000 flowers, depending upon the cultivar, and they reach full bloom in as little as 25–30 days after initiation. The strong dominance of the terminal bud prevents lateral buds from emerging. Surrounding the terminal bud are smaller buds (subapical buds), which are morphologically lateral buds compressed into the apical position, subtending the terminal bud. Apical dominance disappears with the destruction of the apical inflorescence (Nakasone *et al.*, 1955).

The panicle consists of perfect hermaphrodite and male flowers. Perfect flowers consist of four or five calyx lobes and as many free petals (Fig. 9.1). The centre of the flower is occupied by a circular disc (nectary) divided into four or five segments. The one-celled ovary with one obliquely protruding stigma is attached to this disc. There is usually only one functional stamen, the other four being reduced to abortive structures, called staminodes. The male flower is essentially the same, except for the absence of an ovary. Petals are pinkish white, with several yellow ridges on the inner surface. The percentage of perfect flowers on a panicle ranges from 1.25 to 81.0%, with a strong varietal difference in flower-sex expression (Table 9.1). Late-season panicles and those formed in the interior of the tree usually have a larger number of perfect flowers (Adlan, 1965). The percentage of perfect flowers is also higher in the apical zone of the panicle than in the basal and central portions.

A satisfactory crop of fruit is obtained if only a small percentage of the flowers are pollinated. Lack of fruit set is attributed to: (i) lack of fertile pollen; (ii) poor pollen-tube growth; (iii) failure of ovule fertilization; (iv) failure of pistil or ovules to develop; (v) abortions of embryo sac, embryo or endosperm; (vi) anthracnose disease; and (vii) other physical factors (Schaffer *et al.*, 1994). Anthracnose disease is a major problem, if rainfall occurs during flowering, and will eventually affect the whole panicle, leading to flower parts and young fruits being shed.

Table 9.1. Hermaphroditic and male flowers on three cultivars grown in Hawaii (Adlan, 1965).

Sex	Cultivar			
	Fairchild	Zill	No. 9	Total
Hermaphrodite	396	787	556	1629
Male	1660	1153	104	2917
Total	2056	1940	660	4546
% hermaphrodite	19.20	40.50	84.24	

Pollination and Fruit Set

Natural pollination

About 60% of the flowers open before 6 a.m. and the rest open during the day. Anther dehiscence occurs within 1 h after anthesis, with a maximum between 9 a.m. and noon. Ninety per cent of the pollen is viable. Stigmas are receptive from 1 day before anthesis up to 2 days after, with the day of anthesis being the optimum. Insects, including bees, wasps and flies, are the principal pollinating agents, as indicated by the sticky pollen, secretion of nectar, colourful corolla and scent of the flowers. The pollen grains fall on the base of the ovary and the nectary discs, rather than the stigma. Mango cultivars are usually considered to be self-fruitful, but self-incompatibility has been reported. Self-pollination produces 0–1.68% set, while cross-pollination produces 6.4–23.4% set, but there are clear cultivar differences. The use of pesticides during flowering can affect the pollination process, including pollen germination and fertilization.

Polyembryony

Mango cultivars in some regions have a high degree of polyembryony, the condition in which several genetically identical embryos develop from the nucellar tissue of the ovary and often suppress zygotic embryo development (Fig. 9.5). Generally, subtropical Indian cultivars and their derivatives, characterized by roundish, colourful fruit and susceptibility to anthracnose, are largely monoembryonic (Fig. 9.5), while cultivars classified in the Indo-Chinese group of South-east Asia, including those in the Philippines, are largely polyembryonic (Campbell, 1961), are relatively resistant to anthracnose and often lack bright colour. These polyembryous fruit are typically more elongated than round.

Polyembryonic cultivars are generally believed to achieve higher fruit set, since apomictic embryos take over the functions of the aborted zygotic embryos and allow normal fruit development. Zygote abortion in monoembryonic cultivars stops further fruit development and usually leads to fruit drop. However, monoembryonic cultivars can produce heavier yields in some years when mass yields are compared.

A small number of fruit with embryo-aborted ovaries do develop into commercial-sized fruit. Seedless fruit are reported in 'Harders', 'Tommy Atkins' and 'Momi-K' (Galan Sauco, 1979), due possibly to embryo abortion taking place after growth-promoting substances required for fruit development have been produced in the endosperm.

Floral induction and fruit-set control by chemicals

Biennial bearing and postset fruit drop are major problems in commercial mango production. Differences in the trends of various biochemical

Fig. 9.5. Monoembryony (left) and polyembryony (right) in mango seedlings.

constituents in shoots of regular- and irregular-bearing cultivars under different climatic conditions are not consistent. Low temperatures, especially in the subtropics, have the greatest influence on flower induction, which can be enhanced by water stress and differs with cultivar. The subtropical Florida cv. 'Tommy Atkins' does poorly in more tropical climates, while the tropical cultivars may have less demanding temperature requirements. Much research has been directed at overcoming the dependence upon environmental signals for flower initiation using different cultivars, different management strategies and chemical sprays (Table 9.2).

Ethephon increases flowering and fruiting in 'off' years in the biennial-bearing trees. Flowering can also be influenced by potassium nitrate (KNO_3) sprayed on trees in the tropics, but apparently not in the subtropics. Young shoots (1.25 months from bud emergence) of the Philippine cv. 'Carabao' and 'Pahutan' can be induced to flower by spraying 10 and 40 g L^{-1} of KNO_3, respectively. 'Carabao' flowered in 11 days, while 'Pahutan' required 20 days. Older 'Carabao' shoots (8.5 months from bud emergence) sprayed with 10–160 g L^{-1} KNO_3 induced 100% flowering (Bondad and Apostol, 1979).

Table 9.2. Chemicals for manipulating flowering or increasing fruit set and retention in mango (Galan Sauco, 1996).

Area	Chemical	Application time	Dosage (foliar sprays)	Action
Subtropics	GA_3	Prior to flower differentiation (beginning of winter)	100 mg L^{-1}	Delays flowering (repeated use during winter will eliminate flowering)
	Cyclohexamide	Flowering	0.25 g L^{-1}	Destroys apical panicles
	Dinoseb	Flowering	0.5 ml L^{-1}	Destroys apical panicles
	Pentachlorophenol	Flowering	5.0 g L^{-1}	Destroys apical panicles
	Ethephon	Full-flower stage	800 mg L^{-1}	Destroys apical panicles
	Hydrogen cyanide	Beginning of flowering season	0.4% (cv. 'Haden') 0.6% (cv. 'Keitt')	Destroys apical panicles
Tropics	NH_4NO_3	End of autumn	20 g L^{-1}	Interrupts flower dormancy
	KNO_3		40 g L^{-1}	
	Paclobutrazol	Any time during fruit-bud stage or even at the end of autumn	2.5–5.0 g.(a.i.) tree^{-1} (soil drench)	Reduction of vegetative growth and flower induction; increases fruit set
General	Polyamines Spermine Putrescine	Prior to anthesis or at full bloom	10^{-1} mol L^{-1} 10^{-4} mol L^{-1}	Increase fruit set and fruit retention
	KNO_3	Full bloom	2–4%	

NH_4NO_3, ammonium nitrate; KNO_3, potassium nitrate.

Flower induction is uniform, with unsprayed controls remaining vegetative. Genetic differences among seedling trees and between cultivars show variations in response to the chemical used, ammonium nitrate versus KNO_3, with the nitrate ion being the effective ion (Nagao and Nishina, 1993). Soil applications and sprays of the growth retardant paclobutrazol (gibberellic acid (GA) inhibitor) can stimulate early and more efficient flowering under marginally inductive temperature conditions in the tropics (Table 9.2).

Both the number of hermaphroditic flowers per panicle and the fruit-bearing potential increase following application of 100 p.p.m. of naphthalene acetic acid (NAA). Deblossoming of terminal panicles, either manually or with chemicals (Table 9.2), delays flowering and can increase yield. The stimulation of auxiliary panicles also occurs if early panicles are damaged by low temperature in the subtropics. Sprays of GA can delay flowering and fruit maturation by 2 weeks, but concentration and timing of application are critical if yield is not to be reduced (Turnball *et al.*, 1996).

Fruit

Fruit morphology

The fruit of the mango is a drupe of variable size and shape, ranging in weight from a few grams to more than 1 kg. It is fleshy, flattened, rounded or elongated in shape. A number of basic forms of the major morphological characteristics (Fig. 9.6) are used in describing the fruit (Pope, 1929).

Growth and development

Fruit growth showed a simple sigmoidal growth curve in terms of length, thickness, mass and volume against days from anthesis (Fig. 9.7). Immature fruit skin is green or purplish and, upon ripening, becomes yellow, yellowish red, reddish or purplish red. The peel (exocarp) is thick and the flesh (mesocarp) of ripe fruit is yellow or orange-yellow and juicy. The pericarp can be separated into exocarp, mesocarp and endocarp at about 14 days after anthesis. There is a period 9–14 weeks after fruit set when growth rate decreases, and this is associated with hardening of the endocarp and accumulation of starch and sugars (Fig. 9.6). The endocarp is hard, with fibres that may extend into the flesh. The period from fruit set to maturity depends upon cultivar and climate and can range from 10 to 28 weeks. The cv. 'Saigon' grown in a hot climate is ready for harvest in 12–13 weeks. In cool areas where mean temperatures fall below 20°C, maturation is delayed by up to 4 weeks.

A network of latiferous canals and secretory ducts anastomoses in all directions in the exocarp and mesocarp (Joel, 1978). Cultivars with a poorly developed duct system are more susceptible to the Mediterranean fruit fly

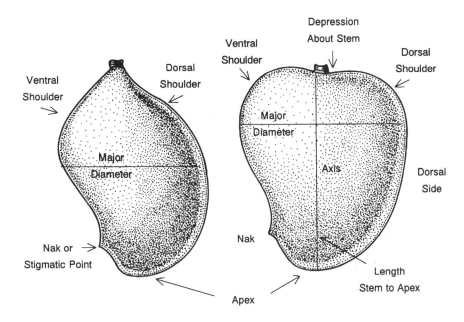

Fig. 9.6. Morphological characteristics of the mango fruit (Pope, 1929).

(*Ceratitis capitata* Wied.). The latiferous canals in mango begin to disintegrate during ripening and the fruit becomes susceptible to fruit flies.

CULTIVAR DEVELOPMENT

Genetics, Cytogenetics and Breeding

Cytological studies of seven species, *M. indica, M. sylvatica, Mangifera caloneura, M. zeylanica, M. caesia, M. foetida* and *M. odorata,* showed all to have chromosome number of 2n = 40 (Mukherjee, 1950). *Mangifera zeylanica* and *M. odorata* have been successfully crossed, indicating the possibility of interspecific crosses without the problem of interspecific sterility. This is supported by the uniformity of chromosome morphology. Although mango hybridization has been carried out, very few cultivars have been developed by controlled pollination. Hence, little genetic information is available.

Problems in Breeding

Breeding is difficult because of the small number of seeds obtained, complex nature of panicle and flower, excessive fruit drop, long life cycle and heterozygosity of the crop (Knight and Schnell, 1993). The objectives are to develop regular bearing, dwarf tree size, extended cropping season,

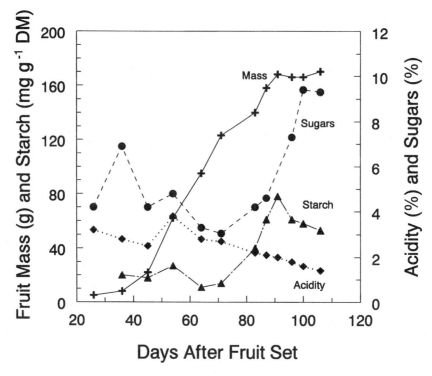

Fig. 9.7. Changes in developing 'Dashehari' fruit mass, starch and total sugars (after Tandon and Kalra, 1983) and 'Nam Dok Mai' titratable acidity (Kasantikul, D., 1983, cited by Mendoza and Wills, 1984). DM, dry matter.

good cropping in the wet tropics, attractive fruit of good size (300–500 g), freedom from internal breakdown and good keeping and eating quality, without fibre. Active breeding programmes are carried out in many countries in the world, including India, the USA, Israel and Australia (Galan Sauco, 1996).

More recent objectives of mango improvement include resistance to diseases, such as anthracnose and powdery mildew, both devastating to fruit yields, and tolerance to environmental conditions, such as soil salinity and cold temperatures. There is some variation in the degree of fruit susceptibility to the anthracnose disease. Genetic transformation of mango is possible, using *Agrobacterium* as the transformant.

Cultivars

The most important commercial cultivars are derived from selections among open-pollinated seedling populations. Many excellent cultivars have been developed through introduction and selection by cooperation between researchers, growers, nurserymen and hobbyists. Seedlings can be produced by mass cross-pollination of a collection of mixed cultivars selected for known desirable characteristics among mono- and polyembryonic races (Whiley *et al.*, 1993). Seeds from only mono-embryonic cultivars as the female parent are planted for later evaluation. Polyembryonic lines are difficult to use as the female parent, as there is no assurance that a sexual seedling will be produced.

There are hundreds of named mango cultivars throughout the tropics and subtropics (Table 9.3). Each country or region usually has its own selected cultivars (Table 9.4). It is without doubt that India, with over 1000 named cultivars and with different sections of the vast country growing specific cultivars, has provided most of the germ-plasm for cultivar development in the Western world. Florida has developed a large number of cultivars, using mainly Indian cultivars, which have shown adaptability to a wide geographical region. 'Haden' was a major cultivar in the early mango industry in Florida. It may be a parent of other cultivars, including 'Irwin' and 'Lippens'. Other well-known Florida cultivars include 'Keitt', 'Sensation' and 'Tommy Atkins' (Campbell and Campbell, 1992). These cultivars have been well accepted in the Americas, the Canary Islands and many African countries.

Isozyme analysis has been used to verify or refute parentage of mango cultivars (Degani *et al.*, 1990). Isozymic banding has shown that 'Haden' appears to be a seedling of 'Mulgoba' and 'Zill' a seedling of 'Haden'. 'Mulgoba' has the *ab* phenotype and 'Keitt' the *cc* phenotype while 'Haden' shows an *aa* phenotype, 'Carabao' *bb* and 'Edward' *ac*. Newer molecular-biology techniques are being used to unscramble the parentage of mangoes and to determine the extent of genetic diversity.

Hawaiian cultivars, such as 'Pope', 'Gouveia', 'Momi-K', 'Ah Ping', 'Harders' and 'Joe Welch', are all derived from monoembryonic parents. The recently released improved cultivars are 'Rapoza' and 'Exel', the former being a seedling of 'Irwin' and having large fruit with excellent eating quality. It is a late bearer, with fruit maturing from August through October, greatly extending the season (Hamilton *et al.*, 1992). 'Exel' was also selected from a population of 'Irwin' seedlings for its high quality, attractive fruit and regular bearing habits, with fruit weight ranging from 400 to 500 g and with 18% total soluble solids. Another desirable feature of the fruit is the thin, flat seed, which results in more than 90% edible flesh (Ito *et al.*, 1992).

There is a need to standardize criteria for evaluating seedling selections and germ-plasm collections. Rating scales for important characteristics, such as size, shape, colour, firmness, fibre content, flavour, disease resistance and yield, have been developed. Characteristics are rated on 1–5 or 1–10 scales (least desirable to most desirable) on merely descriptive terms (Knight, 1985). Preferences for cultivars will vary in different regions of the world.

Table 9.3. Some mango cultivars.

Cultivar	Origin	Poly/mono	Fruit shape	Skin colour	Fruit maturation	Fibre	Eating quality	Storage	Anthracnose susceptibility
Alphonso	India	Mono	Ovate oblique	Yellow	Mid	Low	Excellent		Low
Baptiste	Haiti	Poly	Oval	Bright yellow	Mid	None	Fair to good		
Carabao	The Philippines	Poly	Long and slender	Greenish to bright yellow	Early	None	Good to excellent	Poor	Low
Haden	Mulgoba India	Mono	Oval	Red blush or yellow	Mid	Abundant	Good to excellent		Susceptible
Irwin	Lippens Haden	Mono	Ovate	Red	Mid	None	Good	Poor	Low
Keitt	Mulgoba	Mono	Oval	Red blush on green	Late	Little	Good to excellent	Good	Moderate
Kensington	?	Poly	Oblong ovate	Yellow pink blush	Mid	Low	Excellent		Low
Langna	India	Mono	Oblong	Pink blush or greenish yellow	Mid	Abundant	Fair to good		
Manila	The Philippines	Poly	Long and slender	Bright yellow	Mid	Fibrous	Good to very good		Very susceptible
Mulgoba	India	Mono	Oval to ovate	Yellow pink blush	Mid to late	Little	Good to excellent		
Neelum	India	Mono	Ovate	Bright yellow	Late	None	Good to excellent		Susceptible
Nom Doc Mai	Thailand	Poly	Long slender	Bright yellow	Mid	None	Excellent		Susceptible
Pairi	India	Mono	Round to oblong	Orange red	Mid	Little	Good		Moderate
Saigon	Vietnam	Poly	Oval to ovate	Yellow	Mid	None	Good to excellent		Low
Tommy Atkins	Haden	Mono	Oval to oblong	Dark red or orange yellow	Mid	Moderate	Fair	Good	Resistant
Tong Dam	Thailand	Poly	Oblong to long	Greenish yellow	Early	None	Good to excellent		Susceptible
Zill	Haden	Mono	Oval to ovate	Dark red to crimson or yellow	Mid	None	Good to excellent	Poor	Susceptible

Mono, monoembryonic; poly, polyembryonic.

Table 9.4. Producing countries, selected cultivars and main marketing season.

Country	Selected cultivar	Marketing season
Australia	Kensington Pride, Keitt, Kent, Palmer, Irwin	October–March
Brazil	Haden, Tommy Atkins, Kent, Keitt, Palmer, Bourbon, Espada, Itamarco, Maco, Rosa, Carlota	October–February
India	Alphonso, Banganpalli, Dashehari, Bangalora, Langra, Mulgoa, Neelum, Pairi	April–July
Indonesia	Arumanis, Dodol, Gedong, Golek, Cengkir	September–January
Israel	Keitt, Tommy Atkins, Kent, Maya, Haden	July–August
Malaysia	Harumanis, Golek, Maha 65, MA 200 (Malgoa)	June–August
Mexico	Haden, Manila, Esmeralda, Kent, Keitt, Tommy Atkins, Jan Dyke, Palmer	April–October
The Philippines	Carabao, Pico, Julie	June–September
South Africa	Peach, Zill, Fascell, Sensation, Tommy Atkins, Keitt	November–January
Spain	Tommy Atkins, Keitt, Lippens, Osteen	July–August
Taiwan	Irwin, Yellow No. 1, Haden	July–October
Thailand	Nan Dok Mai, Rad, Tongdum, Okrong	March–May
USA – Florida	Keitt, Irwin, Tommy Atkins, Kent, Van Dyke, Palmer	July–August

CULTURAL PRACTICES

Propagation

Sexual

Mango seeds should be planted while still fresh, as they lose their viability in a few weeks. Uniform germination can be achieved if the kernel is removed from the hard endocarp. The germinated seedlings are separated and transplanted singly into polyethylene bags (c. 17.8 cm × 12.7 cm × 30.5 cm) and allowed to grow under shelter (30% shade). Monoembryonic seeds from breeding programmes can be planted directly into the polyethylene bags (Fig. 9.5). Cultivars of the polyembryonic race may be propagated by seed and still retain the parental characteristics but, generally, even these are grafted to take advantage of the earlier production and shorter, stockier trees, avoiding the juvenile characteristics of seedling trees.

Asexual

Cultivars are propagated vegetatively by such methods as grafting, budding or air layering. Researchers have sought rootstocks with special attributes, such as dwarfing habit and tolerance to high pH and saline conditions. Two Indian hybrid cultivars, 'Mallika' ('Neelum' × 'Dashehari') and 'Amrapali' ('Dashehari' × 'Neelum'), have been reported to exhibit distinctly dwarfish characteristics in terms of trunk circumference, tree height and canopy diameter in Brazil (Pinto and Sharma, 1984). Tolerance to saline soils has been identified and is being used as rootstock (cv. 13–1) commercially in Israel, where calcareous soil and saline irrigation water pose serious problems (Gazit and Kadman, 1980).

Seedlings can be grafted in 6–8 months if fertilized and irrigated regularly. Grafting is usually some form of side-graft or cleft graft, with the side-wedge method being most frequently used and successful. Only a well-matured terminal and the section below it should be used as scion wood; immature wood can lead to failure. Budding methods permit the use of much younger rootstocks. Buds may be prepared in advance by removing the leaves of mature terminal wood. Removal of the leaves and apical bud destroy apical dominance and allow axillary buds to begin to swell in 1–2 weeks. Air layering and inarching are used in some areas to rapidly generate experimental plants. However, these time-consuming methods are appropriate when only a few plants are needed. Cultivars propagated by grafting and inarching grow faster than those propagated by stooling and air layering. Grafting at a low height produces a spreading tree, while grafting high up on the stock is non-spreading.

Field Preparation

Land preparation is similar to that with other tree crops. Deep ripping may be necessary to break up any hard subsoil layer.

Transplanting and Spacing

Transplanting should be done just before or early in the wet season if no irrigation is available. Often, organic matter and frequently P fertilizer are added to the planting hole (0.6 m × 0.6 m × 0.6 m) before planting. No shade is required after transplanting. Spacing is largely dependent upon environment and the vigour of the cultivar. Various patterns, with spacing from 7 to 15 m by 7 to 15 m, are recommended. High-density planting (3 m × 2.5 m) has been tested with grafted trees, and, while individual tree yields are low, total yield is higher, although pruning is essential (Ram and Sirohi, 1991).

Irrigation

Generally, young transplants require about 20–30 L water every 4–5 days for about 2–3 months during establishment. For the remainder of the first year, rates may be increased to 40–50 L at 7- to 10-day intervals. During the second year, rates are increased to about 100–150 L per 10 days. More may be necessary during particularly dry periods. In the third year, rates of 200–300 L tree^{-1} at 15-day intervals may be adequate. Rates may be decreased or increased and the intervals shortened or lengthened, in accordance with soil type and the amounts and periods of rainfall.

Four- to five-year-old grafted trees can begin to bear fruit, and cultural practices should be adjusted to reflect this change. For bearing trees, especially in the tropics, it is desirable to have a 3–4-month dry period prior to flowering for reliable production to reduce vegetative flush growth. This is easily accomplished in areas where this period coincides with the natural dry period. In dry areas, irrigation is desirable after the inflorescence appears and flowers begin to set fruit and during the first 4–6 weeks of fruit development, when cell division is most rapid. Four-year-old trees may require around 400–500 L water tree^{-1} at 2-week intervals. Irrigation is completely cut off as the fruit approach maturity, since dry conditions favour higher sugar content. Heavy irrigation is resumed immediately after harvest to encourage new vegetative growth (Fig. 9.4). In the monsoon tropics, termination of harvest coincides with the rainy season, so irrigation is usually unnecessary. As soon as a major vegetative flush occurs, reduction in soil moisture content is desirable to mature the new flush. In modern orchards in the subtropics, localized irrigation is applied. The choice of irrigation equipment and its management should be based on capacity and efficiency of water delivery, as much as on the usual economic issues. Irrigation should be programmed in accordance with the phenological cycle (Fig. 9.4) to achieve maximum yield.

Pruning

Some pruning is necessary at transplanting time or shortly thereafter to train the tree to a single trunk. All branches from the base to approximately 1 m high are removed. Thereafter, three to four branches are allowed to grow from the leader at different heights, the branches are approximately 25–30 cm apart and facing in different directions to form the framework of the tree. After the trees have been adequately trained, annual pruning is not usually practised. Mango trees normally make very dense growth and, occasionally, light thinning of branches will become necessary to facilitate light penetration, air movement, penetration of pesticides, removal of dead and diseased branches and some control over tree height. The tallest branches are cut back at the fork (point of origin) of the branch. 'Water sprouts' and overlapping branches may be removed annually. In Florida, pruning is done with machines called 'hedgers' and 'toppers' by cutting back the height and width of the trees to about 4 m. This is done annually immediately after harvest (Campbell, 1988). These trees have two or three vegetative flushes before becoming dormant at the onset of winter and then produce flowers in the following spring.

Pruning becomes a major endeavour when trees are allowed to outgrow the space provided. Yields are reduced and harvesting becomes difficult and uneconomical. Removal of alternate trees appears only to aggravate management problems later, as remaining trees will grow even larger. Drastic pruning of large trees to about 2 m to develop new tops on old trunks may cause approximately 3 years' loss of income. Yearly pruning to about 3 m to control tree height is practised in Taiwan, without significant impact on year-to-year production (Plate 19). The fruit are bagged to reduce disease and insect damage (Plate 20) and fruit thinned to match fruit load to tree size.

In cultivars with biennial (irregular) flowering, shoots that have flowered can be removed after harvest, leaving only shoots that may flower next time. Flowering shoots that do not set fruit should also be removed soon after flowering. Removal of apical buds after each flushing cycle to increase the number of terminal shoots that could flower can lead to better fruiting and limits tree size (Oosthuyse and Jacobs, 1995). The number of terminal shoots, however, is not always related to the number of fruit set (Fig. 9.8). There is a strong correlation between number of terminal shoots and fruit number for the cv. 'Sensation', which retains a higher number of fruit per shoot than for 'Kent', indicating that other factors are involved. Deblossoming of the terminal inflorescences can lead to inflorescence development from axillary buds, a 20–30-day later harvest and higher yields (Chang and Lion, 1987).

Fertilization

One of the basic considerations for fertilization amount and time of application is the growth and flowering cycles of the tree (Fig. 9.4). During the first 3 years, approximately 113–227 g per tree of a complete nitrogen (N), P and potassium (K) (NPK) fertilizer is applied three times a year. From the fourth year, trees are considered mature, as they will begin to produce commercial yields, and fertilizer is applied twice a year. One application is made when the first inflorescence begins to appear and the second immediately after harvest to promote new vegetative flushes. Proper placement is important in ground application as the highest feeder-root density is approximately 90–175 cm away from the trunk to a depth of 20 cm. Irrigation is necessary whenever fertilizer is applied.

The major mango-growing countries have usually developed their own fertilizer ratios and amount to be applied. Complete fertilizers, with an oxide ratio of approximately 15: 15: 15, are usually recommended in Hawaii (Yee, 1958). In Florida, a good crop can be obtained by providing 1.4–1.8 kg of N and K tree^{-1} year^{-1} to mature trees (Malo, 1976). Using the triple-15 fertilizer, an application of 4 kg per tree in February–March provides 0.6 kg each of N and K. A second application immediately after harvest of 6 kg tree^{-1} provides

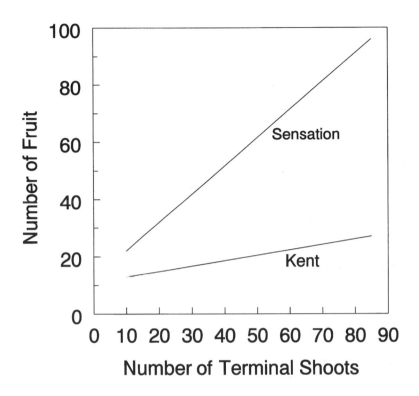

Fig. 9.8. Relationship between the number of terminal shoots on 2-year-old 'Sensation' and 'Kent' trees and number of fruit retained. 'Sensation': $y = 0.98x + 12.8$, $r^2 = 0.86$, $P < 0.001$; 'Kent': $y = 0.18x + 11.8$, $r^2 = 0.31$, $P < 0.005$ (after Oosthuyse and Jacobs, 1995).

0.9 kg each of N and K, for a total of 1.5 kg each of N and K for the year. Nitrogen and K at these levels are reported to enhance total yields, external colour and sugar content, while N in excess of 1.8 kg can produce unfavourable effects on external colour and flesh firmness (Young and Koo, 1974). When leaf N levels are greater than 1.5%, reduced fruit size is expected, with increased incidence of 'soft nose' and soft brown rot and a decrease in the number of days to ripen. Critical levels for various nutrients in mango leaves have been developed.

Fruit Yields

Fruit yields vary with cultivars used, climatic and edaphic conditions of the production site, cultural practices and other factors, such as diseases and insect pests. Yields over many years exhibit a sigmoidal curve, initially with low yields, increasing more rapidly and then dropping off as trees become crowded. The period of maximum production depends upon tree growth rate; a rapidly growing cultivar is more likely to show decreasing yields earlier due to crowding. Mango yield studies over a sufficient number of years involving replicated plantings are relatively rare, due to time and cost.

In Puerto Rico, researchers determined yield potential, year-to-year consistency of production, estimates of incremental increase in yields over age of trees, fruit size and tree growth of 16 cultivars in their first 6 crop years. The adjusted cultivar mean yield can be separated into three yield groups, with no significant differences between cultivars within each group (Pennock *et al.*, 1972). The high-yielding group order was 'Ruby', 'Sensation', 'Eldon', 'Lippens' and 'Irwin', an intermediate group 'Earlygold', 'Keitt', 'Parvin', 'Zill', 'Haden' and 'Palmer', and a low-yield group of 'Pillsbury', 'Kent', 'Edward', 'Santaella' and 'Jacquelin'. Cultivar consistency of bearing also gave three different groupings having different cultivar make-up: 'Edwards', 'Zill', 'Pillsbury', 'Ruby', 'Lippens' and 'Irwin' as regular bearers; 'Sensation', 'Santaella', 'Parvin', 'Earlygold' and 'Jacquelin' with intermediate consistency; and 'Kent', 'Eldon', 'Palmer', 'Haden', and 'Keitt' being highly inconsistent in yield (Fig. 9.9). The consistent bearing cultivars with high yields were 'Ruby', 'Lippens' and 'Irwin'. 'Edwards', 'Zill' and 'Pillsbury' show regular bearing habit but consistently produce low yields.

A mango orchard on the south coast of Puerto Rico under similar conditions to those of the above experimental site would be expected to begin production about 5 years after field planting, with a first crop of around 2.27 kg tree^{-1}, increasing to about 12.7 kg tree^{-1} year^{-1} during the next 5 years. On a per-hectare basis with 173 trees ha^{-1} (57.8 m^2 spacing), the initial yield of 5-year-old trees would be 393 kg ha^{-1}, with a yearly increase of 2.2 Mt ha^{-1} (Pennock *et al.*, 1972). Yields obtained in Florida, Mexico, Central and South America show that, in the fifth year after planting, 0.9 Mt ha^{-1} can be expected, increasing to 1.7 Mt ha^{-1} in year 6, 3.5 Mt ha^{-1} in year 7, 5.2 Mt ha^{-1} in year 8 and 8.7 Mt ha^{-1} in year 9. Mature trees can yield 10–30 Mt ha^{-1} with an average of 22–25 Mt ha^{-1} in the subtropics. In the tropics, commercial yields of 10 Mt ha^{-1} are expected from high quality cultivars.

Pest Management

Diseases

Anthracnose (*Colletotrichum gloeosporioides*) is perhaps the most important disease of the mango in almost all production areas (Table 9.5), as it attacks leaves, flowering panicles and fruit (Plate 21). Yields are drastically reduced when the inflorescence is attacked. This disease is especially serious in areas with high humidity and frequent light showers during the flowering period. In Hawaii, where rainfall coincides with the flowering season, almost the entire

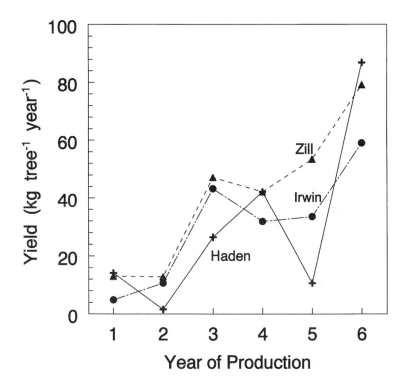

Fig. 9.9. Year-to-year variation in yield of three cultivars with different bearing habits (Pennock *et al.*, 1972).

production of inflorescence can be destroyed. In the Ryukyu Islands of Japan (26°N), a crop of mango can be produced only by constructing polyethylene shelters over the trees to protect the inflorescence from the frequent light showers falling during the flowering season. For this reason, trees are kept at 1.8–2.4 m in height. In Australia, the cultivars 'Carabao', 'Keitt', 'Tommy Atkins' and 'Zill' have been identified as possessing tolerance to anthracnose (Whiley and Saranah, 1981). Powdery mildew on leaves, stems and inflorescences can also become serious, especially under drier conditions (Johnson and Coates, 1993).

A disease that has been reported from a number of areas and is very serious in India, Pakistan and Egypt is the mango inflorescence malformation. Malformation occurs on vegetative shoots and flowering panicles. Panicles become short, branchy and compacted and produce solely male flowers (Shawky *et al.*, 1980). Lower temperatures during panicle development are associated with a higher incidence. It is caused by *Fusarium moniliformes*, with mites playing a minor role. A highly significant reduction in malformation is achieved, with an increase in yield, by spraying with GA_3 and NAA, singly and in mixtures, with a second application 2 weeks later. Pruning all malformed

Table 9.5. Some important diseases, disorders and conditions of mango.

Common name	Organism	Parts affected, symptoms	Region or country
Anthracnose	*Colletotrichum gloeosporioides*	Inflorescence, leaf black spots, lesions on fruit	Universal
Powdery mildew	*Oidium mangiferae*	Leaves, inflorescence; whitish grey powdery spores	Universal
Mango scab	*Elsinoe mangiferae*	Blossoms, leaves, twigs, fruit; greyish brown spore masses, cracked tissues	Widespread
Bacterial black spot	*Xanthomonas campestris*	Leaves, stem and fruit small watery lesions that coalesce, causing necrotic spots 2–3 mm diameter with irregular borders	Subtropical cold areas
Inflorescence malformation	*Fusarium moniliforme*	Inflorescence compact, sex shift to maleness; no fruit set; vegetative shoot compact with small leaves	Egypt, India, Pakistan, Africa, the Americas
Stem-end rot*	*Diplodia natalensis, Fusarium, Alternaria, Cladosporium*	Fruit; blackening of stem end of fruit, skin and pulp, overmature fruit; rotting	Widespread
Sooty mould (not pathogenic)	*Capnodium* sp., *Meliola* sp.	Leaves, stem, fruit; sooty black appearance; symptom of scale insects, mites and mango-hopper damage	Universal
Internal breakdown (jelly seed; soft nose)	Unknown	Fruit; pulp breakdown	Florida

*Some similarity in symptoms; further studies required for clarification.

vegetative and floral flushes, followed by spraying with copper oxychloride (4 g L⁻¹), is more effective than pruning alone (Azzouz *et al.*, 1989), reduces the percentage of malformed panicles and increases the yield in the following season.

Two other diseases with unidentified causes can sometimes cause serious loss, although little is reported in the literature. The first affects the bark at the base of the trunk, with increased gummosis on the upper trunk and large limbs. Wilting of branches occurs, followed by death, particularly to trees around 8 years old. It is especially serious in Nayarit and Sinaloa States in Mexico and has been reported from Colima State, Mexico, and in Guatemala. Both *Fusarium* and *Phytophthora* are suspected but have not been confirmed. The other condition is a disease of the fruit, with no puncture wounds or insects being observed. The external symptom is a black spot 0.5–1.5 cm in diameter, usually located at the nak of the fruit. When the blackened area is cut, there is a hollow 'tunnel' leading to the seed cavity. The seed and seed cavity are blackened. When the fruit are cut open, many show the peduncular vascular system leading to the endocarp to be disintegrated (Nakasone, 1979). It was observed to be extremely serious in orchards of 'Tommy Atkins' in Nayarit and on 'Kent' in Campeche, Mexico, and subsequently reported to occur in the Canary Islands and in Guatemala, where it is called 'pepita negra'. Lakshminarayana *et al.* (1985) reported a preharvest stem-end disease of 'Tommy Atkins' in Mexico similar to that reported, except that no mention was made of the black spot on the nak of the fruit. Preliminary studies isolated a mixture of six species of *Fusarium* and produced more diseased fruit with typical symptoms than any other single isolate. Bacterial black spot (*Xanthomonas campestris*) of the mango appears to be a relatively serious disease in South Africa and Australia. It has also been seen in Hainan, China. Powdery mildew (*Oidium mangiferae*) is a common disease of the dry subtropics.

Insect pests

Mediterranean and Oriental fruit flies are spread widely throughout the world. Other flies are confined to specific regions, but together they constitute a major problem, particularly to the export trade (Table 9.6). Irradiation, hot water and vapour heat treatment have been developed to meet disinfestation requirements in importing countries.

The mango-seed weevil has been a major deterrent to mango export. It is an Old World pest, but is found now in some parts of the New World: St Lucia, Guadeloupe and Martinique in the Caribbean region (Pollard and Alleyne, 1986). Field sanitation, chemical sprays, host-plant resistance, pest-free zones, fruit culling, using X-ray technology, and irradiation are possible solutions to this problem, unless markets are found in temperate countries that do not require disinfestation. All cultivars are susceptible, with some cultivars showing lower seed infestations of 17% to a high of 86%. The weevil

Table 9.6. Some important insect pests of mango.

Common name	Organism	Parts affected, symptoms	Region or country
Mexican fruit fly	Anastrepha ludens	Larva damage to fruit	Caribbean, Mexico
South American fruit fly	Anastrepha fraterculus	Larva damage to fruit	Americas
Caribbean fruit fly	Anastrepha suspensa	Larva damage to fruit	Caribbean, Florida
Queensland fruit fly	Batrocera tryoni	Larva damage to fruit	Australia
Mediterranean fruit fly	Ceratitis capitata	Larva damage to fruit	Widespread
Marula fruit fly	Ceratitis cosyra	Larva damage to fruit	Africa
Natal fruit fly	Ceratitis rosa	Larva damage to fruit	Africa
Oriental fruit fly	Dacus dorsalis	Larva damage to fruit	Asia, Hawaii, the Philippines
Mango seed weevil	Sternochetus mangiferae	Seed	India, Hawaii, the Philippines, South Africa, South-east Asia, Oceania, Caribbean
Mango blossom midge	Erosomyia indica Dasineura mangifera	Sucking sap from floral parts	India, Hawaii
Mango hopper	Idioscopus sp.	Sucking sap from flowering shoots	The Philippines, India, Africa, Oceania, Americas
Red-banded thrips	Selenothrips rubrocinctus	Sucking on underside of young leaves	Widespread
Coconut bug	Pseudotheraptus wayi	Sucking sap from young fruit, watery spot on fruit, fruit drop	Africa

deposits its eggs on the surface in the sinus region of small green fruit in the lower 2 m of the tree. Upon hatching the larva burrows through the soft pulp into the seed, goes through the pupal stages feeding on the developing seed and then finally develops into an adult weevil (Plate 22). When the fruit ripens and pulp decomposes, the adult beetles bore their way out of the endocarp and enter diapause in cracks and crevices on the tree until the next season. Since there is no external evidence of infestation and the fruit normally remains edible, the consumer is unaware of their presence. The weevil can affect the appearance of the flesh, if the mature weevil burrows out of mature fruit, causing it to decay from the seed outward, which hastens ripening and may even cause premature fruit drop.

The mango hopper is a serious pest in India, the Philippines and some other areas. The hopper sucks the sap from flowering stems, causing them to wilt. In serious cases, most of the panicles are damaged. Mango flowers are destroyed by four mango-blossom midge species and others attack the leaves (Table 9.6). One of these midges is a serious pest of mango flowers throughout the state of Hawaii, with several cultivars having 91% of the buds infested, leading to perfect-flower abortion. Eradication by chemical sprays is not feasible, due to the wide distribution and the large size of the trees, and biological control is not an alternative as no predators have been reported in the native habitats.

Other insect pests, such as scale insects, thrips and red spider mites, are found almost universally but they are relatively easily controlled by natural enemies and by chemical sprays and do not usually pose any problems.

Weed Management

Weed control is essential during orchard establishment. Young trees can be grown under clean cultivation or sod. Frequently intercropping is practised during mango establishment with papaya, pineapple or vegetables. Canopy closure of maturing trees prevents weed growth.

HARVESTING AND POSTHARVEST HANDLING

Harvesting

Harvest maturity is determined by using criteria such as changes in colour, fullness of cheeks and hardened endocarp. The most reliable indicator of maturity is when the endocarp has hardened and there is a yellowing of the flesh near the seed; however, this is a destructive test. Immature fruit do not ripen to full flavour and should not be harvested. Fruit-set dates can be established as an index for harvesting. The fruit-set date for each tree is determined when the panicles shows a high percentage of initial fruit set. An old recommendation for judging harvesting date was that, when the first fruit began to drop, the crop was ready to pick. The mango is harvested by hand wherever reachable from the ground or from ladders and by the use of a long pole with a cloth bag to hold two or three large fruit attached at the tip.

Postharvest Treatments

Any form of bruising should be avoided during harvesting and transporting to the packing house. The use of shallow lug boxes minimizes bruising. At the packing house, fruit are usually placed in a water-bath or hand-washed to remove the stem sap from the surface of the fruit. Sap removal is essential to prevent sap burn and should be done within 24 h (Lovey *et al.*, 1992). Fruit anthracnose can be controlled by dipping into hot water (52°C, 5 min). A combination of hot water and a fungicide or chlorine may also be used. When the fungicide is added to the hot water, the temperature can be reduced slightly (Akamine, 1976).

Grade standards are usually based upon size, colour and freedom from injury and defects. Other requirements include full development, freedom from stains and firmness. Since the fruit is easily bruised, fruit is packed in single- or double-layer cartons with adequate protective material or use of trays. In the local markets, mangoes are frequently packed in bamboo baskets.

At ambient temperatures, shelf-life of this climacteric fruit is short: 7–14 days to fully ripe. Precooling to 10–13°C is beneficial during hot weather or when shipping is delayed. Fully ripe fruit can be stored at 8–10°C. The length of shelf-life varies markedly with cultivar, maturity at harvest, injury, calcium (Ca) sprays and exposure to ethylene. A dip in 4–6% calcium chloride can significantly increase shelf-life of some cultivars, with the response varying with season, field-management practices and soil type. Fruit of 'Keitt', 'Tommy Atkins' and 'Muska' from successive harvests show an increasing rate of ripening changes during the 21 days' storage period at 12°C, suggesting a decrease in storage potential as the season progressed.

Mango is a climacteric fruit and ethylene can be used to reduce the time till ripening commences. A treatment of 100 p.p.m. for 24–48 h at 25°C and 90% relative humidity (RH) is adequate. Acetylene generated from calcium carbide and ethephon can also be used. Skin colour is also enhanced by ethylene treatment by increasing degreening. The best ripening temperature range is from 21 to 24°C. At high temperatures of 32°C, ripening can be retarded.

Controlled atmospheres have been tested on mangoes and indicate some possibilities; storage in atmospheres of 5% oxygen (O_2) and 5% carbon dioxide (CO_2) is possible for 20 days, while off-flavours and skin discoloration occur at 1% O_2 or high CO_2 (15%). Cultivar differences in response have been reported and the extension in shelf-life may not be commercially viable. Modified-atmosphere storage using plastic bags or wraps and waxing shows some delay in ripening. Off flavours have been reported with some wraps and waxes that delayed ripening. Waxes are widely and successfully used commercially on mango to reduce water loss.

Fruit flies pose a problem in marketing. Following the banning of the fumigant ethylene dibromide, alternatives, such as cold and heat treatments, have been developed. Irradiation (250 Gy), immersion of fruit in 46.1°C water for 65 min, vapour heat and forced hot-air treatments have potential for different cultivars and fruit flies. Treated fruit can show fresh-mass loss, development of trace amounts of peel pitting (small, slightly sunken, shrivelled areas on the skin at bruised areas) and reduced shelf-life, with no differences in total soluble solids and peel colour at the soft-ripe stage. The heat treated fruit show less severe incidence of stem-end rot and anthracnose.

Postharvest disorders include chilling injury, sap burn, internal breakdown and bumpy tissue (Wainwright and Burbage, 1989). Chilling injury is a storage disorder that occurs at temperatures below 12.5°C, the extent of the injury being dependent upon the storage temperature and duration: at 0°C injury occurs in 4 days, at 5°C in 8 days and at 10°C in 12 days. The symptoms include skin scald, failure to ripen and increased disease susceptibility. Sap burn caused by fruit skin contact with sap exuded from the cut or broken pedicel reduces consumer acceptance because of the browning and blackening of the skin after lenticel penetration. The Australian cultivar 'Kensington' is very susceptible (Lovey *et al.*, 1992), while 'Irwin' is less susceptible. The sap component in 'Kensington' thought to cause the burn is the major non-aqueous terpene component, terpinolene; it can also burn 'Irwin', but in 'Irwin' the predominant terpene (6.8%) is car-3-ene. This sap is present in the latiferous ducts of the fruit and is not interconnected with the stem ducts. The latex is under some pressure and when the pedicel is broken can shoot 300 mm or more. Harvesting with the stem attached, draining with the pedicel down and washing are effective.

One fruit disorder, occurring especially in 'Alphonso' in India, is referred to as 'internal breakdown', 'spongy tissue' or 'soft tissue'. The lower half of the fruit is most affected and it may be related to preharvest heat stress. 'Soft nose' in Florida is serious, with high Ca inhibiting the disorder and high N increasing the disorder (Malo and Campbell, 1978). The 'jelly-seed' disorder is more widespread. The disease usually appears during the initial stages of maturity, with a loss of firmness of the pulp near the endocarp, which becomes jelly-like and translucent with advancing ripeness (Plate 23). The disorder does not develop after harvest. An open cavity may develop in the pulp at the stem end prior to pulp breakdown. 'Tommy Atkins' has been reported to be especially susceptible to this disorder, although 'Kent', 'Irwin', 'Sensation', 'Carabao', 'Alohouron' and a few other commercial cultivars of importance are also susceptible (Campbell, 1988). No pathogenic organism has been detected. The only recourse for this disorder is to harvest the fruit at the mature-green stage, before any colour break occurs on the skin.

Certain cultivars are particularly susceptible to lumpy tissue, which is not evident in green fruit but develops during ripening. The mesocarp contains white starchy lumps and the fruit surface develops indentations. Aetiology is unknown. It has been reported from Thailand and the Philippines. Internal fruit necrosis first appears as a brown area in the mesocarp and endocarp of rapidly growing fruit. This later extends to the skin and a brown-black gummy exudation occurs. These areas then collapse and are surrounded by corky tissue. This non-pathological disorder has been associated with boron deficiency.

Marketing

Consumer preferences vary, with the US market apparently preferring large-sized and highly coloured mango, colour being the more important. Mango marketing has taken on international dimensions, with major marketing centres around the world. Canada and the USA are the major markets in North America, while the major European Community (EC) markets are located in the UK, France, Germany and the Netherlands. In the Far East, Japan, Hong Kong and Singapore are lucrative markets for producing countries, such as the Philippines, Malaysia, Thailand and Pakistan. The Philippine and Mexican mangoes dominate the Japanese market. The major supplier for Hong Kong is the Philippines.

Countries supplying the North American and European markets are Mexico, Ecuador, Brazil, Peru, Venezuela in the Americas, Haiti, Jamaica and other Caribbean-island countries and Ivory Coast, South Africa and Mali in Africa. Egypt and Israel are small producers but are looking at the EC market windows. Increases in demand will largely depend upon increasing consumer familiarity with the fruit, quality and price structure. By virtue of geographical location in the southern hemisphere, South Africa, Brazil, Peru and Australia are able to supply mangoes to the North American, European and Asian markets during their winter months, thus making mango a fruit that is available year-round (Table 9.4).

UTILIZATION

The mango is rapidly becoming one of the leading trade crops in the tropics and subtropics. As postharvest-handling techniques and shipping technology have improved, consumer demand has increased. The fruit is 60–75% flesh, 11–18% skin and 14–22% seed, depending upon cultivar, with the flesh being *c.* 20% dry matter. Most of the mangoes produced are marketed in the fresh state for consumption as a dessert fruit. Fruit can be eaten green, and this practice is very popular in Thailand and the Philippines, with some starchy and crispy cultivars being preferred, such as 'Khieo Sawoey' in Thailand. Fruit may simply be peeled and sliced. Diced pieces may be added to salads and fruit cocktails. People consume mango simply because of its pleasant taste and flavour without much thought about the content of minerals, vitamins, lipids and amino acids. However, the mango is a good to excellent source of provitamin A and is considered a fair source of vitamin C (Table 9.7), although this varies greatly among cultivars, with a range between a low of 5 mg and as high as 142 mg 100 g^{-1} of fresh material (Wenkam, 1990).

Considerable amounts of fruit are processed into various products, such as jellies, jams, marmalades, pulp, juice and canned slices, throughout the world. Green mangoes make excellent chutney. Canned mango slices have been processed in India since before 1925 (Hayes, 1966). Canned mango and dehydrated slices are important export products in the Philippines.

Table 9.7. Proximate analysis of mango (in 100 g edible portion) (Wenkam, 1990).

Nutrient	Units	Haden	Pirie
Proximate			
Water	g	84.12	79.97
Energy	kJ	234	301
Protein	g	0.39	0.55
Lipids (fat)	g	0.02	0.20
Carbohydrate	g	15.05	18.91
Fibre	g	0.54	0.70
Ash	g	0.42	0.37
Minerals			
Calcium	mg	8	6
Iron	mg	0.16	0.16
Magnesium	mg	12	12
Phosphorus	mg	10	15
Potassium	mg	159	126
Sodium	mg	0	3
Zinc	mg	–	–
Copper	mg	–	–
Manganese	mg	–	–
Vitamins			
Ascorbic acid	mg	15.10	15.00
Thiamine	mg	0.041	0.081
Riboflavin	mg	0.057	0.061
Niacin	mg	0.300	0.460
Pantothenic acid	mg	–	–
Vitamin A	IU	3813	4735
Vitamin B$_6$	mg	–	–
Vitamin B$_{12}$	μg	–	–

FURTHER READING

Cull, B.W. (1991) Mango crop management. *Acta Horticulturae* 291, 154–173.

Johnson, G.I. and Coates, L.M. (1993) Postharvest diseases of mango. *Postharvest News and Information* 4, 27N–34N.

Litz, R.E. (ed.) (1997) *The Mango: Botany, Production and Uses.* CAB International, Wallingford, UK, 592 pp.

Mendoza, D.B. and Wills, R.B.H. (eds) (1984) *Mango: Fruit Development, Postharvest Physiology and Marketing in ASEAN.* ASEAN Food Handling Bureau, Kuala Lumpur, Malaysia, 111 pp.

Ridgeway, E. (ed.) (1989) *Mango Pests and Disorders.* Information Series QI89007, Queensland Department of Primary Industries, Brisbane, Australia.

Schaffer, B., Whiley, A.W. and Crane, J.H. (1994) Mango. In: Schaffer, B. and Anderson, P.C. (eds) *Handbook of Environmental Physiology of Fruit Crops*, Vol. II, *Subtropical and Tropical Crops.* CRC Press, Boca Raton, Florida, pp. 165–197.

Wainwright, H. and Burbage, M.B. (1989) Physiological disorders in mango (*Mangifera indica* L.) fruit. *Journal of Horticultural Science* 64, 125–135.

10

PAPAYA

BOTANY

Introduction

The *Caricaceae* are a small family of dicotyledonous plants with four genera: three of tropical American origin (*Carica, Jarilla, Jacaratia*) and one, *Cylicomorpha*, from equatorial Africa. *Caricaceae* species have been variously classified in families such as *Cucurbitaceae*, *Passifloraceae*, *Bixaceae* and *Papayaceae*. Approximately 71 species have been described, although Badillo (1993) reduced the number to 30 species with the following distribution: *Carica* 21 species, *Cylicomorpha* two species, *Jacaratia* six species and *Jarilla* one species.

Papaya (*Carica papaya* L.) is the most important economic species of the 21 species in *Carica*. Common names include papaya, papaw or pawpaw, papayer (French), melonenbaum (German), lechosa (Spanish), mamao, mamoeiro (Portuguese) and mugua (Chinese). *Carica* species are dioecious, except for the monoecious *Carica monoica* (Desf.) and some *Carica pubescens* and the polygamous *C. papaya*. Most species are herbaceous, single-stemmed and erect. The fruit are normally dry and lack the juicy flesh of *C. papaya* or *Carica heilbornii-badillo* var. *pentagona*. *C. heilbornii-badillo* var. *pentagona*, called 'babaco', is of subtropical origin, found at 2000–3000 m in Ecuador. The fruit sets parthenocarpically and is intermediate in size, juicy and somewhat acid in taste. Plants are propagated by stem cuttings.

Origin and Distribution

All *Carica* species are native to tropical America. Isozyme analysis indicates that *C. papaya* is only distantly related to the other *Carica* spp. The greatest diversity exists in the Yucatan–San Ignacio-Peter–Rio Motagua area of Central America, with the wild population having greater diversity than domesticated populations (Morshidi, 1996).

239

The accounts of eighteenth century travellers and botanists indicated that seeds of papaya were taken from the Caribbean to Malacca and on to India (Storey, 1941). From Malacca or the Philippines, distribution continued throughout Asia and to the South Pacific region. Don Francisco Marin, a Spanish explorer and horticulturist, is credited with the introduction of papaya into Hawaii from the Marquesas Islands during the early 1800s. Papaya is now grown in all tropical countries and in many subtropical regions of the world. Early distribution over wide regions was enhanced by abundance of seeds in the fruit and their long viability.

ECOLOGY

Major commercial production of papaya is found primarily between 23°N and S latitudes. Humans have extended cultivation into regions as far as 32°N and S. At these latitudes, papayas may be best grown in well-protected areas at sea level. In Hawaii, at 19–22°N, papaya is grown at sea level and up to 300 m elevation.

Soil

Papayas are grown in a variety of soil types, with the most essential requirement being drainage; poor drainage leads to the development of root rots. A porous loam or sandy loam soil is preferred. In Hawaii, the crop is frequently grown on rocky, volcanic soil called a'a, composed of porous lava with some organic matter and excellent drainage, the planting holes are filled with soil prior to planting.

Papaya grow well at a soil pH between 5.0 and 7.0, with the range between 5.5 and 6.5 being more desirable (Awada *et al.*, 1975). At pH levels below 5.0, seedling growth is poor and mortality is high. In soils with a pH range of 5.0–5.5, lime applications can increase growth and yield.

Papaya has been ranked from extremely sensitive to moderately tolerant to salt stress, germination and early seedling growth being the most sensitive stages. It is probably moderately salt sensitive at other growth stages.

Climate

Rainfall

Papayas grow well and produce substantial yields without supplementary irrigation if there is a minimum monthly precipitation of approximately 100 mm. Such an ideal distribution rarely occurs, even in Hawaii, where rainfall is considered well distributed. Most tropical areas with monsoon-type climates have well-defined wet and dry seasons. Successful production therefore depends upon the availability of supplemental irrigation during the dry period.

A minimum relative humidity of 66% has been reported for papaya's optimum growth. Stomatal opening is controlled by humidity and, as relative leaf-water content is not affected by drought stress, stomatal closure maintains leaf-water status, allowing rapid return of gas-exchange flux and growth upon rewatering (Marler, 1994). Drought frequently leads to the rapid shedding of older leaves and poor fruit set. Within 5 days of flooding, abscission of fully exposed leaves is preceded by chlorosis. Flooding frequently leads to plant death due to root rots, while recovery from non-lethal flooding is slow, due possibly to the low root growth rate in fruiting trees.

Temperature

Optimum temperature for growth is between 21 and 33°C. Papaya is extremely sensitive to frost and, if the temperature falls below 12–14°C for several hours at night, growth and production are severely affected. Dioecious cultivars are better suited to low temperatures (< 20°C), as female trees do not exhibit the sex changes shown by the more sensitive bisexual (hermaphroditic) cultivars to stamen carpellody (Fig. 10.1). Hermaphroditic

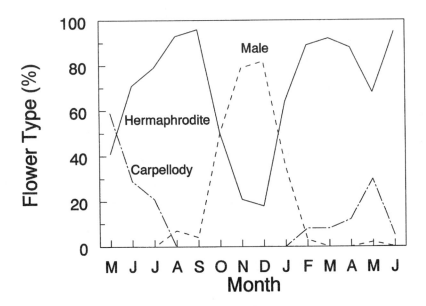

Fig. 10.1. The percentage of true hermaphrodite, hermaphrodite with functional stamens (male) and non-functional pistil and stamen carpellody flowers at different times of the year, on 'Solo' type papaya tree growing in Honolulu, with a mean minimum growing temperature of 21°C and maximum temperature of 27°C (redrawn from Awada, 1958).

cultivars ('Solo' type), grown with minimum temperature less than 17°C, can have 100% carpellodic flowers. At higher temperatures (> 35°C), there is a tendency of bisexual cultivars to form functional male flowers with poorly developed and non-functional female parts. This tendency varies with cultivars and within a cultivar. Net photosynthetic rate also rapidly declines above 30°C.

Temperature during the growing season significantly influences fruit growth and development from the normal 120–150 days. The effect is most pronounced in subtropical areas. In these areas, fruit set does not normally occur in winter and fruit set before the winter can take up to 90 days longer to reach maturity. Fruit developing during cooler parts of the year also have lower total soluble solids. Final fruit size is determined in the first 4–6 weeks of fruit development and temperature plays a dominant role in the process, especially in subtropical areas (Fig. 10.2).

Radiation

Papaya in its wild state is a rapid volunteer in areas where the tree vegetation has been disturbed. The high saturation point above 1000 µmol m^{-2} s^{-1}

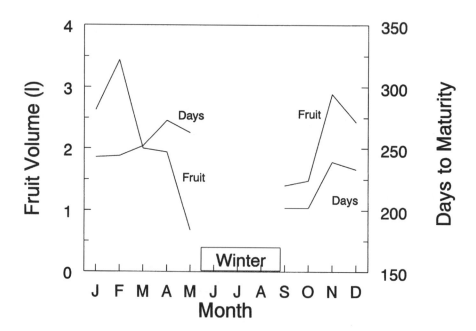

Fig. 10.2. The effect of date of fruit set on final fruit size and days from anthesis to the half ripe fruit stage of plants grown under subtropical condition (redrawn from Kuhne and Allan, 1970). No fruit set occurs during the winter months, due to low temperature.

(photosynthetic photon flux) (*c.* 800 W m^{-2}) supports its rapid development in newly disturbed areas having direct sunlight. When exposed to shade, the plant is shorter, having smaller leaf area, lower stomatal density, increased internode and petiole length and chlorophyll content, and hence it is regarded as a shade-avoiding species. Partial stomatal closure and opening occurs rapidly with cloud-related changes in irradiance, thereby maximizing plant water use efficiency (Clemente and Marler, 1996).

No photoperiodic effects on tree growth, production or sex expression have been reported (Lange, 1961).

Windbreaks

Papaya trees are delicate and require protection from strong winds. The root system is well developed, though relatively shallow and the tree can be uprooted by winds of 64 km h^{-1}, especially if the soil is softened by rain. Even though trees withstand uprooting, considerable damage occurs to the large leaves, leading to flower and young-fruit abscission and low total soluble solids in the more mature fruit on the column. Recovery can take from 4 to 8 weeks.

GENERAL CHARACTERISTICS

Stem

The papaya is a large herbaceous plant, with a single erect stem, which can attain heights up to 9 m, terminating with a crown of large leaves (Fig. 10.3). Although there are occasional lines that produce an abundance of lateral branches, especially during the juvenile period, the monoaxial stem normally grows without branching, unless the growing point is injured. Natural growth of axillary branches develops when the trees are 5–10 years old. The stem is semiwoody and hollow. The bark is smooth, greyish in colour, with large, prominent leaf scars. When the stem is wounded, a thin milky sap oozes from the wound.

After transplanting, shoot growth is initially slow, although considerable root growth is taking place, extending well beyond the canopy line (Fig. 10.4). Stem growth is then rapid up to flowering, increasing in circumference up to 2 mm per day. Growth rate peaks at flowering and then declines as the tree starts bearing (Fig. 10.5). The rate of stem growth is influenced by nitrogen (N) and phosphorus (P) supply, irrigation and temperature. Root growth declines dramatically as flower initiation occurs, continuing at a very low rate during flowering and fruiting (Fig. 10.4). These findings suggest that it is the need to continue stem growth and new leaf and flower formation that decides final yield. Hence, fertilization and irrigation are crucial in the management of this crop.

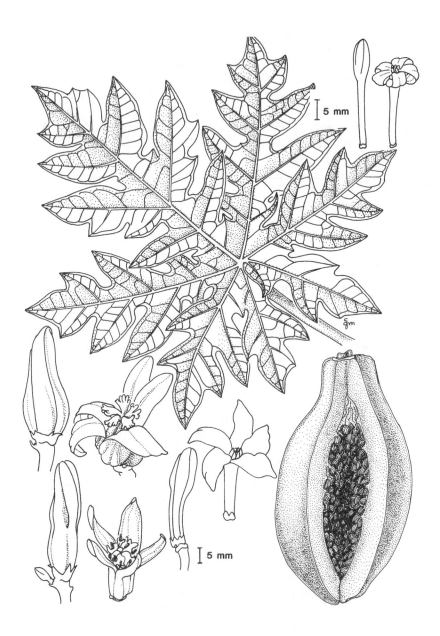

Fig. 10.3. Papaya leaf, female, hermaphrodite and male flower types and fruit. The male flower in the bottom centre is found in hermaphrodite trees that become sterile during hot weather, while the smaller male flowers in the upper right are true males. The unopened female flower is pear-shaped, while the hermaphrodite flower is tubular, with a constriction halfway up (lower left).

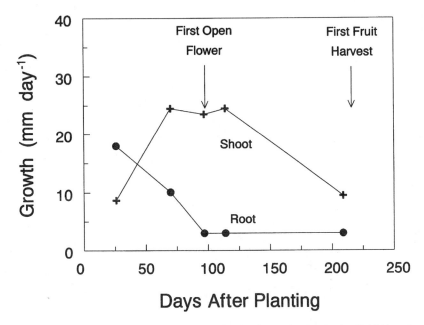

Fig. 10.4. The growth rate of papaya root and stem from planting in the field, showing root growth-rate reduction at the start of flowering and shoot growth as fruit mature (after Marler and Discekici, 1996).

Leaves

The cluster of leaves at the apex and along the upper part of the stem make up the foliage of the tree. New leaves are constantly formed at the apex and old leaves senesce and fall. Leaves are palmately lobed, with prominent venation and can measure 40–50 cm or more in diameter and have an individual leaf area of 1625 cm^2, with c. 15 mature leaves per plant. In the tropics, new leaves appear at a rate of two to three a week; in Hawaii the rate is 2.4 per week during the cool season and up to 3.0 in the warm season. Petioles are cylindrical, hollow and 60–90 cm long, depending upon the cultivar. The most recently matured leaf's fresh weight (c. tenth leaf from 2.4 cm juvenile leaf) can vary from c. 50 g to 170 g. The leaf-petiole dry mass increases at a rapid rate until flowering and then increases more slowly, peaking after fruit bearing starts (Fig. 10.5).

Inflorescence and Flowers

Flowers are borne on modified cymose inflorescences, which appear in the axils of the leaves. The type of inflorescence depends upon the sex of the tree.

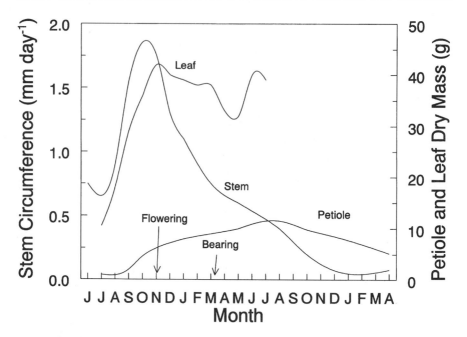

Fig. 10.5. Growth rate of papaya stem circumference, dry weight of the most recently matured leaf petiole (redrawn from Awada, 1977) and the seventeenth leaf from the 2.4 cm juvenile leaf (redrawn from Shoji *et al.*, 1958) under dry-land conditions of Puna, Hawaii. The leaf dry-weight reduction in April is attributed to low rainfall in the previous 2 months.

In staminate trees, flowers are sessile and are produced in clusters on long pendulent racemes 60–90 cm long. The individual flower is tubular, with ten stamens in two series of five attached to the throat of the corolla tube, and lacks a pistil (Fig. 10.3). There may be two races of male trees, one that rarely produces fruit and the other that fruits during cool months from normal or teratological hermaphroditic flowers. The female tree produces only female flowers, on short 4–6 cm peduncles, with a functional pistil devoid of stamens. The five petals are separate but are inconspicuously fused together at the very base of the ovary. The hermaphroditic form is between the two unisexual flower types and exhibits numerous deviations. The basic flower on a short peduncle is characterized by an elongated pistil, with five stigmatic rays and five petals, which are fused to about two-thirds of their length, forming a corolla tube. There are ten stamens in two series of fives attached to the throat of the corolla tube (Fig. 10.3). The pistil is normally five-carpellate and elongated, oval or pyriform.

Problems in pollination, fruit set and production are intimately associated with sex expression resulting from genotype–environment interactions. Cultivar and environmental differences have produced a wide array of modified forms so the number and types of modifications have varied in reports by various researchers. Storey (1958) summarized and simplified the classification into eight categories and this has been generally accepted by papaya breeders (Table 10.1). For practical purposes, the eight flower types are classified into three primary groups: (i) male; (ii) hermaphrodite; and (iii) female. The widest variations in sex types are found in hermaphrodites and

Table 10.1. Types of papaya flowers (Storey, 1958).

Types	Tree	Flower	Description
Staminate	M	M	Typical unisexual flower on long peduncles
Teratological staminate	M	M	Found on sex-reversing male tree, with some degree of carpel initiation and development. A number of hair-like processes – vestigial carpels – at base
Reduced elongata	MF	M	Modified normal elongata flower differs from staminate flowers in having a thicker and stiffer corolla tube, abortion of pistils and reduced ovary size and number of carpels. More frequent during warm periods and late summer, and can last from 1–2 weeks to 6 months, depending upon cultivar and temperature
Elongata (normal type)	MF	MF	Elongata refers to the shape of the pistil terminating in fine stigmata lobes; develops into pyriform or cylindrical fruit, five laterally fused carpels. Petals fused two-thirds length
Carpelloid elongata	MF	F	Transformation of the inner series of stamens into carpel-like structure. Numerous types with different number of stamens becoming carpelloid and degree of carpellody, from slight to developing locules with functional stigma. Fruit to varying degrees misshaped
Pentandria	MF	F	Normal hermaphrodite type, modified unisexual pistillate flower, through stepwise stamen transformation to carpels, with loss of the original carpels. Short corolla tube, only five stamens of the outer whorl on long filaments globose and furrowed pistil. From five to ten carpels.
Carpelloid pentandria	MF	F	The stamens of the outer whorl become carpelloid. Carpellodic forms in carious stage, especially under cool conditions. All five stamens fully carpelloid and fuse laterally, with abortion of original carpels, flowers resemble pistillate flowers – pseudo-pistillate
Pistillate	F	F	Unisexual flowers larger than MF flower, lack stamens. Form stable and unchanged by environment

males. The appearance of a large number of modified forms occurs in progenies from appropriate hybridization when grown under a temperature regime, which may range from around 13 to 32°C. Therefore, recovery of extreme sex modifications may not be seen in a breeding programme unless conducted with heterozygous parental types in an area with wide fluctuations in seasonal temperatures. For example, stamen carpellody is expressed under cool temperatures, with increasing severity at lower temperatures in the c. 40 days before anthesis. Instead of ten stamens in a double whorl, there are only five stamens, with the other five fixed to the normal carpels. The fruit that develop from this carpellody are severely misshapen and unmarketable (Fig. 10.6). The incidence of carpellody also declines with increasing plant age and may be related to internode length (Chan, 1985). Female sterility occurs at warm temperatures, again with increasing severity at higher temperatures in the c. 40 days before anthesis. Excessive N and moisture also favour stamen carpellody, while plant stress, such as N deficiency and moisture stress, influences female sterility (Awada and Ikeda, 1957).

The floral primordia are laid down at a rate of about one new flower in each leaf every 2–3 days (Arkle and Nakasone, 1984). Stamen differentiation begins 50–56 days before anthesis and is completed by the 35 days prior to anthesis. Differentiation of the ovaries begins 42–50 days prior to anthesis and is completed within the 28 days before anthesis. Flower primordia initiation to anthesis ranges from c. 46 days in Hawaii to 80 days in India, with the wide discrepancy being due to temperature.

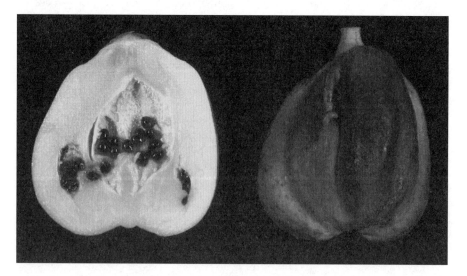

Fig. 10.6. Stamen carpellody induced by environmental conditions (temperature, water stress, fertilization) on young hermaphrodite trees significantly alters fruit shape. The mild forms are sometimes referred to as 'cat face'.

Pollination and Fruit Set

In mixed planting of pistillate and hermaphroditic trees or in purely hermaph-roditic stands, no pollination problems are experienced. Problems occur when dioecious cultivars are planted with an inadequate number of male trees. In Australia, the ratio of female to male recommended is 8 : 1. However, one male tree per 15–20 female trees provides adequate wind pollination, if male trees are located appropriately with respect to prevailing winds.

In normal bisexual lines, the anthers extend directly over the stigmatic rays, ensuring automatic self-pollination, even when flowers are bagged to prevent contamination. However, in some cultivars and breeding lines, the anthers, with short filaments or adnation of filaments at lower positions on the corolla neck, may be 5–8 mm below the stigma. Pollen should be placed upon the stigma before bagging. Self-incompatibility in cultivars is relatively rare, although there are isolated cases when controlled self-pollinations are made.

Fruit set is no problem under open pollination in a well-managed orchard. On bisexual trees, it is most common for the terminal flower to set while the laterals abscise. Under favourable conditions, one or two laterals may be set and only persist for 2–3 weeks or remain to produce undersized fruit, which crowd the fruit column. This fruit crowding leads to fruit compaction and misshaped fruit, and thinning may be practised. Annual fruit set depends upon the length of the female-sterility period in hot weather and, with one fruit per leaf axil, the range is 85–95%.

Fruit

The fruit superficially resembles a melon, being spherical, pyriform, oval or elongated in shape. Fruit from female trees are spherical and those from hermaphroditic trees can show diverse shapes depending upon modifying factors affecting flower morphology during ontogeny. Fruit size ranges from 255 g to 6.8 kg, with flesh thickness from 1.5 to 4 cm. The fruit is normally composed of five carpels, united to form a central cavity containing the seeds (Fig. 10.3). The fruit-seed cavity can be star-shaped to round. Placentation is parietal, with the seeds attached by a 0.5–1 mm stalk. The seeds are dark grey to black when mature and enclosed in a sarcotesta. Parthenocarpy in papaya is rare. Seedless fruit or fruit with very low seeds can be produced on female trees and are generally smaller in size.

In gynaecia less than 1 mm, all tissue is meristematic; later, the outer layer of the epidermis increases in size, while the subepidermal layer continues to divide, both anticlinally and periclinally. The central parenchyma of the pericarp increases in size and divides, with the placenta forming opposite the marginal vascular bundles. Fruit growth shows two major phases. The first lasts about 80 days postanthesis (Fig. 10.7), with a greater increase in dry weight occurring just before fruit maturity. Mesocarp growth parallels seed

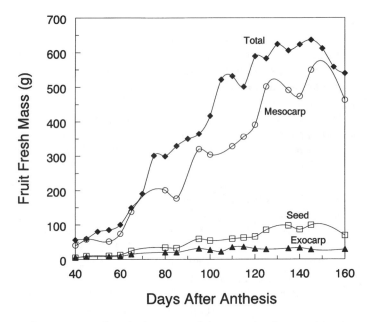

Fig. 10.7. The increase in fruit fresh mass and the mass for skin, seeds and mesocarp during fruit development (redrawn from Qiu *et al.*, 1995).

and total fruit growth. Flesh colour is white in immature fruit to a pale orange-yellow, salmon-pink or red, depending upon cultivar, in ripe fruit (Table 10.2). Total fruit starch declines from 0.4% to less than 0.1% during the first 80 days of fruit development. Sugars, however, do not begin to accumulate until 110 days from anthesis during the last 28–42 days of fruit development (Fig. 10.8). Flesh total soluble solids can be as low as 5% up to 19%. Fruit development takes 150–164 days, which is extended another 14–21 days in

Table 10.2. Preference of fruit characteristics in some countries.

Country	Fruit mass (g)	Shape	Flesh colour	Sex type
Australia	800–1000	Cylindrical, spherical	Yellow	Dioecious
Cook Islands	500–1134	Pyriform	Yellow	Hermaphrodite
Fiji	397–510	Pyriform	Red, yellow	Hermaphrodite
Hawaii	397–510	Pyriform, oval	Yellow, red	Hermaphrodite
Mexico	1360–5443	Spherical, elongata	Yellow, red	Dioecious, hermaphrodite
South Africa	1000–1500	Oval	Yellow	Dioecious
Caribbean Islands	500–4000	Round, oval, pyriform, elongata	Red, yellow	Dioecious, hermaphrodite

Hawaii in the colder months (Paull, 1993). In subtropical areas, such as South Africa, it can vary from 190 to 270 days (Fig. 10.2).

Green fruit contains an abundance of milky latex, which contains the protease papain. The pericarp consists of a network of lacticifers, which develop close to the vascular bundles and anastomose profusely throughout the fruit (Roth and Clausnitzer, 1972). This latex can be under pressure and spurt out to a distance of 20 cm when the skin is pricked. Commercially, the skin is scarified to induce latex flow, which is allowed to dry and then collected to be later processed into papain. Laticifers collapse as the fruit ripens and there is little or no latex at the fully ripe stage.

CULTIVAR DEVELOPMENT

Cytogenetics and Genetics

The species in the genus *Carica* possess nine pairs of chromosomes, and meiosis is normal in the three sex types. Cytological studies have been unsuccessful in showing the presence of a heteromorphic pair of sex chromosomes. However, one bivalent shows precocious separation at anaphase I in the male

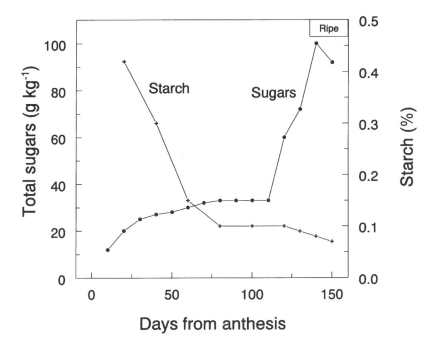

Fig. 10.8. Changes in fruit starch and total sugar of developing 'Solo' papaya fruit. Note the dramatic increase in sugars during the last phase of fruit development (redrawn from Chan *et al.*, 1979; Chittiraichelvan and Shanmugavelu, 1979).

papaya (Storey, 1953). Variation in papaya sex expression cannot be associated with any visible morphological difference between autosomes and the precociously separating pair, which might be the sex chromosomes.

Genetic studies and examination of seeds from fruit of males and hermaphrodites have shown that the Mendelian factors (m = female, M_1 = male and M = hermaphrodite) are allelomorphic and that the sex forms represented by double dominant alleles, M_1M_1, M_1M_2 and M_2M_2, are not recovered, due to some form of lethality. A cross between female and male (mm × M_1m) produced females and males with a 1 : 1 ratio. Selfing a hermaphrodite or crossing two hermaphrodites (M_2m) produced females (mm) and hermaphrodites with a ratio of 1 : 2. Crossing female with a hermaphrodite (mm × M_2m) produced females and hermaphrodites with a ratio of 1 : 1. A hermaphrodite (M_2m) crossed with a male (M_1m) produced 1 mm : 1 M_2m : 1 M_1m. Occasionally, when male trees produced bisexual flowers, these could be self-pollinated, resulting in one female (mm) and two males (M_1m) (Hofmeyer, 1938; Storey, 1938). These data indicate a complex of closely linked genes in different segments located at identical regions of a chromosome and which code for secondary characteristics, such as peduncle length and number of flower per node. Long peduncles (p) are confined to males (Mp) and short peduncles are common to females (mp) and hermaphrodites (mp); the A gene (l) enforces heterozygosity to males and hermaphrodites and prevents crossing over in the sex-differential segment. A suppressor of androecium (sa) completely suppresses androecium development when homozygous, as in the female; it is not completely recessive to the normal allele, as it exerts partial suppression on males and hermaphrodites under certain environmental conditions. Another suppressor gene of gynoecium (sg), when homozygous, exerts complete suppression of gynoecium development in strongly female-sterile males; most hermaphrodites and males appear to be heterozygous for this gene and sex expression is pushed towards maleness during the warmer months of the year (Storey, 1953). Two additional independent sets of factors may interact with the two suppressor genes to modify sex expression in males and hermaphrodites under certain environmental pressures (Fig. 10.1). One set of factors induces seasonal changes from female fertility to sterility and back. The other set influences expression of stamen carpellody. Both factors are limited to males and hermaphrodites, but females may be carriers of these factors. An initial genetic linkage map of papaya indicates a locus for sex.

Stamen carpellody and female sterility may involve a number of gene pairs involving three loci, two for carpellody (c/c, c/c) and one for sterility (s/s). Their normal alleles (+) are partially or completely dominant. The s/+ combination is normal. In c/+, the + is partially dominant over 'c' so there will be some carpellody. The 's' factor, whether homozygous or not (s/–), is epistatic over the 'c' allele when the carpellody factors are heterozygous. The phenotype would be normal but unstable, depending upon environmental conditions. If either of the 'c' factors is homozygous (c/c), the combined strength of the two alleles at one locus can overcome epistasis, thus exhibiting carpellody (Arkle, 1973). The high heritability for carpellody of stamens does not lend itself to phenotypic selection, due to the interaction between genotype, plant age and the environment (Chan, 1985).

A factor that has not been explained satisfactorily is the production of a highly uniform hermaphroditic line with minimum carpellody and sterility through continuous selection and inbreeding, within a specific microclimatic location. When seeds of a uniform line are planted in another environment with higher or lower or a wider range of temperatures, carpellody and/or sterility can again be expressed strongly. Whether these modifying factors are further influenced by other genes associated with adaptation to specific physiological changes brought about by temperature changes is unclear. It appears that there are several independent sets of factors, with interactions among them and which are influenced strongly by environmental conditions.

Studies of mutant genes in the papaya have largely been unsuccessful attempts to find sex-linked vegetative characters that might differentiate sex at the seedling stage (Table 10.3). Yellow flower colour (Y) is dominant to white (y), purple stem and petiole colour (P) is dominant to green (p) and grey seeds (B) are dominant to black (b), and these are all sex-linked (Hofmeyer, 1941). Purple stem and petiole colour is loosely linked, with approximately 41% crossover value. The crossover value is about 32% between sex and seed-coat colour; grey seeds at planting would therefore give an increased number of female trees. Non-sex linked genes include yellow flesh dominant over red flesh and yellow skin colour recessive to normal green.

Fruit shape is determined by flower type and, within a single sex type, different shapes are exhibited. Hermaphroditic fruit of 'Waimanalo' produces a roundish fruit with a short neck, while 'Kapoho' produces a more cylindrical or

Table 10.3. Reported mutant papaya genes.

Symbol	Description
a	Albino plants recessive to green plants
d	Dwarfness recessive to tall; excessive branching in juvenile stage
dp	Diminutive plant recessive to large plant; short, slender trunk and petioles, small leaves, flowers and fruit
cp	Crippled leaves recessive to normal leaf; appearance resemble symptoms of PRSV
rg	Rugose leaf recessive to smooth leaf; puckering of blade; petioles short, obliquely upright
w	Wavy leaf recessive to normal leaf
r	Red flesh recessive to yellow
Y	Yellow flower colour dominant to white
P	Purple stem and petiole dominant to green; colour intensity may be affected by modifying genes
B	Grey seed-coat dominant to black seed-coat

PRSV, papaya ringspot virus.

pyriform fruit. Fruit size and weight are quantitatively inherited, with F_1 fruit falling near the arithmetic means of the two inbred parental lines. Tree height is largely determined by internode length and is controlled by multiple factors. The height of initial flowering is controlled by the number of nodes from the cotyledonary node to the first flowering node and can range from 15 to 50. The old 'Solo' type papaya produces initial flowers after approximately 50 nodes, with some later hybrids initiating flowering after producing 30–35 nodes (Nakasone and Storey, 1955). Yield and fruit total soluble solids are extremely variable, due to the significant genotypic–environment interaction (Chan, 1985).

Breeding

Some papaya breeding objectives are common to all regions, with specific objectives in different localities due to climatic differences, consumer preferences, desired sex types and export markets (Table 10.4). These objectives have changed little in the last 80 years. Desirable tree characteristics are tree vigour, low and precocious fruiting, minimum expression of stamen carpellody and female sterility, if hermaphrodites are preferred, resistance to diseases and insect pests and yielding ability. Universally desired fruit characteristics are smooth skin, free from blemishes, firm fruit with thick flesh, round seed cavity, absence of internal lumps and long shelf-life. The desired fruit size, shape and flesh colour vary with production regions (Table 10.5). An extreme example is a preference for a heavy musky papaya aroma in South-east Asia, which would not be suitable in Western markets. Preference

Table 10.4. Ideals in papaya breeding have not changed much since the following were proposed; additional criteria have related to disease resistance (Higgins and Holt, 1914).

	Character
Tree	Vigour
	Early and low fruiting – wide variation exists
	Freedom from branching habits
Fruit	Productive but not compact fruiting
	Small size for table use
	Large size for animal feed and papain production
	Uniformity in shape, symmetry and smoothness
	Uniformity in ripening
	Colouring before softening
	Colour of flesh – yellow, pink or red
	Easily separable without scraping flesh
	Flavour – not easily described but easily recognizable
	Keeping quality
Other	Papain yield

Table 10.5. Some papaya 'cultivars' reported in the literature, many are variable and not true cultivars.

Country		Cultivar	Sex type	Flesh colour
Australia		Improved Petersen	Dioecious	Yellow
		Guinea Gold	Hermaphrodite	Yellow
		Sunnybank	Dioecious	Yellow
		Arline/57	Dioecious	Yellow
America	Mexico	Verde	–	–
		Gialla	–	–
		Cera	–	–
		Chincona	–	–
	USA – Florida	Cariflora	Dioecious	Yellow
		Betty	Dioecious	Yellow
		Homestead	Dioecious	Yellow
	USA – Hawaii	Kapoho Solo	Hermaphrodite	Yellow
		Sunrise	Hermaphrodite	Red
		Sunset	Hermaphrodite	Red
		Waimanalo	Hermaphrodite	Yellow
	Venezuela	Paraguanera	–	–
		Roja	–	Red
Caribbean	Barbados	Wakefield	–	–
		Graeme 5, and 7	–	–
	Cuba	Maradol	Hermaphrodite	Red
	Trinidad	Santa Cruz Giant	–	–
		Cedro	–	–
	Dominican Republic	Cartagena	Hermaphrodite	Yellow
Asia	India	Coorg Honey Dew	Hermaphrodite	Yellow
		Coimbitor 2	Dioecious	Yellow
	Indonesia	Semangka	Hermaphrodite	Red
		Bangkok	Hermaphrodite	–
		Dampit	Hermaphrodite	–
	The Philippines	Cavite Special	Mixture	–
	Thailand	Sai-nampueng	–	–
		Kek-dum	–	Red
South Africa		Hortus Gold	Dioecious	Yellow
		Kaapmuiden	–	Yellow
		Honey Gold	Dioecious	Yellow

for the small (450–650 g) 'Solo' fruit is increasing in tropical countries, when export is considered. Requirements for processing cultivars are the same, although larger fruit may be more desirable, size and shape being uniform. Flesh-colour preference is usually based upon traditional experience and familiarity with cultivars. Total tree height is not a good criterion to assess tree vigour, stem diameter or girth being a more reliable measure.

In breeding for bisexual types, selections should be evaluated during the cool period for occurrence of stamen carpellody and again during the warm period for female sterility. Trees that show no carpellody in winter may show a high degree of female sterility in summer, or the opposite. Within a specific locality, if temperature fluctuations are not too wide, continual selection and inbreeding can produce lines with minimum carpellody and female sterility. In subtropical climates, where the range of temperatures between summer and winter is large, dioecious cultivars are more suitable, as stamen carpellody and female sterility do not occur in dioecious cultivars. The significant genotype–environment interaction shown by papaya types makes selection essential in the area where cultivation is to occur.

Increased yields may be accomplished by increasing yield per tree, by increasing density per unit area or both. Usually, only the terminal flower sets fruit and others abscise or produce small, unmarketable fruit if set. However, lines have been observed to produce two, three or even four normal-sized fruit on each peduncle. Multiple fruiting is strongly influenced by proper soil fertility. Peduncles must be long enough to prevent overcrowding, which can lead to misshapen fruit. Increasing production by increasing trees per unit area is limited by leaf shading. Petioles of papaya trees are approximately 75–100 cm or even longer and are more or less horizontal, this requires at least 2 m between plants in a row. A mutant 'Solo' line, with short petioles (45–60 cm) that are positioned obliquely upright, is available and trees can be planted at 0.9–1.2 m apart in the row. Short, upright petiole hybrids have been produced, although the yield of the commercial cultivars has not been achieved.

A common papaya disease is the root, stem and fruit rot caused by *Phytophthora palmivora* Butl., which is especially severe during wet seasons. Early studies in Hawaii involved repeated selection for seedling mortality counts and vigour, and replanting each advance generation in the same field resulted in a few highly tolerant lines (Mosqueda-Vazquez *et al.*, 1981). Cultivars showing high tolerance to the *Phytophthora* root rot have also exhibited tolerance to stem canker and fruit rot caused by the same organism. Other fruit diseases that appear to be important enough to warrant breeding efforts are anthracnose and chocolate spot, both caused by strains of *Colletotrichum gloeosporioides*, and stem-end rot, caused by *Phoma caricae-papayae*.

The aphid-transmitted papaya ringspot virus (PRSV) (papaya mosaic virus (PMV) or distortion ringspot virus (DRV)) of different strains often limits commercial production in most papaya-growing areas. Tolerant lines have been created and a dioecious cultivar named 'Cariflora' with strong tolerance to PRSV has been released (Conover *et al.*, 1986). Using this material, several highly tolerant hermaphroditic 'Solo' selections have been made in Hawaii (Zee, 1985). More rapid tolerance is achieved using molecular biology and the transformation of papaya with the virus coat protein of a mild virus strain that confers resistance. Two resistant 'Solo' lines, one with yellow flesh, 'Rainbow', and the other with red flesh, 'Sunup', have been released in Hawaii.

The hermaphroditic papaya is naturally self-pollinating, and continuous inbreeding has not shown inbreeding depression. Also, F_1 hybrids between 'Solo' lines have not shown hybrid vigour, probably due to close genetic relationship, with many genes in common (Hamilton, 1954). This narrowness in germ-plasm has been confirmed for ten Hawaii cultivars and three non-Hawaii cultivars by deoxyribonucleic acid (DNA) analysis at 80% similarity. Heterosis has been observed in the F_1 of crosses between 'Solo' and widely different papaya accessions and between interspecific crosses involving *Carica cauliflora* Jacy. × *C. monoica* and *Carica goudotiana* Trian et Planchon × *C. monoica* (Mekako and Nakasone, 1975).

Cultivars

Wide variability is shown by the papayas grown in various countries and, with a few exceptions, most cannot be classified as cultivars (Table 10.5). Plantings are usually heterogeneous and seeds are obtained from open pollinated fruit from trees having desirable characteristics. A number of horticultural cultivars have been reported to produce relatively uniform progenies. Stabilizing characteristics in dioecious cultivars is more difficult than in hermaphroditic ones, as the genotype of the male with respect to fruit characteristics is unknown. In hermaphroditic cultivars, proper selection and self pollination can stabilize characteristics at a more rapid rate. The Hawaiian 'Solo' type is perhaps one of the few cultivars that has been continuously inbred since its introduction to Hawaii in 1911 from Barbados (Yee, 1970). 'Solo' cultivars, such as 'Kapoho', 'Sunrise', 'Sunset' and 'Waimanalo', have been inbred for many generations and have different characteristics (Table 10.6).

Table 10.6. Fruit and tree characteristics of five Hawaiian cultivars. Data are averages from tests conducted at three locations (winter harvest only).

Characteristics	Cultivar				
	Kapoho	Higgins	Line 8	Waimanalo	Sunrise
Height to first flower (cm)	146.8	96.5	169.2	75.7	91.4
Fruit mass (g)	343.0	368.6	516.0	694.6	425.3
Soluble solids (%)	15.7	15.9	15.0	15.2	14.5
Flesh colour	Yellow	Yellow	Yellow	Yellow	Red
Carpellodic fruit (%)	0	1.0	1.0	0.8	–
Other culls (%)	16.2	11.8	5.2	10.2	–
Marketable (kg tree⁻¹)	17.2	31.3	25.8	39.1	–
Phytophthora resistance[*]	I	S	R	R	I
Virus resistance[†]	IT	IT	S	S	S

[*] R, resistant; I, intermediate; S, susceptible.
[†] IT, intermediate tolerance; S, susceptible.

CULTURAL PRACTICES

Propagation

Papayas are propagated by seeds. Growers select trees with desirable characteristics, from which seeds, generally from open-pollinated hermaphroditic fruit, are saved. This seed produces uniform progenies if from a single cultivar. Fresh seeds are washed and air-dried in the shade. Seeds germinate earlier and with a higher germination percentage if the sarcotesta (gelatinous material covering the seed) is removed during seed washing. Seeds kept at 7–10°C and 50% relative humidity are viable for several years. Approximately 100–150 g of seeds, with 80% germination, are needed to produce the 2000 plants needed for 1 ha.

Seeds may be sown in trays filled with a suitable medium, peat pots or polyethylene bags or directly in the field. Germination occurs in 12–20 days. Seedlings are transplanted into 7.6 cm peat pots or into 10 cm plastic bags at the two-leaf (cotyledonary leaves) stage. Seedlings grown in containers should be hardened gradually in sunlight and field-transplanted around 1.5–2 months after germination at about 20 cm high. In field planting, up to 15–20 seeds are sown in each hole. Upon germination, seedlings are thinned out to leave three to five seedlings to grow to flowering. So many seeds are planted in each hole to allow for uneven germination, birds and field-mice loss. At first flowering, a vigorous plant of the desired sex is kept and the others removed.

Clonal propagation of papayas by grafting, air layering or rooting of side-shoots has been done but does not appear to be feasible for commercial planting. Tissue culture methods have been demonstrated to be feasible.

Field Preparation

Field preparation in many areas is poorly done, due to lack of appropriate equipment or to rough terrain. Subsoiling or ripping down to 50 cm or more is desirable on heavy or compacted soils so that roots can penetrate deeply. Subsoiling provides better drainage if done parallel to the contour lines. Discing, levelling and furrowing follow standard practice. Planting holes 30–45 cm in diameter are best dug with a soil auger attached to a tractor. Raised beds are used if there is a chance of flooding.

Transplanting and Plant Spacing

In rain-fed areas, field planting is done at the beginning of the rainy season if this is not the cool season. Seedlings grown in peat pots are planted directly and buried completely without removing the pot, polyethylene bags are removed. Transplanted plants must be watered soon after planting to settle the soil around the root system. To reduce the height of first fruit, hardened plants may be transplanted at 45° to the vertical, the leaves touching the ground are removed and the plants are allowed to bend upwards to the vertical by subsequent growth.

Spacing between plants and between rows varies widely. Universally practised is the single-row system with between-plant spacing ranging from 1.8 to 3 m and between-row spacing varying from 1.8 m to as much as 3.6 m. The between-row spacing largely depends upon degree of mechanization; a standard tractor requires *c.* 3 m. The most frequently used spacing is 2.0–2.5 m within row × 2.5 m, giving a density of 2000–1600 plants ha^{-1}. A double-row system, with 2 m between a set of rows and 3.5 m between double rows, is also used (Plate 24). Polyethylene mulch over the beds prevents moisture losses and greatly minimizes weed growth within the beds. Organic mulches of grasses, wood shavings, rice hulls or other kinds of material are also beneficial.

Irrigation

Non-irrigated fields in Hawaii are located in a region with 2500–3125 mm of rain, fairly well distributed throughout the year. However, occasional droughts of 2–3 months occur. Low soil moisture tends to shift the sex type to male and results in lower fruit yield, while high moisture levels can lead to excessive production of misshaped carpellodic fruit with rapid tree growth. Dioecious cultivars fare better, unless moisture stress is severe. A monthly minimum rainfall of 100 mm is needed, without supplementary irrigation for some production. Irrigation should replace at least that lost by pan evaporation, and 1.25 times potential pan evaporation is required for maximum yield from mature trees (Fig. 10.9). Young trees may only need about 0.3–0.5 times potential pan evaporation. Good production occurs with 60–90 L tree^{-1} week^{-1} immediately after planting or during the wet season and 120–240 L tree^{-1} week^{-1} during the dry period.

Irrigation may be by flooding between the row spaces by furrows running along both sides of the rows of trees or via microsprinklers, jets or drip. Irrigation intervals of around 10–15 days may be necessary to sustain production, unless this interval is broken by rainfall. Papayas are wind-pollinated and overhead sprinkling reduces pollen dissemination and therefore is not recommended. A drip system must be capable of providing the maximum amount rather than average demand. If a tree requires 56 L day^{-1} but 114 L day^{-1} during the dry season, the delivery system must have the capacity to deliver the maximum requirement. One emitter per tree is sufficient for the first 3–4 months but, thereafter, two emitters, one on each side of the tree, have been found necessary. A single microsprinkler (30 L h^{-1}) can meet requirements.

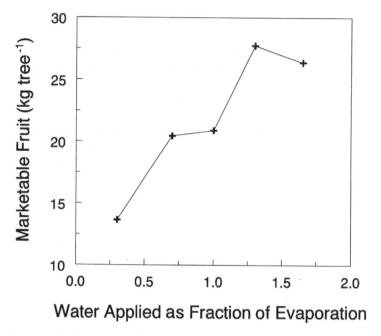

Fig. 10.9. Relation between yield of marketable fruit and the water application as a fraction of potential pan evaporation (redrawn from Awada *et al.*, 1979).

Pruning

As papaya is a monoaxial tree, normal tree pruning is not practised. Side-shoots produced by some cultivars during the juvenile stages or enforced when the plant apex is damaged are removed as early as possible to maintain a single stemmed tree. If the apex is damaged or destroyed, one shoot nearest the terminal is maintained to re-establish the tree. Some lower leaves and senescent leaves and petioles are removed to facilitate harvesting, improve penetration of spray materials, prevent the petiole of senescent leaves from rubbing on maturing fruit and generally to increase marketable yields. No statistical differences are found in total soluble solids, fruit size, number of fruit and marketable yield after a year of leaf pruning when only 15 fully expanded leaves are retained (Ito, 1976). Each mature leaf can support about four developing fruits.

Fertilization

The objective is to balance fertilization to maintain growth and continue fruit production. Diversity of soil types, climatic conditions and practices makes it necessary to develop recommendations for specific areas, based upon soil analysis and preferably foliar analysis. In the sandy loam soils of North Moreton, Australia, a preplant application of 0.5 kg of 5 : 7 : 4 complete fertilizer, 0.25 kg of single superphosphate and 1.0 kg of lime per individual plant (over 1 m^2 area) is recommended. For postplant application, 100 g plant^{-1} of 10 : 2 : 16 complete fertilizer is given at about 2-month intervals, with monthly application of 150–200 g plant^{-1} during summer and autumn. In the second year, 250 g per plant is given at 2–3-month intervals. Because of boron-deficiency problems in this area, 20 g of borax is applied at 3–4-month intervals and 30 g plant^{-1} is given the second year at about 7–8-month intervals.

Studies on plant-tissue analysis of papaya have been made to determine critical levels of N, P and K (Awada *et al.*, 1986). Most of the plant-tissue work was in Puna, Hawaii, where the growing medium is low in NPK and is well drained and plants respond to fertilizer application. Tentative ranges of critical levels for bearing plants of 'Kapoho Solo' are: N = 1.10–1.40%, P = 0.15–0.18% and K = 2.5–3.5%.

Pest Management

Diseases

The importance of frequently reported papaya diseases (Table 10.7) varies, and may be widespread or localized. *Phytophthora* root rot is a major concern in all areas where the same land is used repeatedly for production. Depending upon soil type and rainfall, the replant problem may become very serious within two or three successive plantings, with seedling mortality up to 45% (Nakasone and Aragaki, 1973). Continuous production on the same area should be avoided, with at least a 3-year break under some alternate crop. Alternatively, virgin soil (soil never planted to papaya) is placed into each planting hole in a replant field, allowing the seedling to become established before the roots penetrate beyond the virgin soil when they are little affected by the fungus. Development of cultivars with strong tolerance through breeding is slow but practical, with the cv. 'Waimanalo' being such a product. *Phytophthora* can also cause stem canker (Plate 25) and fruit disease. Seedling damping off immediately after germination is caused by a complex of organisms. The use of a seed treatment, with 2500–5000 µl L^{-1} sodium hypochlorite, and a sterilized germination medium, with good aeration and control of moisture, are effective preventive measures.

Fruit diseases generally occur postharvest. The incidence of some fruit diseases can be minimized by field sanitation (Plate 1) and field application of appropriate fungicides, while others can be reduced by careful handling of fruit, as these organisms become infected through wounds. Anthracnose is a preharvest infection and fruit-surface rot and stem-end rot, a harvest-wound rot, are the two most common postharvest rots of papaya (Table 10.7).

Table 10.7. Some important diseases and disorders of papaya.

Common name	Organism	Parts affected	Distribution
Phytophthora blight	*Phytophthora palmivora*	Root, stem, fruit	Widespread
Pythium root rot	*Pythium aphnidermatum*	Root	Widespread
Damping off	*P. aphanidermatum, Pythium ultimum, P. palmivora, Rhizoctonia* sp.	Seedling stem at soil line	Widespread
Collar rot	*Calonectria crotolariae*	Base of trunk, crown roots	Hawaii, Thailand
Powdery mildew	*Oidium caricae*	Underside of leaves, petioles	Widespread
Alternaria fruit rot	*Alternaria alternata*	Fruit body	Widespread
Blackspot	*Cercospora papayae*	Fruit body	Widespread
Anthracnose chocolate spot	*Colletotrichum gloeosporioides*	Ripe fruit body	Widespread
Soft rot	*Rhizopus stolonifer*	Injured mature fruit at stem end	Widespread
Stem-end rot	*Fusarium solani*	Wounded mature fruit at stem end	Widespread
Phoma rot	*Phoma caricae-papayae*	Stem end, wounded area	Widespread
Phomopsis rot	*Phomopsis* sp.	Stem end, wounded area	Widespread
Stemphyllium rot	*Stemphyllium lycopersici*	Stem end, wounded area	Widespread
Fruit rots	*Botrydiplodia* sp.	Stem end, wounded area	Widespread
	Cladosporium sp.	Fruit body	Widespread
Stem-end rot	*Mycosphaerella* sp.	Stem end	
Yellow crinkle	Tomato bug bud organism	Leaves, flowers	Australia
Papaya ringspot virus	Aphid vectors	Leaves, stems and petioles, fruit ringspots or distorted	Widespread
Bunchy top	Bacteria	Death of growing point	Caribbean, Tropical America
Dieback	Unknown	Yellowing and death of crown leaves	Australia, Israel
Yellow crinkle	Leaf hopper	Yellowing of old leaves; crown leaves show translucent areas	Australia

The papaya ring spot virus (PRSV) has become the limiting factor for commercial production in many areas. This disease has been referred to as papaya mosaic virus and distortion ringspot virus (Gonzalves and Ishii, 1980). The fact that PRSV is suspected to be a number of different viruses is due to the variations in the expressed symptoms associated with different strains. There are no control methods once the plant is infected. Eradication, sanitation methods and isolating the papaya-growing area by a papaya-free buffer zone allows papaya to continue to be grown. An orchard or growing area isolated far enough from disease sources remains uninfected because, by the time aphids can move from the virus source to the healthy orchard, they are non-infective due to the non-persistent characteristic of the virus (Namba, 1985). In Taiwan, large structures (1 ha) covered with 32-mesh screen are used to keep aphids out, along with use of aphid predators in the structure to prevent PRSV infection of plants. Also, curcurbit plants should not be interplanted in papaya orchards, as they may be alternative hosts of the virus. A cross-protection method, using a mild strain of the same virus inoculated into the plant, reduces the establishment rate of the virulent strain, enabling growers to be productive despite high virus pressure. Breeding for resistance or high tolerance has also shown promise (Zee, 1985). Transforming papaya with the viral coat protein has been successful; however, each virus strain has a specific coat protein, so that different coat-protein genes are needed in different areas.

The bunchy-top disease is a serious disease in the Caribbean, including Florida and some Latin American countries. It is caused by a bacterium (Davis *et al.*, 1996) and characterized by stunting, yellowing and stiffening of the leaves, much like the symptoms of PRSV. A major differentiating factor is the lack of latex in the affected parts in bunchy top, while latex flow is present in the parts affected by PRSV. A leaf hopper (*Empoasca* sp.) has been identified as the vector (Adsuar, 1946) and is also a cause of severe insect damage.

A dieback disease, which in some respects resembles PRSV and in others bunchy top, has been reported in southern Queensland and in Costa Rica. The actual causes of this dieback are still unknown, indicating a complex syndrome. It is not viral and no mycoplasma-like bodies have been found. Theories proposed include soil condition, moisture stress, calcium deficiency and climatic conditions.

Insects and mites

Many insects have been reported on papayas, but most are unimportant and the damage is negligible or easily controlled. A few may cause severe problems in localized areas (Table 10.8). Fruit flies are troublesome in the export trade, as papayas may need to receive a disinfestation treatment for fruit-fly eggs and larvae. Fruit of established 'Solo' cultivars are rarely infected at the mature green to colour-turning stage, egg lay becoming a problem when fruit have 25% or more skin yellowing (Seo *et al.*, 1982).

Several mite species damage papaya. One of the most serious is the broad mite (Table 10.8), which attacks the underside of young emerging leaves, causing stunting and distortion. Because of their small size, it is difficult to see

Table 10.8. Some important insects and pests of papaya.

Common name	Scientific name	Parts affected	Distribution
Melon fly	*Bactocera cucurbitae*	Fruit	Widespread
Oriental fruit fly	*Bactocera dorsalis*	Fruit	South-east Asia, the Philippines, western Pacific, Hawaii
	Bactocera melanotus	Fruit	Cook Islands, South Pacific
Mediterranean fruit fly	*Ceratitis capitata*	Fruit	Hawaii, Mexico, Central, South America, Middle East, Africa
Fruit fly	*Toxotrypana curvicaula*	Fruit	American tropics, Florida
American fruit fly	*Anastrepha fraterculus*	Fruit	Subtropical and tropical America
Caribbean fruit fly	*Anastrepha suspensa*	Fruit	Florida, Caribbean
Green peach aphid	*Myzus persicae*	Virus vector	Widespread
Red and black flat mite	*Brevipalus phoenicis*	Fruit	Widespread
Broad mite	*Hemitarsonemus latus*	Emerging leaves, leaves of young seedling	Widespread
Carmine spider mite	*Tetranychus cinnabarinus*	Lower surface of mature leaves	Widespread
Texas citrus mite	*Eutetranychus banksi*	Mature leaves	Widespread
	Eutetranychus orientalis	Mature leaves	Thailand
Citrus red mite	*Panonychus citri*	Mature leaves	Widespread
Leaf hoppers	*Empoasca* spp.	Emerging and first few leaves	Caribbean and Hawaii
Monkeys			Kenya, Barbados
Birds			Caribbean, Hawaii
Bats			Vanuatu, South Pacific

them with the naked eye and damage is usually done by the time their presence is known. Multiplication is rapid and is more prevalent during the cool months (Yee, 1970). The Texas citrus mite, citrus red mite and carmine mite are also found on mature leaves. The latter species is the more serious one, as it reproduces throughout the year on many host plants in all papaya-growing areas. On papaya, the carmine mite is found on the underside of mature leaves. The red and black, flat mite is a major pest, as it causes scarring of the fruit, often causing latex to exude. The affected area on the fruit is greyish in colour, with a corky appearance. Wettable sulphur is effective for most mite species. The carmine mite is somewhat tolerant of sulphur but can be kept under control. In Hawaii, 2.7 kg of 95% wettable sulphur in 380 L of water has been recommended for use on papaya for mite control.

Aphids are important only as vectors of PRSV. Two aphid species, *Aphis gossypii* and *Myzus persicae*, give positive results for the papaya virus. Leaf hoppers (*Empoasca* spp.) can be very important insect pests, causing severe damage. Malathion provides good control if coverage is thorough.

Nematodes, such as the root-knot (*Meloidogyne* sp.) and the reniform (*Rotylenchulus reniformis*) can cause some problems (Hiranpradit, 1983). Preplant soil fumigation is effective.

Weed Management

Hand-weeding, especially around young seedlings, mowing and mulching are often used for weed control. Plastic mulch is effective against annual broad-leaved weeds but ineffective against perennial grasses or sedges (Plate 26). Postemergence herbicides, such as paraquat and glyphosate, are currently widely used. The latter is effective for a wide spectrum of weed species and, as a systemic herbicide, its effects are not visible for a number of days. Diuron is used as a pre-emergence herbicide and can also be used as a postemergence spray by adding a non-ionic surfactant.

A common practice in some areas is to interplant papaya orchards with tree crops, which later becomes the principal crop. Herbicides selected for the principal crop must also be compatible with papaya and be registered for use for both crops.

HARVESTING AND POSTHARVEST HANDLING

Harvesting

Harvesting is simple when fruit can be reached by hand; as trees become taller, some form of harvesting aid, such as poles and ladders, must be used. In Hawaii, most growers use a rubber cup or chisel-shaped metal with a 'V'-shaped notch and attached to long poles, which are placed against the bottom of the fruit or against the peduncle and pushed upward, snapping the fruit at the peduncle. The falling fruit is caught by the picker with the other hand or removed from the cup. The harvested fruit are accumulated in a bucket, tray or cloth picking bag. These methods are possible only with the small 'Solo' fruit. When the container is full, it is emptied into wooden bins or big boxes on field roads. An experienced harvester can harvest from 360 to 450 kg 8-h day^{-1}. Various hydraulic-operated mechanical harvest aids, with a platform for the harvesters and driven by an operator, have been used.

The degree of ripeness for harvesting depends upon distance to markets. Fruit may be one-quarter to one-half ripe for local markets. Fruit to be transported long distances or exported are harvested at colour break to one-quarter ripe, depending on the cultivar's ripening characteristic and season. Colour assessment is based on the judgement of pickers. The Hawaii grade standard requires fruit to have 11.5% total soluble solids, with colour-break stage normally meeting this standard. Green immature fruit do not ripen well and have total soluble-solid values that are frequently less than 10%, giving a bland fruit (Paull *et al.*, 1997).

Postharvest Treatment

The need for postharvest treatments depends upon the importing countries. Overseas markets require disinfestation treatments to eradicate fruit-fly eggs and larvae from the fruit before shipment. The vapour heat method is most frequently used, in which the fruit are heated at greater than 93% relative humidity for about 4 h. The fruit core temperature reaches at least 47.2°C. Total treatment time is about 7 h. Alternative methods include irradiation at a minimum dose of 250 Gy. The low irradiation dose sterilizes the fruit-fly eggs and larvae, preventing them from completing their life cycle. Irradiation at 250 Gy does not provide postharvest disease control and needs to be coupled with a 20 min 49°C hot-water dip. These treatments all put the fruit under stress and can lead to some disorders.

Fruit are packed into cartons in a screened area to prevent recontamination by fruit flies after vapour heat treatment. Defective fruit are culled either before or after disinfestation, or both, and fruit are graded for size and colour and hand-packed into cardboard cartons with a capacity of about 4.5–6.0 kg. Packed cartons are held at 10–12°C to delay ripening before shipping. Some packers preripen fruit for 1–2 days at 22–25°C before cooling and shipping (Paull *et al.*, 1997). Other in-transit storage methods, such as controlled atmospheres, low oxygen storage and hypobaric or low-pressure (LP) storage, have been investigated. They do not offer any major commercial benefit in extending shelf-life and lowered incidence of fruit diseases.

A fungicide–wax combination is often applied prior to packing. Incidence of storage diseases can be reduced by field spraying and proper care in harvesting and handling to avoid wounding and bruising. Skin injury is a major problem and is caused mostly by impact and abrasion during harvesting (Quintana and Paull, 1993). The latter is mainly caused when harvested fruit are being dropped into field bins with rough side-walls and bottom. Careful handling is essential to avoid these unsightly blemishes, which provide invasion sites for postharvest rots.

Marketing

Many of the papayas produced in the tropics are consumed locally, due to long distances to markets and the difficulty in handling the large fruit size. Marketing standards for trade are normally set by agreement between the shipper and the wholesale or retail buyer. In Hawaii, standards for Hawaiian-grown papaya have been developed. Fruit must conform to the 'Solo' type, with similar cultivar characteristics as to size and shape, must be mature, with a definite tinge of yellow at the blossom end, must be clean, well-formed and meet the total-soluble-solid standards, averaging not less than 11.5% for any lot of papayas, provided not more than 5%, by count of fruit in the lot, have soluble solids less than 10.5%. Fruit is sized into the following size classifications: small 284–369 g; medium 369–454 g; large 454–907 g; extra large over 907 g. A major problem with the standard is the difficulty of grading to the same size and skin colour that will ripen together in the carton (Paull *et al.*, 1997). Other requirements include that the fruit must be free from decay, breakdown, internal lumps and other undesirable characteristics and free from various categories of injury, including insect and mechanical injury. These standards apply to fruit marketed in Hawaii and shipped out of state. Postharvest culling may remove up to 40% of the fruit.

UTILIZATION

Nutritionally, the papaya is a good source of calcium (30 mg 100 g^{-1}) and an excellent source of provitamin A and ascorbic acid (Table 10.9). Papayas are consumed fresh as breakfast fruit or dessert or in salads. In Asia, green fruit are served in salads, as a vegetable or made into preserves. Papayas are also processed into various forms such as dehydrated slices, chunks and slices for tropical fruit salads and cocktails, or processed into purée for juices and nectar base, usually frozen and as canned nectar, mixed drinks and jams. Papaya purée is the basis for remanufacturing of many products. It is processed aseptically or frozen, with a yield of *c.* 50%. The flavour of aseptically processed purée is stable during processing and after 6 months in ambient storage, with some colour changes noted.

Papain is a proteolytic enzyme that digests proteins and is used as a meat tenderizer, as digestive medicine in the pharmaceutical industry, in the brewing and tanning industries and in the manufacture of chewing gum.

Table 10.9. Composition of hermaphrodite 'Solo' type papaya (Wenkam, 1990).

		Amount per 100 g edible portion
Proximate		
Water	g	87
Energy	kJ	192
Protein	g	0.39
Lipid	g	0.06
Carbohydrate	g	12.2
Fibre	g	0.58
Ash	g	0.57
Minerals		
Calcium	mg	30
Iron	mg	0.2
Magnesium	mg	21
Phosphorus	mg	12
Potassium	mg	183
Sodium	mg	4
Vitamins		
Ascorbic acid	mg	84
Thiamine	mg	0.03
Riboflavin	mg	0.04
Niacin	mg	0.33
Vitamin A	IU	1093

Papain production is primarily centred in Tanzania and India, where labour is abundant and inexpensive. The latex is obtained from green papaya by making about four surface lancings on the fruit and catching the drippings in cups, yielding 88.5–227 g tree^{-1} of dried latex per year. Approximately 2.27 kg of fresh latex will produce 0.45 kg of dried latex.

In processing papayas, 22% of the waste is seed. The seed-oil content of 33% on a dry-mass basis is considered high when compared with seeds of other fruit. The protein content of 29% on a dry-weight basis is comparable to that of soy bean with 35%. Papaya-seed meal, with 40% crude protein and 50% crude fibres appears to be a potentially rich animal feed. A salad dressing is also made from ground papaya seeds.

FURTHER READING

Alvarez, A.M. and Nishijima, W.T. (1987) Postharvest disease of papaya. *Plant Disease* 71, 681–686.
Marler, T.E. (1994) Papaya. In: Schaffer, B. and Andersen, P.C. (eds) *Handbook of Environmental Physiology of Fruit Crops* Vol. 2, *Subtropical and Tropical Crops*. CRC Press, Boca Raton, Florida, pp. 216–224.

Nakasone, H.Y. (1986) Papaya. In: Morselise, S.P. (ed.) *CRC Handbook of Fruit Set and Development*. CRC Press, Boca Raton, Florida, pp. 277–301.

Paull, R.E., Nishijima, W., Reyes, M. and Cavaletto, C.G. (1997) A review of post-harvest handling and losses during marketing of papaya (*Carica papaya* L). *Postharvest Biol. Technol.* 11, 165–179.

Qiu, Y.X., Nishina, M.S. and Paull, R.E. (1995) Papaya fruit growth, calcium uptake, and fruit ripening. *Journal of the American Society for Horticultural Science* 120, 246–253.

Sankat, C.K. and Maharaj, R. (1997) Papaya. In: Mitra, S.K. (ed.) *Postharvest Physiology and Storage of Tropical Fruits*. CAB International, Wallingford, pp. 167–189.

11

PASSION-FRUIT

BOTANY

Introduction

The passion flower family (*Passifloraceae*) includes 12–18 genera of about 400 species of dicotyledonous, herbaceous or woody vines, usually with axillary tendrils. Erect shrubs and trees are rare. They are native to the tropical and subtropical regions of both hemispheres at medium to high elevation where temperatures are moderate. Only two genera, *Passiflora* and *Tetrapathaea*, are cultivated. The most important genus is *Passiflora*, whose species are mostly vines with axillary tendrils and the fruit is a many-seeded berry. Flowers of many species are conspicuous in their form and colour and are grown for their ornamental value.

Only 50–60 species bear edible fruit and most are unknown outside the areas where these species may have originated or are cultivated by the native people (Table 11.1). A number of species are considered commercial in the sense that fruit of *Passiflora quadrangularis*, *Passiflora ligularis*, *Passiflora laurifolia*, and *Passiflora mollissima* are often found in village markets in Latin American countries. However, only the purple passion-flower fruit (*Passiflora edulis* Sims), the yellow passion-flower fruit (*P. edulis* f. *flavicarpa* Deg.) and hybrids between the two are considered to be of value in inter-national commerce (Table 11.2). Commercially, the fruit is referred to as the passion-fruit and this will be used in this chapter. The common names besides passion-fruit include granadilla (English), markisa or buah susu (Malaysia), linmangkon (Bangkok, Thailand), benchawan (Chiang Mai, Thailand), limangkan (Laos), maracuja, morada (purple) and maraarya amarillo (yellow) (Spanish), maracuya peroba (Portuguese), fruta de pasion, pasionaria (Tagalog, the Philippines) and parcha or marflora (Ilokano, the Philippines).

Table 11.1. Some selected *Passiflora* species, their areas of origin and use (Martin and Nakasone, 1970).

Species	Common names	Area of origin	Ornamental	Fruit
			____Uses____	
P. alata Dryand		Brazil, Peru	x	x
P. antioquiensis Karst	Banana passion-fruit	Colombia		x
P. banksii Benth		Australia	x	
P. caerulea	Blue passion-flower	Brazil to Argentina	x	
P. coccinea Aubl.	Red granadilla	Venezuela to Bolivia	x	x
P. edulis Sims.	Purple granadilla	Brazil		x
P. incarnata L.	Wild passion-flower, maypop, apricot vine	Southern USA	x	x
P. laurifolia L.	Yellow granadilla, water lemon, Jamaica honeysuckle	West Indies to Brazil and Peru	x	x
P. ligularis Juss.	Sweet granadilla	Mexico to Bolivia		x
P. maliformis L.	Sweet calabash, sweet cup, conch apple	West Indies to South America		x
P. mollissima (HBK) Bailey	Banana passion-fruit, curuba	Venezuela to Bolivia		x
P. quadrangularis L.	Giant granadilla	Unknown		x
P. vitifolia HBK		Nicaragua to Venezuela and Peru	x	x

Table 11.2. Some similarities and differences between *P. edulis* Sims and *P. edulis* f. *flavicarpa* Deg. (after Knight, 1980).

Characteristics	*P. edulis*	*P. edulis* f. *flavicarpa*
Ecology	Cooler elevation	Low elevation
Vine	Less vigorous	Vigorous
Leaves	Similar shape, smaller	Larger size
Flowers	Smaller, fragrant, less protandry, anthesis a.m.	Larger, stronger fragrance, stronger protandry, anthesis p.m.
Fruit	Purple, small	Yellow, large
Juice	Mild acid	Acid
Chromosome number	2n = 18	2n = 18
Meiosis	Normal	Normal
Ovules, pollen grains	Fully viable	Fully viable
Compatability	Self-compatible	Self-incompatible

Origin and Distribution

The well-known purple passion-fruit, *P. edulis*, is considered native to southern Brazil and was widely distributed during the nineteenth century to other countries of South America and the Caribbean and into Asia, Africa, India and Australia (Martin and Nakasone, 1970). It was introduced into Hawaii from Australia in 1880 and soon became established in the wild.

The origin of the yellow form, *P. e.* f. *flavicarpa*, is unknown. It may have originated in Australia as a mutant form of *P. edulis* or seeds may have been obtained from tropical America. The hybrid origin of *P. e.* f. *flavicarpa* as a natural cross between *P. edulis* and possibly *P. ligularis* is unlikely, as normal meiosis with fully viable ovules and pollen grains occurs in *P. e.* f. *flavicarpa*. *P. edulis* and *P.e.* f. *flavicarpa* both have n = 18. An interspecific hybrid would exhibit some meiotic irregularities, leading to irregularities in fertility.

ECOLOGY

Climatic preferences of the different species of *Passiflora* vary rather widely, from tropical climates in the lowlands (*P. e.* f. *flavicarpa*) to the highlands with cool periods (*P. edulis*). *P. edulis* appears to be a variable species in Brazil, with forms having different ecological preferences, including those represented by yellow fruit type (Martin and Nakasone, 1970).

Soil

The passion-fruit can tolerate a wide range of soil types, although the vines are highly susceptible to poor drainage and waterlogging. Soil pH may range from 5.5 to 6.8.

Climate

Rainfall

A well-distributed annual rainfall is necessary for passion-fruit culture, especially if supplemental irrigation is not available. However, rainfall must be minimal during the flowering period, as pollen wetted by free moisture bursts open and become non-functional. Furthermore, rain minimizes insect activity and hinders pollination. The yellow passion-fruit has been grown quite successfully in Hawaii with a rainfall of from 1000 to 1500 mm and supplemental irrigation during dry periods.

Moisture stresses less than −1.3 MPa lead to a significant decline in leaf area, flowering and yield (Stanley and Wolstenholme, 1990). When hybrid vines are subjected to water stress, leaf potential recovers within a day of irrigation, leaf growth returning to prestressed levels in about 4 days and net carbon dioxide (CO_2) assimilation in 6 days (Menzel and Simpson, 1994).

Temperature

Different temperature regimes for growing passion-fruit in the tropics can be chosen by elevation above sea level. The economic life of the purple passion-fruit at elevations of about 800 m is 3–4 years, while at elevations between 1200 and 1500 m plants produce reasonable crops for about 8 years. Low temperature (15 day/10°C night) reduces vegetative growth and potential yield, while high temperatures (30/25°C) can prevent flower production (Menzel *et al.*, 1987). Some hybrid cultivars show differences in the optimum temperatures for growth and yield. The cv. 'Lacey' does not flower at 25/20°C, while 'E-23' and 'Purple Gold' have reduced flowering (Menzel and Simpson, 1994). Purple passion-fruit thrives and yields well at night temperatures of 4.5–13°C and day temperatures of 18–30°C. Mature vines of the purple passion-fruit withstand light frost, but are injured at 1–2°C below freezing (Beal and Farlow, 1984). The yellow passion-fruit, *P. e.* f. *flavicarpa,* is tropical in its requirements, exhibiting more vigour and a wider range of adaptability. It grows and flowers well from sea level to 600 m or even higher, depending upon proximity to the Equator (Campbell and Knight, 1987).

Light

Seasonal changes in solar radiation can significantly influence productivity. Lower average irradiance, in the cool season, during the wet season with cloudy weather and with self-shading, reduces plant growth, the number of floral buds and open flowers and vine growth (Fig. 11.1). Short periods (1 out of 4 weeks) of heavy shade significantly reduce flowering and potential yield. Cultivars differ in response; cv. 'Purple Gold' is more precocious at lower light levels than 'Lacey' (Menzel and Simpson, 1994). Shaded vines have more stem and a lower proportion of root. There is an interaction of sunlight with temperature, as no flowering occurs at high temperatures with low irradiance.

Photoperiod

Flowering of yellow passion-fruit was suggested to be photoperiodic (Watson and Bowers, 1965). Artificially induced short days (8 h) prevent vines from flowering and no flowers are produced under natural day lengths of about 11 h or less in Hawaii (21°N). In the absence of light-interruption studies, these results can be interpreted as being due to greater solar radiation or higher temperature (Menzel and Simpson, 1994).

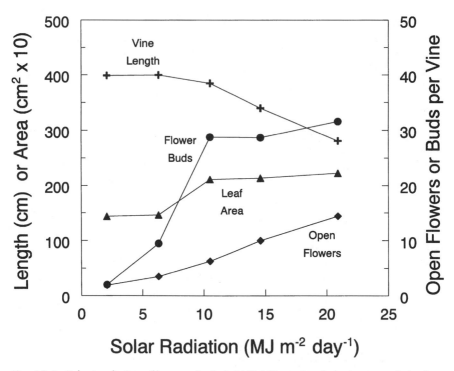

Fig. 11.1. Solar irradiation effect on the hybrid 'E-23' passion-fruit vine growth, leaf area, flower buds and number of open flower buds (redrawn from Menzel and Simpson, 1988).

GENERAL CHARACTERISTICS

Vine

Growth of the vine is essential for continued flowering. The vigorous perennial passion-fruit vine has medium to large toothed leaves (Fig. 11.2). Purple passion-fruit has green tendrils, while the yellow passion-fruit possesses reddish or purplish tendrils. Leaves of the yellow passion-fruit are somewhat larger than those of the purple.

Flowers

The solitary showy flowers (7.5–10 cm diameter) consists of five sepals and five white petals, which form a tubular calyx tube, usually surrounded by a thread-like crown or corona in the centre (Fig. 11.2). The five stamens unite into an elongated stalk bearing the ovary, with three horizontal styles (0.5 cm diameter). The ovary is superior, one-celled and with three parietal placentas.

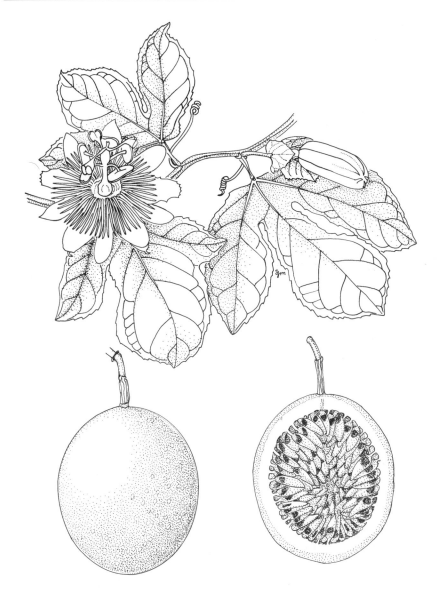

Fig. 11.2. Stem, leaf, flower and fruit of *Passiflora edulis* f. *flavicarpa*.

Flowers of the yellow passion-fruit are larger and more fragrant, with a strong tendency towards the protandrous habit (Table 11.2). Floral anthesis occurs about 40 days after the buds become visible in the yellow passion-fruit (Akamine and Girolami, 1959). The time from bud formation to floral anthesis in *P. quadrangularis* ranges from 17 to 24 days (Haddad and Figueroa, 1972).

The name 'passion-flower' was given by early European travellers, to whom it represented the passion of Christ. The ten sepals and petals represented the ten apostles at the Crucifixion; the fringed crown represented the crown of thorns and the five stamens and three styles represented the wounds and nails, respectively. The tendrils are the cords or scourges; the lobed leaves represent the hands of the persecutors. The white symbolized purity and the blue the heavens.

Flower buds are produced at every node of new growth and, theoretically, there should be a succession of fruit set, if pollination is adequate. After four to ten flowers have set fruit, further setting of the remaining flowers ceases, even when hand-pollinated. Fruit set resumes when the initially set fruit begin to mature, this leads to peaks of fruit production with continuous vine growth (Fig. 11.3). The alternation of fruit setting and cessation of setting leaving fruitless spaces on long vines, could be a physiological mechanism to balance the number of fruit to leaves.

Pollination and Fruit Set

The flower is large (Fig. 11.2), with a strong fragrance, attractive colours, abundance of nectar and large sticky pollen, all being conducive to insect pollination. The most frequent visitors are carpenter bees, *Xylocopa sonorina*, honey-bees, *Apis mellifera*, and various flies in the order *Diptera*. The relative abundance of species varies with locality, with different species of carpenter

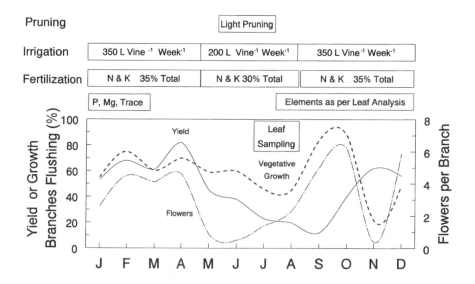

Fig. 11.3. The growth of hybrid passion-fruit vines, flowering and yield in relation to pruning, irrigation and fertilization (redrawn from Menzel *et al.*, 1989, 1993). Mean of 2 years' data from southern Queensland, Australia. N, nitrogen; K, potassium; P, phosphorus; Mg, magnesium.

and honey-bee being reported as the principal pollinators in most countries. Fruit set is about 70% by natural pollination while hand-pollination fruit set can be 100%. Hand-pollinated fruit are larger, heavier and higher in juice content, due to the increase in seed number (Akamine and Girolami, 1959).

Style curvature is crucial for passion-fruit flower pollination and fruit set. When the flowers open, the styles are upright, gradually recurving during the day, to *c.* 55° in the first hour after anthesis and 85° by 6 p.m. Styles then revert to an average of about 20° curvature at flower-closing time (Hardin, 1986). The most effective time for pollination is after the styles have completely recurved, with the stigmas being receptive only on the day of anthesis. Ruggiero *et al.* (1976) observed three kinds of flowers with respect to style curvature, sometimes on the same vine. Around 71% of the flowers show total curvature (TC), 23% achieve partial curvature (PC) and 6% have upright styles (SC). Flowers with upright styles fail to set fruit with pollen from compatible vines and wither under natural conditions or when hand-pollinated. Pollen of SC flowers effect good fruit set when deposited upon stigmas of TC and PC flowers of different vines, indicating high pollen viability.

The purple passion-fruit is self-compatible, with the yellow passion-fruit often requiring cross-pollination. Upon pollination, stigmas must be kept dry for at least 2 h, the time necessary for pollen germination (Akamine and Girolami, 1959). Pollen bursts and becomes useless upon contact with free water, germinated pollen not being affected. Since flowers of the yellow passion-fruit open during the afternoon, rain or irrigation by overhead sprinklers during the afternoon can greatly reduce fruit set.

Fruit

Fruit can range from 6 to 8 cm in diameter by 7 cm in length in the yellow passion-flower and 3.5 to 7 cm (diameter) and 4 to 9 cm for the purple, both having a hard exocarp (Fig. 11.2). Both fruit are round to oval in shape. The mesocarp is a thin green layer below the exocarp and the endocarp is white, this forms the 3–6-mm-thick shell. Each seed has a hard black testa and is surrounded by a juicy edible aril. The seeds are attached by peg-like funiculi to the endocarp. The yellow passion-fruit has up to 350 seeds. The yellow to orange juicy aril pulp of the purple passion-fruit is very aromatic, having a very desirable flavour. Total soluble solids usually exceed 15% and total titratable acidity ranges from around 2.5 to 4.0% (Table 11.3), depending upon the season. The yellow passion-fruit pulp is more acid than the purple passion-fruit (Seale and Sherman, 1960).

The purple and yellow passion-fruit take 60–90 days to mature (Fig. 11.4) and 65–85 days for the giant granadilla. Growth follows a sigmoid growth curve, reaching maximum size in about 21 days, when sclerification leads to a hardening of the shell. Subsequent fruit-mass increase occurs at a slower rate and reaches a maximum at about 50 days for purple and 60 days for yellow

Table 11.3. Pulp characteristics of *P. edulis, P. edulis* f. *flavicarpa* and some hybrids (after Beal and Farlow, 1984).

	Fruit wt (g)	No. of seeds	% Pulp	TSS (%)	Titratable acidity (%)	Total volatile esters (p.p.m.)
P. edulis	36	122	49	15.3	2.4	159
P. edulis f.*flavicarpa*	72	50	37	15.3	3.4	83
3−1	−	−	46	14.5	3.2	122
Other hybrids	−	−	35−68	13.4−16.1	2.3−3.6	49−200

TSS, total soluble solids.

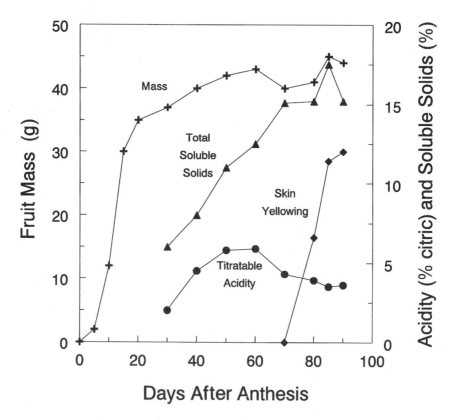

Fig. 11.4. Change in purple passion-fruit fruit mass, skin yellowing, total soluble solids and titratable acidity during growth and development (redrawn from Shiomi *et al.*, 1996).

from anthesis. The increase in sugars, with a concomitant decline in starch, occurs during this last phase. When maximum mass is achieved, titratable acidity begins to decline as maturity approaches. There is a positive correlation between number of seeds that develop and the fresh-fruit mass and juice content. Juice content reaches a maximum on the vine and declines due to dehydration after abscission (Shyy *et al.*, 1987).

CULTIVAR DEVELOPMENT

Genetics and Cytogenetics

On the basis of established chromosome numbers of a relatively few species, the $2n = 18$ group is the largest, with all of the horticulturally important species and hybrids (Storey, 1950). A basic number of $x = 3$ provides an explanation of the origin of *Passiflora* species. The horticultural species are hexaploids having fairly high interspecific compatibility and regular meiosis.

Few genetic studies have been done on passion-fruit. Purple tendril colour is dominant over green. Fruit-shell colour is controlled by a single pair of genes lacking dominance, with three colour types being recovered in the F_2 (Nakasone *et al.*, 1967). Inheritance data for crown rot indicate a simple dominant gene for resistance to *Fusarium oxysporum* f. *passiflorae*, with *P. e.* f. *flavicarpa* being the resistant parent. The yellow passion-fruit has also been found to be resistant to a similar disease in South Africa, attributed to *Phytophthora nicotianae* var. *parasitica*. The purple passion-fruit is susceptible to passion-fruit woodiness virus (PWV), while some yellow passion-fruit lines are tolerant.

Breeding and Selection

In establishing a passion-fruit breeding programme, it is highly desirable to develop a collection of as many species as possible in at least two ecological zones, at sea level and at a higher elevation. The breeding objectives in Hawaii were: (i) oval-shaped fruit (10% more juice than in round fruit); (ii) flavour of purple passion-fruit and acidity, fruit size, juice recovery and vine vigour of the yellow passion-fruit; (iii) bright tangerine-coloured juice with high total soluble solids; (iv) high degree of self- and cross-compatibility and good fruit set; (v) resistant to wilting disease and *Alternaria* brown-spot disease; and (vi) resistant to broad mite and other insects. For other regions, breeding for cold tolerance and resistance to PWV, *Fusarium* and *Phytophthora* crown or collar rot and other problems must be added to the set of objectives (Farlow *et al.*, 1984). Where the purple passion-fruit is the primary crop, rootstock breeding for resistance to crown and root rots is a major objective.

In Australia, lines produced from *P. edulis* × *Passiflora incarnata* have shown improved cold-hardiness and tolerance to the severe strain of PWV. The original

F_1 hybrid and selections from the F_2 and F_3 populations backcrossed to *P. edulis* hybrids and to *P. e. f. flavicarpa* show promise as improved rootstock (Farlow *et al.*, 1984). In South Africa, *Passiflora caerulea* is tolerant to both *F. o. passiflorae* and *Phytophthora parasitica* and hence is used as a rootstock (Grech and Frean, 1986). Australia has developed purple hybrids ('3–1', 'E–23', 'Lacy' and 'Purple Gold') that are maintained by grafting upon seedlings of the yellow passion-fruit.

Breeding through interspecific hybridization has not been explored adequately in any sustained programme. Numerous hybrids have been made and the most frequent cross is between *P. edulis* and its form, *P. e. f. flavicarpa*. This cross can be accomplished only if *P. edulis* is used as the female parent, as the reciprocal is strongly incompatible. The latter condition has discouraged the development of clonal cultivars in the yellow passion-fruit. A cross between *P. e. f. flavicarpa* and *Passiflora alata* produces fertile hybrids with high-quality fruit. *P. e. f. flavicarpa* can be transformed using *Agrobacterium tumefaciens*, offering the possibility of incorporating genes for virus resistance and other desirable traits (Manders *et al.*, 1994).

Cultivars

Many growers plant selected seeds or use purple passion-fruit vines grafted on to *Fusarium*-resistant lines of the yellow passion-fruit. The higher acidity of the yellow passion-fruit, along with the need to develop disease-resistant cultivars, has led to the development of hybrids (Table 11.3). The selections and hybrids (between *P. edulis* and *P. e. f. flavicarpa*) have focused on winter and summer cropping and tolerance to passion-fruit mosaic virus (PMV), nematodes, *Alternaria* spot and *Fusarium* wilt. In Australia, two or three hybrids are grown to spread production peaks, with 'E-23' and 'Purple Gold' being most widely used. In Hawaii, the more recent release is 'Noel's Special', a yellow passion-fruit with an unusually bright orange-coloured juice and tolerance to *Alternaria* brown spot. The round fruit averages 90 g in weight, yielding 43–56% juice by weight. Total soluble solids range from 15 to 19.8% (Ito, 1978). There are a number of selections in Brazil, including 'Muico', 'Peroba' and 'Pintado' (all purple) and the yellow 'Miram' and 'Grande'. In Colombia and Venezuela, the yellow passion-fruit 'Hawaiiana' is used along with some Brazilian selections. Hybrids have been developed and released in Taiwan, with 'Tainung No. 1' being one of the most common.

CULTURAL PRACTICES

Propagation

Passiflora species can be propagated quite readily by seeds, cuttings or air layers or by grafting upon a selected seedling rootstock. Where the yellow passion-fruit is favoured, plants are produced exclusively from seed, both for commercial production and for use as rootstocks, due to its vigour, its resistance to root and stem rots and nematodes and its self-incompatible nature.

Seeds are washed to remove the pulp, dried in the shade and sown immediately or stored at about 10–13°C for future use. Seeds stored at room temperature for 3 months give better than 85% germination. Seeds germinate in about 2 weeks, although germination can extend over 2–3 months, due to seed-coat dormancy. Cracking the seed-coat increases germination, while scarifying with sandpaper or fermenting seeds with wall-degrading enzymes does not increase germination. However, seed cracking is feasible only for small quantities of seeds. Twelve-hour alternating between 20 and 30°C also increases germination over a constant 30°C.

When the purple passion-fruit or its hybrids are the desired cultivars, plants are propagated by grafting upon seedlings of the yellow passion-fruit. Grafted vines are more vigorous than their seedling counterparts and also have longer lifespans (Bester, 1980). Grafting is done at a height of approximately 50–55 cm above ground to prevent the scion from coming into contact with soil.

Transplanting and Spacing

Seedlings at the two- to four-leaf stage are transplanted into individual plastic bags 75–100 mm in diameter and about 150 mm deep, grown in semishade for 1–2 months and then gradually provided with more sunlight. Seedlings are considered ready for field transplanting when they have attained heights of 25–50 cm and have been hardened in full sunlight for 1–2 months. For grafted vines, the scion portion should have grown about 25 cm and hardened. The practice of applying fertilizer in the planting holes varies widely, from no fertilizer to 1 kg of superphosphate in South Africa. In Hawaii and Australia, fertilizer (60–114 g of 10–5–20) and/or manure are incorporated in a circular area (*c.* 0.8 m diameter) at each planting site.

The trellis is the principal initial cost of production. There are a number of different trellis types, each having variations in height, number of strands and placement of wires, length of cross-arms, if arms are used, spacing of posts and method of construction. The two most commonly used types are the 'grape' or 'fence' trellis, with one to several wires strung parallel, one below the other, on upright posts, and the 'T'- or 'cross'-type trellis, with three strands of wires, one running on top of the posts and the other two attached at the ends of each cross-arm (Plate 27). With taller posts, two to three strands of wires are used, each spaced about 0.6–0.91 m apart. In the cross type, the cross-arm is 1.22–1.52 m long and is placed 0.46–0.60 m from the top of the post.

The cross-type trellis was found in Hawaii to give higher yields over the 'grape'-type trellis. A higher trellis provides a longer time for vines to trail from the wires to the ground. There is less piling of vines at the top, due to better spreading of growth on the three strands of wires. The high trellis and spreading of vines by the cross-arms allows greater exposure of the vine to sunlight (Anon., 1972).

Minimum row spacing for grape trellises should be about 3 m to permit mechanization. If the cross-type trellis with a 1.0 m bar is used, there should be a distance of 2 m between the ends of bars on the parallel trellis of the adjacent row. Plant spacing within the trellis row is dependent upon the type of passion-fruit being grown, too high density leading to lower yield (Fig. 11.5). In cooler subtropical areas using the purple passion-fruit or its hybrids, plants may be spaced 2.5–3 m apart, due to less vigorous growth. In-row spacing may be up to 5 m if very vigorous growth occurs. For the yellow passion-fruit in warm areas, plants can be initially spaced at 2.5 m and, after the first harvest year, every other vine is thinned to provide a permanent spacing of 4.8 m (694 plants ha^{-1}). Taking advantage of the high yields in the first crop year may be well worth the added cost of the additional plants.

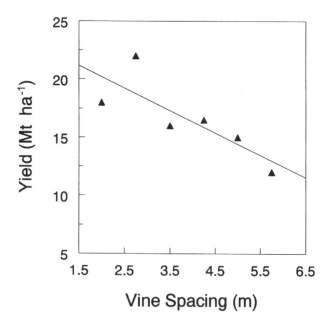

Fig. 11.5. Vine spacing significantly affects yield of yellow passion-fruit during its third year at Viamo, Rio Grande do Sul, Brazil (redrawn from Manica et al., 1985).

Pruning and Training

Training

Planting holes are made 0.6–1.0 m away from the posts. In Hawaii, seedlings 45–60 cm tall are topped at time of transplanting to force lateral shoots to develop, of which two to four leaders are allowed to grow. One to two of the lateral shoots are trained to grow in one direction and the remaining shoots to grow in the other direction of the trellis. These shoots may be tied to stakes placed in correct positions. Bamboo stakes with short branches and about the height of the wires provide excellent support for the vines. Tendrils wrap around the bamboo branches, so tying of the vines to the stakes is unnecessary. All other trailing branches are removed. A single leader is preferred in some systems and is trained along the wire in one direction.

Pruning

Passion-fruit vines produce flowers on the current season's growth, and practices that encourage new lateral growths increase flowering (Fig. 11.3). Pruning, besides encouraging new growth, allows sunlight and air movement into the vine, minimizes disease conditions and allows maximum penetration and coverage of insecticides and fungicides (Bowers and Dedolph, 1959). Occasional pruning is needed, although the degree of pruning varies. If not pruned, laterals begin growing on the ground, the vines tend to form a tangled mass on the wires and only the new outside growth bears fruit.

Pruning of old growth can limit the current crop, while removal of new growth restricts future cropping. The following general conclusions are drawn:

1. Light, selective pruning, particularly at the end of the annual cycle of production, enhances new growth and maintains high yields the following year. This consists of removing all vines about to reach the ground or which are growing on the ground. Vines are cut near the trellis wires but a few nodes away from the main stems. Vines hanging only halfway to the ground are left.

2. Long vines should not be thrown over the trellis as this only increases entanglement on the trellis and depresses yield.

3. Some vine growth on top of the entangled tops of the trellis may be pruned, as fruit produced on these vines is more apt to be lost in the maze of vine entanglement, especially when the planting is 2 or 3 years old. Pruned vines on the trellis should be left to dry in place, as attempts to remove the cut vines damage uncut vines.

4. Cross-arm trellises 2.4–3.0 m high, although more costly initially, offer advantages, such as better spread of vines and greater height for vines to grow and reach the ground (Plate 27). It is also easier to remove old vines hanging down from the wires.

Irrigation

Adequate soil moisture is required to sustain vegetative growth and production (Fig. 11.3). No floral buds are initiated under dry conditions (Fig. 11.6), as vine extension and growth are curtailed. Production is generally associated with higher rainfall in the 2 months before flowering. In Australia, 300–400 L water vine^{-1} week^{-1} is required during the summer.

Fertilization

Recommendations vary widely, with all directed at encouraging new growth throughout the season. Since all flowers occur on new growth, there is a high demand for nitrogen (N); leaf N levels higher than 4.5% are recommended (Fig. 11.7). In Hawaii, mineral analysis of fruit indicates that to produce each 1000 kg of fruit ha^{-1} requires 33 kg ha^{-1} of fertilizer based on a 10–5–20 formulation, assuming 100% uptake efficiency by the fruit; a more realistic uptake value is 50%. Therefore, at least 66 kg ha^{-1} of 10–5–20 fertilizer per

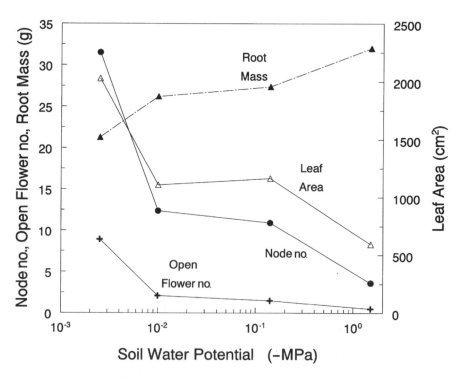

Fig. 11.6. Effect of different moisture stress applied for 10 weeks on the number of nodes produced, leaf area, open-flower number and root dry mass (redrawn from Menzel *et al.*, 1986).

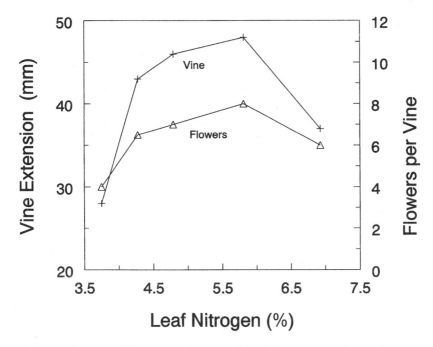

Fig. 11.7. Effect of leaf nitrogen on vine growth and flower production of 'E-23' passion-fruit (redrawn from Menzel *et al.*, 1989).

1000 kg fruit is necessary. Based on 667 plants ha⁻¹ (3 m × 5 m spacing), each plant should receive 4.4 kg of 10–5–20 fertilizer. Deficiency symptoms have been described for N, phosphorus and potassium (NPK) and the microelements iron (Fe), boron (B), zinc (Zn), magnesium (Mg), calcium (Ca) and sulphur (S) (Abanto and Muller, 1976, 1977a,b).

Time of fertilizer application depends upon climatic condition, general appearance and vine performance. In the first year in Hawaii, about 113 g of 10–5–20 fertilizer is placed in the hole at planting and the same amount applied every 6–8 weeks. In the second year, when vines become productive, application should be based on vine growth. Where temperatures allow year-round growth, 5.4 kg of a 10–5–10 fertilizer per plant per year in four applications is recommended for the yellow passion-fruit in Malaysia. Similarly, four applications of 15–4–11 at 500 g per mature vine per application, alternating with four applications of urea at 460 g per vine per application, are recommended in Queensland for purple passion-fruit and its hybrids (Menzel *et al.*, 1993).

Yield

Passion-fruit yields vary with climate, species of passion-fruit, cultivation practices, type of trellis, presence of diseases and the abundance of appropriate pollinating agents. Among the two most commonly cultivated passion-fruit, the yellow passion-fruit, *P. e.* f. *flavicarpa,* has the highest yield potential, with 44,800 kg ha^{-1} year^{-1} in Hawaii and 41,800–61,900 kg ha^{-1} year^{-1} at the Malaysian Agricultural Research and Development Institute (MARDI) research station in Sungai Baging, Malaysia (Chai, 1979). In commercial orchards, a more realistic yield is between 20,000 and 30,000 kg ha^{-1} year^{-1}. The purple passion-fruit is less vigorous and yields between 5000 and 10,000 kg ha^{-1} year^{-1} can be obtained in Australia, with the hybrids having yields up to 25,000 kg ha^{-1} year^{-1} (Beal and Farlow, 1984). The lifespan of the purple passion-fruit is considered to be from 3 to 4 years (6–8 years in cooler areas).

Pest Management

Diseases

Passion-fruit has very few serious diseases, with the extent of these diseases depending upon the species selected and the growing environment (Table 11.4). *Alternaria* brown-spot disease has been reported from various passion-fruit regions of the tropics and subtropics on both the purple and the yellow passion-fruit. Fruit of the yellow passion-fruit in high-rainfall areas can be 66–98% infected and most of the leaves defoliated. The reddish brown to brown, sunken spots, 1.3–5 cm in diameter, are easily identified. Lines with high tolerance when grown in low- to moderate-rainfall areas, combined with

Table 11.4. Some diseases of passion-fruit.

Common name	Organism	Parts affected	Region
Anthracnose	*Colletotrichum gloeosporioides*	Leaves, fruit	Probably universal
Brown spot	*Alternaria passiflorae, Alternaria tenuis, Alternaria tomato*	Fruit, leaves	Universal
Alternata spot	*Alternaria alternata*		Australia
Fusarium wilt	*Fusarium oxysporum* f. *passiflorae*	Roots, crown	Probably universal
Phytophthora blight	*Phytophthora nicotianae* var. *parasitica, Phytophthora cinnamomi*	Leaves, stem, fruit, root	Probably universal
Woodiness virus (insect vectors)	Long, flexuous virus, 750 nm long	Stem, leaves, fruit	Australia, South Africa, Malaysia, Panama, South America, Caribbean area

a good fungicidal spray programme, can produce up to 80–90% marketable fruit (Nakasone *et al.*, 1973).

Fusarium wilt can cause devastating losses on the purple passion-fruit, with a very rapid onset of 24–48 h. Symptoms include browning of the vascular system of roots, crown and stem. *Phytophthora* blight is serious where the purple passion-fruit is grown. It affects the vines, causing defoliation and fruit rots. The disease may completely girdle the stem, and causes root rot in soils with poor drainage.

The PWV is an almost universal, aphid-transmitted disease of both the purple and the yellow passion-fruit. It is serious in Australia, South Africa, Malaysia, Taiwan, Panama and the Dominican Republic. Infected plants have leaves with a light green mosaic pattern and yellow speckling and crinkling and the fruit is misshapen with a woody, thick rind and shrunken cavities (McCarthy, 1982). It is caused by a long flexuous virus, 750 nm in length, belonging to the potyvirus group, which is transmitted by the aphid *Myzus persicae*, as well as mechanically by grafting or by contaminated tools. The virus has been purified and characterized (Jan and Yeh, 1995). There is no evidence of seed transmission. Hawaii has been kept free of PWV, largely by a strict embargo placed on the importation of vegetative parts.

Insects

The most troublesome pests are fruit flies (Table 11.5), as they are difficult to control even using a well-executed spray programme. The Oriental and melon flies can be particularly injurious to passion-fruit grown at low elevations in some areas. *Dacus tryoni*, the Queensland fruit fly, occurs only in Australia and has caused severe injuries at times on purple passion-fruit and its hybrids. Most fruit-fly damage occurs on young passion-fruit. Such fruit will shrivel and drop. The larvae develop better in soft, immature fruit than in mature ones (Anon., 1972). Fruit-fly ovipositing can also cause crater-like scars on mature fruit.

Mites are particularly prevalent in areas of low rainfall and during the dry season, damaging and defoliating the vines; they are effectively controlled with sulphur sprays. Other pests, such as the barnacle scale (*Ceroplastes cistudiformis*) and the red-banded thrips (*Selenothrip rubrocinctus*), occasionally build up to serious proportions and cause defoliation.

HARVESTING AND POSTHARVEST HANDLING

Harvesting and Handling

Hand-harvesting is the most costly operation in passion-fruit culture and can account for approximately 40–50% of the variable costs (Chapman *et al.*, 1978). Normally, fruit are allowed to ripen on the vine and abscise. Fallen fruit are gathered once or twice per week, depending upon the quantity of fruit.

Table 11.5. Some insect pests of passion-fruit.

Common name	Organism	Parts affected	Region
Mediterranean fruit fly	*Ceratitis capitata*	Mostly fruit	Hawaii, Central America, Panama, Mexico
Melon fly	*Dacus cucurbitae*	Mostly fruit, also vegetative parts	Hawaii, Panama, other areas
Oriental fruit fly	*Dacus dorsalis*	Fruit	Hawaii, Asia, parts of Africa, western Pacific, Malaysia
Queensland fruit fly	*Dacus tryoni*	Mostly fruit	Australia
Caribbean fruit fly	*Anastrepha* sp.	Fruit	Caribbean, American tropics
Borer	*Cryptorhynchus* sp.	Root, stem, branches	Panama
Red banded thrips	*Selenothrip rubrocinctus*	Leaves	Panama, Hawaii
Red spider mite	*Brevipalpus papayenis*	Vegetative parts	Universal
Red and black flat mite	*Brevipalpus phoenicis*	Bark, diebacks, fruit	Universal
Carmine mite	*Tetranychus cinnabarinus*	Mature leaves	Universal
Broadmite	*Hemitarsonemus latus*	Young terminal leaves	Universal

During rainy periods, fruit are gathered more frequently and, if sunburn is a problem in summer, fruit may need to be gathered daily.

Vineyard layout and management, such as locations of access roads, line lengths, row widths, trellis types, field grading, trellis orientation and pruning before vines trail on the ground, influence harvesting efficiency. Trellis posts aligned on the raised part of the bed, with ground sloping away on both sides of the posts allows falling fruit to roll to the side, concentrating the fruit outside the canopy and making it easier to gather the fruit. Harvesting using hand-raking and a machine that straddles the fruit and sucks them up has been developed. Alternatively, fruit are picked up from a mown sward with minimal damage. In spite of the fruit's thick skin, it is very susceptible to mechanical damage. Mechanical harvesting aids cause damage and make the fruit generally only suitable for processing.

Postharvest Treatments

Fruit for the fresh market are carefully handled in small 5–10 kg fibreboard cartons. Marketing standards for the fresh-fruit market in Australia require half- to full-ripe fruit, not less than 35% pulp and larger than 4 cm in diameter. Diseased or badly blemished fruit are culled. Dark purple fruit with 120–140 fruit per carton attract best prices. Most growers wax their fruit to extend shelf-life and secure higher prices (Anon., 1984). Purple passion-fruit can be kept for up to 4–5 weeks with little loss of mass at 5°C and 80–90% humidity (Fig. 11.8). The yellow passion-fruit can only be stored for about 1 week at 5–7.5°C. Ethylene can be used to enhance skin-colour development of mature fruit without affecting soluble solids or juice pH.

UTILIZATION

Passion-fruit is valued more for its unique flavour and aroma than for its nutritional value. According to Wenkam (1990), the yellow and purple passion-fruit are good sources of provitamin A, niacin, riboflavin and ascorbic acid (Table 11.6). In processing, passion-fruit's characteristic juice flavour and aroma must be retained. Vines that produce fruit consistently with off flavour are easily recognized and must be eliminated from the planting. Pulp from *Alternaria* brown-spot-infected fruit can seriously affect juice flavour. The yellow passion-fruit has an average juice yield of 30–33%, while the purple

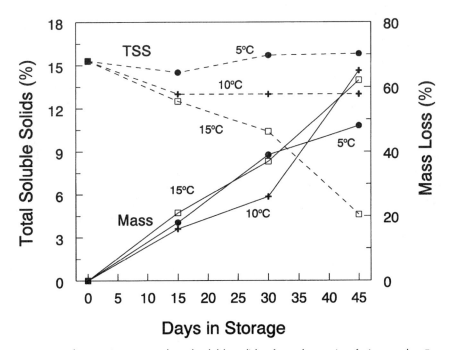

Fig. 11.8. Changes in mass and total soluble solids of purple passion-fruit stored at 5, 10 and 15°C (redrawn from Arjona *et al.*, 1992).

Table 11.6. Composition of edible portion of passion-fruit (Wenkam, 1990).

		P. edulis	*P. edulis* f. *flavicarpa*
Proximate			
Moisture	g	85.6	84.9
Energy	kJ	213	222
Protein	g	0.39	0.67
Fat	g	0.05	0.18
Carbohydrate	g	13.60	13.72
Fibre	g	0.04	0.17
Minerals			
Calcium	mg	3.6	3.8
Phosphorus	mg	12.5	24.6
Iron	mg	0.24	0.36
Potassium	mg	–	–
Vitamins			
Ascorbic acid	mg	29.8	20.0
Thiamine	mg	Trace	Trace
Riboflavin	mg	0.131	0.101
Niacin	mg	1.46	2.24
Vitamin A	IU	717	2410

has a yield of 45–50%. The extracted juice is quick-frozen as purée and kept frozen for later processing of finished products. The juice can be pasteurized (85°C, 1 min) prior to freezing. However, the flavour constituents of passion-fruit are extremely sensitive to heat treatments and the pasteurization process can cause the loss of 35% of the volatile components. The effect of heat is minimized when passion-fruit juice is diluted or blended with other juices (Seale and Sherman, 1960).

The strong, pleasing flavour of passion-fruit permits its use as a pure juice or as a constituent in various frozen and heat-processed punches. It adds an excellent flavour to other products, such as pies, cakes, sauces, salads and sherbets. The fresh juice and concentrates are refreshing mixers with alcoholic beverages, such as gin, vodka and rum. Other products include tropical fruit cocktail, passion-fruit sherbet and ice, and jelly and jam combinations (Seale and Sherman, 1960).

Passion-fruit processing produces large quantities of waste in the form of rind and seed, which create a disposal problem. The wastes are composed of 51% rind and 11% seed by mass for the yellow passion-fruit. Dehydrated rind has been fed experimentally in rations to swine and dairy cattle, with good results.

FURTHER READING

Bora, P.S. and Narain, N. (1997) Passion Fruit. In: Mitra, S.K. (ed.) *Postharvest Physiology and Storage of Tropical and Subtropical Fruits.* CAB International, Wallingford, UK, pp. 375–386.

Martin, F.W. and Nakasone, H.Y. (1970) The edible species of *Passiflora. Econ. Bot.* 24, 333–343.

Menzel, C.M., Winks, C.W. and Simpson, D.R. (1989) Passionfruit in Queensland. 3. Orchard management. *Queensland Agricultural Journal* 115, 155–164.

Menzel, C.M. and Simpson, D.R. (1994) Passionfruit. In. Schaffer, B. and Anderson, P.C. (eds) *Handbook of Environmental Physiology of Fruit Crops,* Vol II, *Subtropical and Tropical Crops.* CRC Press, Boca Raton, Florida, pp. 225–241.

Sao José, A.R., Ferreira, F.R. and Vaz, R.L. (1991) *A Cultura do Maracuja no Brasil.* Fundacao de Estudos e Pesquisas em Agronomia, Medicuna Veterinaria e Zootecnica (FUNEP), Jaboticabal, Brazil, 247 pp.

Winks, C.W., Menzel, C.M. and Simpson, D.R. (1988) Passionfruit in Queensland. 2. Botany and cultivars. *Queensland Agricultural Journal* 114, 217–224.

12

PINEAPPLE

BOTANY

Introduction

Pineapple, a member of the *Bromeliaceae* or bromeliad family, is one of about 45 genera and 2000 species in this family. The bromeliad family are of tropical American origin, except for one species, *Pitcairnia felicana* (Aug. Chev.) Harms & Mildbr., a native to tropical West Africa (Collins, 1960). Plants are herbaceous or shrubby and classified as epiphytic or terrestrial. Some bromeliads, including pineapple, are grown for their leaf fibres, and many more are grown as ornamentals. *Ananas comosus* (L) Merr., pineapple, piña, ananas or abacaxi (Portuguese), is the only species grown commercially for its fruit (Fig. 12.1).

Origin and Distribution

The pineapple was first seen by Europeans when Columbus and his men landed on the island of Guadelupe during the second voyage in 1493. They called the fruit 'piña' because of its resemblance to the pine cone. Early explorations by botanists in South America indicated the area of origin to be south-eastern Brazil, Paraguay and northern Argentina, because of the abundance of the wild species. Based on materials collected in South America, Leal and Antoni (1980) proposed an area further north, between 10°N and S latitudes and 55–75°W longitude. This general area includes north-western and eastern Brazil, all of Colombia and Guyana and most of Venezuela.

At the time of Columbus's arrival, the pineapple was already widely distributed throughout most of tropical America. The antiquity of this fruit even at that time is evidenced by the presence of distinct types (Plate 28), all of which were nearly or completely seedless. Its wide use as food, wine and medicine at the time of Columbus's arrival in the Americas and the absence of recognizable wild progenitors of the cultivated pineapple are further evidence of pineapple's antiquity (Collins, 1948).

Fig. 12.1. Pineapple flower and fruit.

Distribution of pineapple from the Americas is attributed to Spanish and Portuguese explorers and was aided by the resistance of crowns and slips to desiccation. Pineapple was introduced into Africa at an early date and reached southern India by 1550. Before the end of the sixteenth century, it had become established in China, Java and the Philippines (Collins, 1949). Today, the pineapple is found in almost all the tropical and subtropical areas of the world and has become one of the leading tropical fruits in international commerce.

ECOLOGY

Major areas of commercial cultivation are found between 30°N and S latitudes, with some areas considered marginal for various reasons (Bartholomew and Malezieux, 1994). Minor plantings extend pineapple production to subtropical areas with mild climates beyond 30°N and S latitudes and even under protective shelters. Pineapple cultivars show considerable variation in their plant growth and fruit size when grown in different environments (Chan and Lee, 1985), the variation in cultivar response being greater in less favourable environments.

Soil

Pineapple can be grown in a wide variety of soil types, with drainage and aeration being crucial. It is found cultivated in the peat soils of Malaysia, the black sandy loam of Oaxaca, Mexico, and the volcanic soils of Hawaii. Collins (1960) defines an ideal soil type for pineapple as volcanic or sandy loam, with good drainage to prevent waterlogging and root diseases.

   In Hawaii, it is grown primarily on red soils classified in the Molokai series, Typic Torrox subgroup in the order Oxisol. These red soil types are derived from basaltic rocks and volcanic-ash alluvium and are high in oxides of iron (Fe), aluminium and, in some cases, manganese (McCall, 1975). These soils are fine-textured with clay-size particles that are not sticky and allow good drainage. Iron chlorosis associated with manganese occurs in these soils with pH 4.5 or higher. Generally, a pH range of 4.5–5 is considered best for pineapple (Collins, 1960).

Climate

Rainfall

The pineapple is an obligate crassulacean acid metabolism (CAM) plant, with a xerophytic characteristic that enables it to withstand long periods of drought. The leaves have a water-storage parenchyma that serves as a reservoir of moisture during drought. The leaves are covered with trichomes and a highly cutinized upper epidermis. The stomata are small and are located in furrows on the underside of the leaf. At the early-inflorescence development stage, 38% of water use occurs at night when the stomata are open. At midday, no water loss is detected. Water-use efficiency is 3.3 times greater for pineapple than for wheat (Bartholomew and Malezieux, 1994). Pineapples are produced under a wide range of rainfall, from 600 mm to over 3500 mm annually, with the optimum for good commercial cultivation being from 1000 to 1500 mm. Where pineapple is grown in Hawaii, rainfall ranges from 510 to 2540 mm, with an average of 1190 mm annually. Pan evaporation in the major growing area of Wahiawa, Hawaii, is 1850 mm year^{-1} (5 mm day^{-1}). Even if annual rainfall closely approximates the optimal range, poor distribution can result in periods of serious drought. In Loma Bonita and Acayucan (Mexico), 89 and 82%, respectively, of the rainfall falls from June to November.

Despite the xerophytic characteristics of pineapple, growth is adversely affected by prolonged dry periods. Plants may not attain the desirable size by the scheduled time for chemical flower induction. However, most areas of the world where pineapple is cultivated have fairly high humidity, which can reduce the impact of drought. At higher elevations, night air is cooled enough to produce heavy dew, which condenses on the leaves and drains into the heart of the plant and into the soil, providing supplemental moisture that can be absorbed by roots in the leaf axils.

Temperature

Temperature is probably the most important factor in pineapple cultivation. It is difficult to provide an ideal range of temperatures, inasmuch as different plant activities are best achieved at different sets of temperatures. The rate of growth and development are positively correlated with temperature up to 29°C. In the cool season, growth is delayed, leaves are narrow, rigid and shorter, the number of slips is higher, fruit are smaller, with prominent eyes and the flesh is opaque, higher in acidity and lower in sugars (Bartholomew and Malzieux, 1994). Optimum soil temperature for root elongation is around 29°C (Fig. 12.2), with 32°C optimum for leaf elongation and 20–30°C for fresh weight (Bartholomew and Kadzimin, 1977). The pineapple does not tolerate frost and even night temperatures of 7–10°C for a few hours for several weeks during winter can result in leaf-tip necrosis. Such low temperatures can also cause fruit injury. A range of desirable minimum and maximum temperatures would be 15–20°C and 25–32°C, respectively. The optimum is close to 30°C day and 20°C night. The interaction of solar radiation and temperature significantly affect the days from forcing to harvest (Fig. 12.3). Further away from the equator, the amplitude of the difference between seasons increases and number of days to harvest increases.

Selecting proper elevation is a means of achieving a desirable range of temperature conditions. Generally, close to the equator, pineapples are grown at higher elevations. In Mexico, the elevation at around 18°N is between 25 and 200 m. This contrasts with 1555 m at 14.5°N in Central America. In the eastern Cape and border area (34°S) in South Africa, pineapple cultivation goes up to 700 m, while the maximum elevation in Hawaii (21°N) is about 670 m. In South Africa, Florida and southern

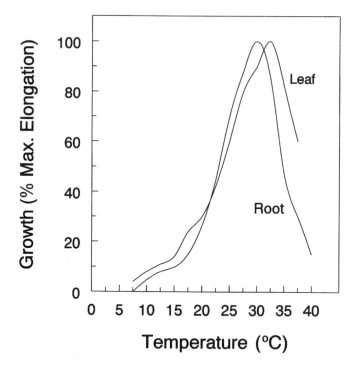

Fig. 12.2. Effect of temperature on leaf and root growth (redrawn from Sanford, 1962).

Queensland, minimum temperatures occasionally approach or reach freezing-point for short durations, while in Hawaii winter temperatures rarely fall below 10°C. At 670 m elevation in Hawaii, plants are smaller, leaves shorter, fruitlets are more pointed, flesh colour is pale yellow, flavour is poor and acid is higher (Collins, 1960).

Cool night temperatures stimulate floral induction if the plant is of a minimum size. Cool temperatures also increase the responsiveness to flower-inducing compounds (Bartholomew and Malezieux, 1994) and shorten the time to floral induction under a short-day regime. High temperatures (> 28°C) make chemical flower induction more difficult, with the percentage of plants forced decreasing linearly with increasing temperature (Glennie, 1981). In Mexico, fruit produced for processing between June to August generally have low acidity and Brix. This lower quality is due to the combined effect of high temperature, excessive rain and an increased number of cloudy days.

The 'Smooth Cayenne' group is the more productive in tropical conditions, while the 'Queen' group is grown mainly in subtropical areas. 'Smooth Cayenne' is most sensitive to low-temperature-induced internal browning (chilling injury, blackheart), while 'Red Spanish' is less sensitive.

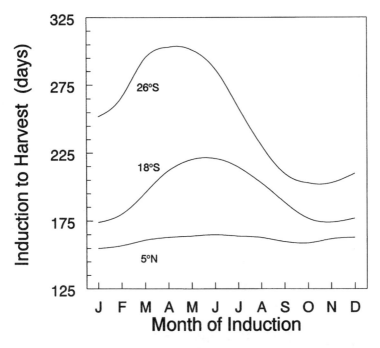

Fig. 12.3. Effect of latitude as an indicator of solar radiation and temperature on the time from forcing to harvest; 26°S is Nambour, Australia (redrawn from Wassman, 1990), 18°S is Madagascar (redrawn from Moreau and Moreuil, 1976) and 5°N is Ivory Coast (redrawn from Malezieux *et al.*, 1994).

Sunlight

Fruit weight is significantly correlated with mean irradiance from planting to harvest (Fig. 12.4). The lower fruit mass with lower irradiance is due to a lower plant mass at forcing. Fruit acidity in the month before harvest declines with solar radiation levels, with no significant effect on total soluble solids (TSS). Cloudy days reduce pineapple growth and result in smaller plants and smaller fruit, with higher acid and lower sugar content. A rule of thumb is that yield decreases about 10% for each 20% decrease in solar radiation. In a pineapple canopy, about 95% of light interception occurs at a leaf-area index (LAI) of about 5.0, due to relatively erect leaf orientations. Shading at higher planting densities leads to a linear decrease in individual fruit weight and a curvilinear increase in total yield.

Intense sunlight, particularly during fruit maturation, can cause sun-scalding of fruit, with the 'Queen' group being more susceptible than 'Smooth Cayenne' (Broadley *et al.*, 1993). The damaging effect can be prevented by shading the fruit or spraying a reflective coating. Fruit covered with newspapers or the gathering and tying of the longest leaves over the fruit is observed in Okinawa and Taiwan (Plate 29). In Mexico, newspapers and weeds are used to cover the fruit, with the principal method being painting the

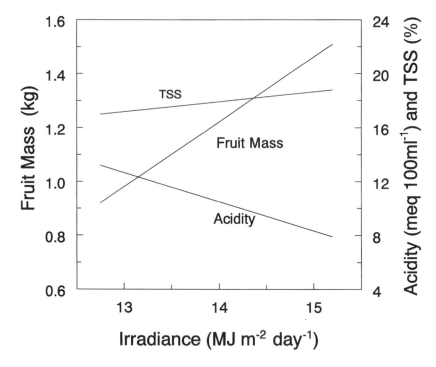

Fig. 12.4. Effect of mean irradiance from planting to harvest on average fruit mass, mean irradiation 2 weeks prior to harvest on fruit-juice titratable acidity and total soluble solids (TSS) of 'Smooth Cayenne' pineapple fruit grown in Côte d'Ivoire over 6 and 10 years, respectively (redrawn from Bartholomew and Malezieux, 1994).

side exposed to the afternoon sun with a thick lime paste. In Australia, some growers spray a water–lime suspension containing a sticker on the fruit.

Photoperiod

'Smooth Cayenne' is known to flower naturally at any time of the year, depending upon the planting material and time of planting (Gowing, 1961). Thus, short days are not necessary for flowering. In Mexico, before the advent of artificial floral initiation, a relatively high percentage of late summer and early autumn sucker plantings flowered in the following July to August, producing winter and spring fruit. Pineapple is a quantitative, but not an obligate, short-day plant, which does not require diurnal variations in temperature to flower (Friend and Lydon, 1978). Plants grown under 8-h days all flowered, 69% flowered under a 10-h, 53% under a 12-h and 30% under a 16-h day. Interruption of the dark period by illumination suppresses flowering. Low temperatures may enhance the effects of short days with a small diurnal temperature range, cool nights being more effective. Some minimum plant size is necessary before natural flowering will occur.

GENERAL CHARACTERISTICS

Stem

The stem of a mature plant is 30–35 cm long and club-shaped, with the thickest diameter being 6.5–7.5 cm below the apex (Fig. 12.1). Internodes are short, ranging from 1 to 10 mm with the middle region of the stem having the longer internodes. There is an axillary bud at each node (Krauss, 1949). These axillary buds can produce either slips or suckers, which are used as planting material. True slips are shoots produced in the axils of the floral bracts on the peduncle (Fig. 12.5), while suckers are shoots from axillary buds on the peduncle that bear the ratoon crop. Suckers can grow continuously until harvest or until flowering occurs and, therefore, are potentially larger than slips but they also have a wide range of sizes. Slips can weigh from 250 to 450

Fig. 12.5. Pineapple fruit showing multiple crowns and slips at the base of the fruit.

g and crowns from 100 to 350 g. Sliplets are produced by treating the plant with chloroflurenol and vary widely in number and size.

Roots

Adventitious roots are formed from preformed root primordia in the stem. Krauss (1948) separates the roots into 'soil' roots, which develop at the base of the stem and form the underground root system, and 'axillary' roots, which form above the soil surface in the leaf axils. The underground root system is very dense but shallow, extending to a depth of about 15 cm. The roots of pineapple plants grown in deep, loose, fertile soil free from parasitic organisms can grow downward more than 50 cm and extend 1.83 m beyond the plant within a year's time.

The axillary roots will absorb both water and nutrients, facilitating the utilization of foliarly applied nutrients. The main absorptive area of the root is in the unlignified white tissue of the root tip. Roots without white tissues at the root tips are not actively growing and are inefficient in absorbing water; the tips may have been lost because of root rot and nematode feeding.

Leaves

Leaves are produced in a spiral around the stem in a tight rosette. A dormant axillary bud is found at the midpoint of each leaf on the main stem. By counting from a leaf or bud at a selected node on the stem upwards in a spiral, the thirteenth leaf or bud will be directly above the initial leaf or bud and five turns around the stem will have been made, resulting in a 5/13 phyllotaxy (Bartholomew and Kadzimin, 1977).

The number of leaves increases regularly, with an average of five or six leaves per month. Old leaves do not abscise and a mature plant may have about 70–80 active leaves. The most frequently mentioned developmental index for carbohydrate and nutrient status is the whorl of leaves designated and described as the D leaves by Sideris and Krauss (1936) and first used by Nightingale (1942). Usually, the D leaves are the fourth whorl from the base of the plant and not more than three leaves would qualify as D leaves at any one time. These are the longest (80–100 cm) and the youngest, nearly physiologically mature set of leaves. In non-fruiting plants, the leaf base of these leaves is only slightly broader than the green blade. In 'Smooth Cayenne' a small group of marginal spines is found near the leaf base and near the tip. Also, a short area of spines are formed on new growth of the leaves after temporary cessation of leaf growth due to adverse conditions. There is always a slight constriction at the point where leaf expansion has been interrupted.

There are a number of structures in pineapple plants and other bromeliads that contribute to their strong resistance to moisture stress. The position and trough shape of the leaves and the presence of trichomes and stomata located in furrows beneath trichomes on the underside of the leaf are features that enhance drought resistance. Long multicellular trichomes are found in abundance on the lower (abaxial) side of the leaf and fewer on the upper (adaxial) side. It is generally believed that they absorb moisture and nutrient solution and reduce water loss through the stomata by forming a dense covering over the stomata.

A unique internal feature of bromeliads, including the pineapple leaf, is a water-storage tissue that is colourless and translucent. This tissue can be identified with the naked eye in the adaxial part of transverse sections of the leaf and contrasts with the chlorophyllous mesophyll tissue beneath. The width of both tissues in leaf cross-section varies with age and environmental factors. In the pineapple-growing district of Wahiawa, Hawaii, the water-storage tissue of a turgid, fully developed leaf from a field-grown plant occupies about half the leaf cross-section at the middle of the blade. This tissue becomes narrower as moisture stress increases, and it is absent towards the tip and margins.

The pineapple plant uses CAM, a system of metabolism characterized by large diurnal fluctuations in organic acids (malic and citric acids) and an inverted pattern of gas exchange, in which carbon dioxide (CO_2) is fixed at night and transpiration is suspended during the daytime (Bartholomew and Kadzimin, 1977). Pineapple-leaf stomata are closed during much of the day (9 a.m.–2 p.m.) and are open from around 2 p.m. and throughout the night. These opening and closing phases are closely related to intercellular-space CO_2 levels. During the night, gas exchange takes place, with uptake of CO_2, which is fixed as malic acid, thus increasing leaf acidity. Shortly after sunrise, malic acid moves from the vacuole to the cytoplasm, is decarboxylated and loses a molecule of CO_2; the low intercellular CO_2 level begins to build up and the stomata close. In the morning to early afternoon hours, malic acid is fixed and reduced to carbohydrate by normal Calvin-cycle (C_3) photosynthesis. As the malic acid level declines, the intercellular CO_2 also decreases and the stomata open. During the remainder of the afternoon hours, CO_2 is assimilated from the atmosphere by conventional C_3 photosynthesis. Normal night opening of pineapple stomata requires sunlight on the previous day.

Inflorescence and Flower

The first sign of floral initiation, whether natural or induced, is a rapid increase in diameter of the apical meristem and, 5–6 days after this change has taken place, the peduncle begins to elongate. It continues to elongate as the inflorescence develops. There are 100–200 flowers per inflorescence, and at anthesis one to several flowers open each day, beginning at the base of the inflorescence, over a period of 3–4 weeks (Okimoto, 1948). When inflorescence development is initiated, the phyllotaxy changes from one of 5/13 for the leaf to that of 8/21 in the inflorescence. Production of many small leaves just below the base of the inflorescence marks this change. When flower production ceases, the apical meristem again reverts to the 5/13 phyllotaxy, with a transition area of short leafy bracts, followed by the growth and development of the crown (Fig. 12.1).

The individual flower is hermaphroditic, with six stamens in two whorls of three and a three-carpellate, inferior ovary with numerous ovules (Fig. 12.1). The ovules and pollen grains are functional, but seeds are not normally formed, as the 'Smooth Cayenne' is strongly self-incompatible.

Fruit

The fruit, a sorosis, is a syncarp of fused fruitlets from inferior ovaries (Fig. 12.1). Fruitlets develop from flowers that do not abscise. The style, stamens and petals wither, with remaining floral parts developing into the fruitlet (Okimoto, 1948). Yeast and bacteria can enter through the nectary gland and mature fruit is not sterile (Rohrbach and Apt, 1986). In the normal fruit, there are eight gently sloping rows and 13 shorter, steeper rows of flowers. The number of fruitlets can be estimated by counting the fruitlets on the long, gently sloping spiral and multiplying by eight, the number of spirals (Bartholomew and Paull, 1986). It takes approximately 4 months from end of last open flower to fruit maturity and the total time from floral initiation to harvest takes between 6 and 7 months (Fig. 12.6). Temperature significantly accelerates or delays development.

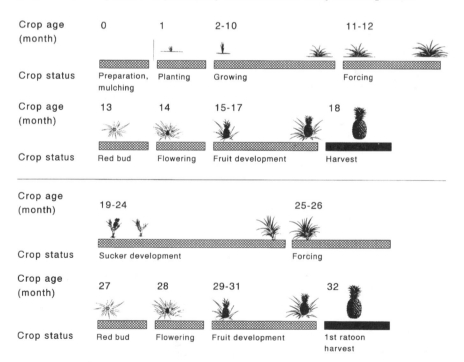

Fig. 12.6. The growth cycle of a pineapple crop to harvesting the first ratoon crop in relation to the various stages of development and crop age (Rohrbach, undated).

Cell division is completed prior to anthesis, with all further development being the result of cell enlargement (Okimoto, 1948). Fruit mass increases about 20-fold from the time of flowering until maturation. Fruit development studies have shown that fruit weight and its components (core, fruitlets, the collective flesh, fruit shell) increased in a sigmoidal fashion after the inflorescence is initiated (Fig. 12.7). Crown growth increases about 30–45 days after fruit growth has commenced. The crown has been reported to have no direct effect on the development of the fruit, although crown removal early in fruiting leads to greater fruit weight. The most marked changes in flesh composition occur in the 3–7 weeks prior to and at the half-yellow shell-colour stage. Just prior to this stage, fruit translucence can start to develop, and translucent development continues after harvest. Titratable acidity declines and TSS gradually increases, with a more rapid increase in the last 6 weeks, as the fruit approaches the full-ripe stage. Fruit sugars continue to increase through to senescence, unless the fruit is harvested.

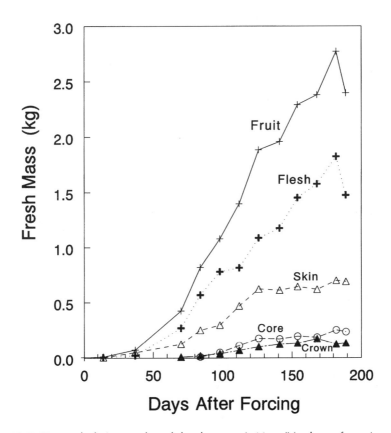

Fig. 12.7. Pineapple fruit growth and development in Hawaii (redrawn from A. Hepton, personal communication).

CULTIVAR DEVELOPMENT

Cytogenetics and Genetics

Information on cytogenetics and genetics come from early studies conducted at the former Pineapple Research Institute of Hawaii. Chromosome counts of 'Cayenne', 'Queen', 'Spiny Samoa', 'Ruby', 'Pernambuco', 'Spiny Guatemala', an F_1 hybrid between 'Cayenne' and an unknown wild type from Brazil and *Bromelia pinquin* L. showed the pineapple cultivars to possess an n = 25 chromosome number and *B. pinquin* n = 48, with no irregularities in meiosis (Collins, 1960). Several triploid plants with 75 chromosomes are found among the F_1 hybrids. The triploids appear to be products of the conjugation between an unreduced 'Cayenne' egg cell with 50 chromosomes and a normal haploid pollen of the Brazilian wild type. The commercial cultivar 'Cabezona' is a natural triploid with 75 chromosomes (Collins, 1933).

Collins and Kerns (1938) described about 30 heritable mutant forms in 'Smooth Cayenne' in fields, most being undesirable. One desirable processing mutant form is for elongated fruit. Collar-of-slips, a condition in which excessive numbers of slips are either attached directly to fruitlets at the base of the fruit or massed around its base, occurs in successive vegetative generations (Fig. 12.5). These plants are continuously rogued out. This character is dominant and occurs in a heterozygous state in 'Smooth Cayenne'. A spiny leaf is due to a homozygous recessive gene and the smooth leaf condition of 'Smooth Cayenne' is carried as a heterozygous dominant. However, it is unstable and mutates frequently to the spiny-leaf type; other spiny conditions also occur. Self-incompatibility in pineapple is an advantage in commercial production of seedless fruit and in making certain types of cross-pollinations. Pollen shows good viability for several cultivars but not for the triploid 'Cabezona'. A single *S* locus, with multiple alleles, and gametophytic control of pollen phenotype are involved in self-incompatibility (Brewbaker and Gorrez, 1967), in which pollen-tube growth beyond the upper third of the stylar canal is inhibited.

Breeding and Cultivars

Breeding

Pineapple breeding and selection objectives vary with the locality, but almost always emphasize disease and insect resistance (Leal and Coppens d'Eeckenbrugge, 1996). More recently, development of cultivars for fresh-fruit consumption has been a major focus. Populations have been produced from crosses to allow selection of improved types. This selection involves constant roguing of undesirable mutations and selection of superior types. Mutation continues to occur within selected clones, so the effects of selection are not permanent and, without continued roguing and selection, commercial fields can revert to conditions of unselected field population.

In Hawaii, advanced hybrid clones derived from many years of breeding are now in advanced stages of testing by individual plantations. In Australia, sources of germ-plasm consist of selections from Queensland 'Cayenne' clones, hybrid populations, meristem-cultured plants from known clones and introductions from other countries (Winks and Glennie, 1981). Hybrid selections were derived from 'Cayenne' crosses with rough-leaf types, such as 'Queen', 'Ripley Queen', 'MacGregor', 'Alexandra' and 'Collard'. 'Singapore Spanish', a spineless plant with good fruit quality, has been widely used in their breeding programme. Malaysia has similarly used 'Smooth Cayenne' and 'Singapore Spanish' in their hybrid-breeding programme (Chan and Lee, 1985). Breeding programmes in Puerto Rico, Brazil, Taiwan, the Philippines and South Africa maintain the major cultivars in their germ-plasm collections.

'Smooth Cayenne' is highly susceptible to mealy-bug wilt, although a few clones have shown variable resistance, which is transmitted to seedling progenies. Other cultivars and species have shown some resistance: 'Red Spanish', 'Pernambuco', 'Queen', *Ananas ananasoides*, *Ananas bracteatus*, *Pseudananas sagenarius* and some hybrids involving 'Cayenne' and resistant cultivars (Collins, 1960). In Hawaii, hybrid No. 59-656 is resistant to *Phytophthora cinnamomi* and *Phytophthora parasitica* root- and heart-rot diseases and it is being tested in Australia (Winks and Glennie, 1981).

Concomitant with breeding for disease and insect resistance have been the attempts to develop cultivars suitable for fresh-fruit export (Paull, 1997). A suitable clone should have a high yield, high sugar, a good balance of sugars to acids, high ascorbic acid and an appealing flavour. These must be incorporated in a clone having the desired disease resistance, fruit shape and weight.

Cultivars

Collins (1960) developed a botanical key to the genera and species and the major characteristics of five 'Cayenne' clones for selection of appropriate parents. A later taxonomic key identified 18 commercial clones, with vernacular names and descriptions (Grazia *et al.*, 1980); this included a new group, 'Maipure', composed of smooth-leaved clones. Clones in this group are primarily grown in South America. The major fruit and leaf characteristics of the principal clones can be placed in five phenotypic groups (Table 12.1). 'Monte Lirio' and 'Perolera', formerly unclassified, are placed in the 'Maipure' group. 'Cayenne' and 'Maipure' smooth-leaved groups differ, in that leaves of the former group exhibit a few spines near the leaf tip while the latter group shows leaf piping (leaf margins) with a greyish streak due to folding over of the lower epidermis on to the upper leaf surface (Collins, 1960). Pineapple isozyme variation indicates five genetically diverse groups that do not perfectly

Table 12.1. Characteristics of pineapple groups and clones (after Leal and Soule, 1977).

Characteristic	Group				
	Spanish	Queen	Abacaxi	Cayenne	Maipure
Leaves	Spiny	Spiny	Spiny	Smooth	Smooth
Fruit					
Weight (kg)	0.9–1.8	0.5–1.1	1.4	2.3	0.8–2.5
Shape	Globose	Conical	Conical	Cylindrical	Cylindrical
Colour of skin	Large, deep eyes; O, R	Deep eyes; Y	Y	Flat eyes; O	Y to OR
Colour of flesh	Pale Y to W	Deep Y	Pale Y to W	Pale Y to Y	W to deep Y
Core	Large	Small	Small	Medium	Small–medium
Taste	Spicy acid, fibrous	Sweeter, less acid, low fibre	Sweet, tender, juicy	Sweet, mildly acid, low fibre, juicy	Sweeter than 'C', fibrous, tender, very juicy
Market					
Canning	F	F	F	VG	F
Fresh					
Local	G	G	G	G	G
Export	VG	G	F to P	F	F to P
Disease problems	Gummosis, wilt resistance	More resistant than 'Cayenne'	Resistant	Mealy-bug wilt	Unknown
Clones	Red Spanish	Queen	Abacaxi	Smooth Cayenne	Maipure
	Singapore Spanish	MacGregor	Abakka	Cayenne Lisse	Perolera
	Green Selangor	Natal	Sugar Loaf	Smooth Guatemalan	Lebrija
	Castilla	Ripley	Papelon	Typhone	Monte Lirio
	PRI-67	Alexandria	Venezolara	St Michael	Abacaxi
	Cabezona		Amarella	Esmeralda	Rondon

VG, very good; G, good; F, fair; P, poor; Y, yellow; O, orange; OR, orange; R, red; W, white; C, Cayenne.

match these phenotypic groupings (Loison-Cabot, 1992; Aradhya *et al.*, 1994). A brief discussion of the five horticultural groups follows.

'CAYENNE' GROUP

'Smooth Cayenne' is the standard for processing and for the fresh-fruit trade, because of its cylindrical shape, shallow eyes, yellow flesh colour, mild acid taste and high yields. In most areas, 'Smooth Cayenne' constitutes a mixture of clones, due to new introductions from mutations, lack of roguing and various other sources. Local selections are mostly known by their areas of origin, such as 'Sarawak' in Malaysia. 'Champaka' is a selection of 'Smooth Cayenne' originating in India and widely grown in Hawaii. The group is susceptible to mealy-bug wilt and nematodes.

'QUEEN' GROUP

This group generally produces smaller plants and fruit with spiny, shorter leaves than the 'Cayenne' group. The 'Queen' is grown in South Africa, Australia and India for the fresh-fruit market. The 'Z-Queen' or 'James Queen' is reported to be a mutant of 'Natal Queen' and is a natural tetraploid.

'SPANISH' GROUP

The plants are generally small to medium, spiny-leaved, vigorous and resistant to the mealy-bug wilt, but susceptible to gummosis, caused by the larvae of the *Batrachedra* moth. It is acceptable for the fresh-fruit market but not favoured for canning due to deep eyes and poor flesh colour. The 'Red Spanish' or 'Espanola roja' is the major cultivar in the Caribbean region. The 'Singapore Spanish', or 'Singapore Canning', and 'Nanas Merah' are the principal canning pineapples in west Malaysia because of their adaptability to peat soil. The flesh has a bright yellow colour. Other Malaysian cultivars are 'Masmerah', a spineless type with large fruit, and 'Nanas Jabor', a 'Cayenne'–'Spanish' hybrid that is susceptible to fruit marbling and cork spot. 'Cabezona', a natural triploid, is an exception, having large plants and fruit weighing 4.5–6.5 kg. It is grown primarily in the Tabasco State of Mexico and a small area of Puerto Rico, where local consumers prefer the larger fruit. The Puerto Rico clone 'PR 1-67' is suspected to be a hybrid between the 'Red Spanish' and 'Smooth Cayenne', as these were the only clones grown in adjacent fields. The fruit has light yellow flesh, with adequate sugar, is resistant to gummosis, is fairly tolerant to mealy-bug wilt, has good slip production and has good shipping qualities.

'ABACAXI' GROUP

This group is grown mostly in Latin America and in the Caribbean region. Py *et al.* (1987) called this the 'Pernambuco' group. The fruit is not considered suitable for canning or for fresh-fruit export, but the juicy, sweet flavour of the fruit is favoured in the local markets. 'Perola', 'Pernambuco', 'Eleuthera' and 'Abacaxi' are the principal clones in Brazil, along the eastern Espirito Santo in the south through Bahia and Pernambuco to Paraibo.

This group is cultivated in Central and South America as fresh fruit for the local markets. These clones may be of interest to breeders in the western hemisphere, as they constitute a gene pool of adapted forms almost unused in breeding programmes.

CULTURAL PRACTICES

Planting Material and Propagation

Commercial pineapple production utilizes crowns, slips and suckers (shoots) and sliplets produced by growth-regulator treatment at the time of flower induction. It is important that these vegetative materials be kept separate, as mass and possibly growth rate, harvest time and other factors differ (Collins, 1960). Time from planting to harvest is dependent primarily on the mass of the propagule, and crowns produce fruit in 18–24 months (Fig. 12.6), slips in 15–20 months and suckers in 14–17 months. Planting materials are treated with a fungicide and are 'cured' by drying the butt end to prevent rots before planting.

A shortage of planting materials can occur under certain circumstances. The fruit and crown are sold when marketed as fresh fruit and the crown becomes unavailable for planting. 'Smooth Cayenne' produces only one or two suckers per plant and seldom more than three slips (Fig. 12.5). In breeding programmes, adequate numbers are needed for replicated evaluations, and named clones must be multiplied for release. Several methods of asexual propagation can be used. The mature plant can be stripped of leaves, exposing all dormant axillary buds. The stem is then split longitudinally into four or more sections and each is cut transversely into triangular pie-shaped pieces, with at least one axillary bud. After immersion in a fungicide solution, the sections are left to dry for a few days before planting in well-prepared beds with good drainage and aeration. Water is applied sparingly during the first 2 weeks. Axillary buds will grow and produce plantlets, which can be transplanted in 4–6 months. A normal-sized 'Cayenne' plant can produce an average of 25 sectioned plants (Collins, 1960). Planted crowns and shoots with adequate root systems can be split longitudinally into quarters, which forces one to several axillary buds to develop. These plantlets can be transplanted when large enough and the process repeated. In Okinawa, crown leaves planted in sand produce plantlets at the leaf base. Only one plantlet is produced per leaf, at a low rate. As no axillary buds are present on the leaf, shoots must be generated from meristematic callus formed at the leaf-base tissue or from stem tissues adhering to the leaf base. There are 40–70 leaf-bud cuttings from a single crown.

The growth regulators, called morphactins, containing chloroflurenol, induce sliplets. The morphactin induces axillary bud growth after apical dominance has been broken with ethephon. One to 9 weeks after flower induction and before flower differentiation, chloroflurenol (Maintain CF-125; Multiprop R) is applied, and abnormal fruit is produced, along with slip-like plantlets; additional suckers are also produced. There can be a tenfold increase in available planting material.

Numerous 'Smooth Cayenne' plantlets and protocorm-like bodies are produced from shoot tips and axillary buds using tissue-culture methods (Mapes, 1973). About 5000 plantlets can be produced from a single crown and 100,000 plants from a single shoot in 12 months. Excessive somaclonal variation has limited the commercialization of tissue culture to increase planting material, but the limited success indicates that avoiding callus-tissue formation reduces and may eliminate such variation (Drew, 1980).

Field Preparation and Layout

Large plantations usually have heavy equipment to clear virgin land or to prepare replant fields. In virgin and old fields with soil compaction, subsoiling is desirable to break the hard pan so as to improve drainage and soil aeration. Deep ploughing to a depth of around 64 cm is preferred. Old pineapple plants may be chopped, dried and burned, if the field is to be replanted soon after the previous crop, or buried in the ground to rot, if time before replanting permits. Discing to break clumps of soil and improve soil texture is important to improve efficiency of fumigants and planting.

A multiduty machine has been developed in Hawaii, which fumigates the soil for nematodes, applies preplant fertilizer and lays the irrigation drip line and the black polyethylene mulch in a single operation. The black polyethylene mulch (0.05 mm thick and 81 cm wide) helps to prevent rapid escape of fumigants, maintains warmer soil temperatures during the cool season, retains moisture at the soil surface, reduces fertilizer leaching during rainy periods, controls weed growth in the beds and increases yields. In many pineapple-growing areas where plastic mulching is too costly, mulching with straw, grass, sugar-cane bagasse or other available materials is done. Crop rotation for nematode control is a possible practice when economic.

Fumigation

'Smooth Cayenne' is susceptible to nematodes and, if left unchecked, nematodes can devastate the plant. Fumigation of the soil with various volatile chemicals as a preplant treatment for control of root-knot and reniform nematodes is an established practice in Hawaii. In recent years, loss of some fumigants has occurred, due to health and environmental concerns. Nonvolatile nematocides are also applied through drip irrigation systems and this leads to better plant condition, especially in ratoon fields, and significantly increases yields. Ratoon yields are highly dependent upon the health of the plant-crop root system, which can be severely damaged by nematodes. In South Africa, where *Meloidogyne* and *Helicotylenchus* nematodes are serious problems in pineapples, preplant dipping in a systemic nematocide, followed by postplant spraying at monthly intervals for 12 months, increases plant-crop yield by 11 t ha^{-1}.

Planting

Planting is commonly done manually, with the holes dug with simple planting tools of various sizes. Good soil preparation makes it easy to place the crown butt deep enough into the soil. To keep rows aligned, planting cords marked with planting distances are stretched from one end of the field to the other. In Hawaii, planting material is inserted into the ground through the polyethylene mulch at factory marked intervals, using a *c.* 25-cm-long spatula-shaped planting iron. In Queensland, Australia and Mexico, some growers use a planting machine similar to a vegetable transplanter.

Crowns, slips and suckers are planted separately in different fields. These materials differ in size within each morphological group and between groups. Crowns are somewhat more uniform in size than slips and suckers, both of which are larger than crowns. Both slips and suckers have wide size variation, which is dependent upon when they are harvested. Sizing prior to planting is crucial in order to obtain uniform-sized plants at the time of floral forcing in order to ensure a uniform ripening fruit crop (Py *et al.*, 1987). Large suckers have a high tendency for precocious fruiting, especially when sucker size exceeds 600 g. The following size segregations are suggested (Py *et al.*, 1987): small crowns 100–200 g and medium crowns 200–300 g; slips or suckers 200–300 g small, 300–400 g medium and 400–600 g large.

Spacing

Plant spacing or density is one of the major factors affecting individual pineapple fruit size and yields per unit area. Plant density and other cultural practices are directed towards the production of fruit size appropriate to the principal use: the fresh-fruit market or processing. Plant density for processing fruit is generally less than for the fresh-fruit market, as smaller fruits are generally preferred for the fresh market, especially in the USA. Spacing is also dependent upon cultivar. Cultivars of the 'Spanish' group produce smaller plants but have spiny leaves, so wider spacing is used than for smooth-leaved cultivars.

Small growers use the single-row system with relatively wide spacing between plants and between rows, giving low plant densities of 15,000–25,000 plants ha^{-1}. In commercial plantings, the double-row system has been almost universally accepted. Multiple-row systems (three- to five-row beds) are unsuitable, due to difficulties in field maintenance and reduction in fruit size in the inner rows. In the conventional two-row bed system, various spacing regimes are employed to achieve desired plant densities. In Hawaii and other areas where cultivation practices are highly mechanized, changes in bed width are limited by the width of mulch used. A widely used spacing under plantation cultivation is 30 cm between plants in a double row of 60 cm on a bed with from 90 to 120 cm between beds, providing a density of from 44,444 to 58,700 plants ha^{-1}. Planting densities as high as 75,000 plants ha^{-1} are used where smaller fruit are desired. There is a fruit size decrease of about 45 g for each population increase of about 2500 plants ha^{-1}.

Irrigation

The pineapple can survive long periods of water stress under natural conditions. However, yields are low, with poor-quality fruit of unacceptable size. Irrigation of pineapple in the Ivory Coast increased yields by 14–22 Mt ha^{-1}, with a cost equivalent to a yield of 5 Mt ha^{-1}. The potential evapotranspiration of pineapple can reach 4.5 mm day^{-1} and a soil's water-holding capacity rarely exceeds 100 mm, so, without rains, the water supply will be exhausted within 3–4 weeks. Water deficit can be indexed by the relative thickness of the water-storage tissue of the youngest, physiologically mature leaf (D leaf) and by the percentage of white root tips visible on the roots in the soil (Sanford, 1962).

Irrigation allows year-round planting, provides fresh fruit throughout the year, allows plants to attain desirable plant size for chemical flower initiation at scheduled times of the year, allows the use of soils with poor water retention and allows fertilizing even during the dry season. Weed problems, however, are increased by overhead irrigation.

Sprinkler delivery systems, including a self-propelled boom sprayer delivering approximately 205.6 m^3 per irrigation have been used. Drip irrigation has replaced other methods of irrigation in Hawaii, with one tubing orifice for every two plants. The crop cycle is shortened, fertilizers and nematocides are applied through the drip system with increased efficiency and safety and yields are increased. Yield increases have been especially evident in ratoon crops, due to a decrease in plant mortality and the healthier condition of the plants at plant-crop harvest. The healthy condition of the ratoon plants is primarily due to frequent delivery of nematocides.

Fertilization

Various published reports on pineapple nutrition indicate that the quantity of nitrogen (N) ranges from 225 to 350 kg ha^{-1} and of potassium (K) from 225 to

450 kg ha^{-1}. The crop has a low requirement for phosphorus (P), and many reports have omitted the amount applied, but, generally, it is between 150 and 225 kg ha^{-1}, as P or P_2O_5. Some idea of pineapple fertilizer requirements may be obtained by analyses of elements immobilized in the various plant parts (Table 12.2). Large amounts of N are found in the plant, fruit and slip. In ratoon fields, which develop on suckers from the mother plant, nutrients removed by the first fruit crop must be replenished. This amounts to approximately 175 kg N, 27 kg P, 336 kg K, 47 kg calcium (Ca) and 27 kg magnesium (Mg) ha^{-1}.

Pineapple in Hawaii is grown on soils with relatively strong P-fixing capacity and for this reason as much as 1120 to 2240 kg ha^{-1} of rock phosphate is worked into the soil. Ammonium phosphate and potassium sulphate in equal amounts, totalling about 900 kg ha^{-1} may also be applied in planting rows during bed preparation. Urea and potassium sulphate are sprayed on a weekly or biweekly basis from about 3 months after planting. Urea can be applied as a foliar spray at up to 20% without damage if biuret is less than 1%.

Iron fixation is a problem in the Hawaiian pineapple soils due to high manganese content. Ferrous sulphate up to 17 kg ha^{-1} per application is given whenever needed as a foliar spray. Plants with Fe deficiency show yellowing of interveinal areas. However, Fe deficiency also affects absorption of N, via its effects on the root system, leading to a general yellowing of the entire plant. Zinc (Zn) deficiency is usually found in eroded areas and soils with high amounts of coral sand and high pH. Deficiency causes curved and twisted leaves and plants. Apical dominance is lost and plants begin to sucker profusely. Zinc sulphate 0.5% (w/v) applied at 935 L ha^{-1} (0.5%, 100 gals acre^{-1}) is sprayed as needed. Boron deficiency is characterized by chlorosis of young leaves, development of red margins and even death of the apical region, resulting in profuse suckering. Boron deficiency is a common problem in Australia and 1–4 p.p.m. boron is applied with the flower-induction chemicals.

Nutrient needs of pineapple in Hawaii are based on the concept of the crop log, which involves the development of laboratory soil and foliar indices

Table 12.2. Minerals immobilized or removed by pineapple plants at a density of 54,340 plants ha^{-1}

	Amount (kg ha^{-1})				
	N	P	K	Ca	Mg
Plant	437	47	538	134	134
Fruit	135	20	269	33.6	20.2
Slip	40	6.7	67	13.4	6.7
Total	612	73.7	874	181	160.9

(see Chapter 2), visual deficiency symptoms for N, Fe and Zn, and other measurements of growth rate, pathogens, root parasites, moisture stress and weather conditions (Sanford, 1964). A crop log not only indicates deficiency symptoms but also attempts to determine causes of deficiencies. Soil and foliar analyses are both important, as the former shows reserve amounts of soil nutrients and the latter indicates efficiency in absorption. An element may be sufficient in the soil but deficient in the plant, due to causes such as moisture stress, a diseased root system or an elemental imbalance causing unfavourable interactions affecting absorption (Sanford, 1962). Fertilizer is applied on a schedule to meet different needs at different stages of growth (Table 12.3). Foliar applications should cease after forcing to prevent injury to the inflorescence and reduced fruit yields.

Chemical Flower Induction

The accidental discovery in the Azores that smoke from burning organic materials induces premature flowering in greenhouse-cultured pineapples led to the wide practice of burning rubbish around the periphery of the fields in Puerto Rico (Rodriguez, 1932). The active ingredient in the smoke is ethylene gas, with acetylene and calcium carbide also inducing flowering in pineapple (Aldrich and Nakasone, 1975). Forcing plants into flowering allows synchronization of harvest and makes it possible to control harvest dates to meet anticipated fresh-market and cannery needs (Fig. 12.6). Fruit-harvest date can be predicted with good accuracy, with only daily maximum and minimum air temperature needed to calculate fruit heat units (Malezieux *et al.*, 1994). This model can be adjusted, using historical data.

The relatively soluble sodium salt of α-naphthaleneacetic acid (SNA) was the first plant growth regulator used commercially to force flowering in Hawaii (Bartholomew and Criley, 1983). Ethylene and acetylene gases dissolved in water have been used in many pineapple growing regions. The gases require specialized equipment for effective application and safe use. In the warm tropics, SNA has not been effective, while a water solution of acetylene produced from calcium carbide or calcium carbide as granules has been successful.

Ethephon (2-chloroethylphosphonic acid) is probably the most widely used chemical in commercial pineapple production, because of its effectiveness and ease of application. This chemical breaks down to produce ethylene at a neutral pH (Bartholomew and Criley, 1983). The effective ethephon concentration ranges from 500 to 1500 μg L^{-1} with greater amounts required to force flowering in warm than in cool months. At least 90% flowering is obvious 40–60 days after applying about 1000 μg L^{-1} solutions (Fig. 12.6), with 50 kg ha^{-1} urea (Bartholomew, 1977). In some regions, adjusting the solution pH above 7 with sodium borate improves forcing success.

Acetylene and ethylene are less effective when applied during the day, but

Table 12.3. Fertilizer protocol for pineapple planted at 58,710 plants ha⁻¹ applied using a spray valve of 2500 L ha⁻ for different stages of plant development (after Evans et al., 1988).

	Stage of development (months)					
	0	0–3	4–8	9–10	11–12	
Stage	Prior to planting	←	Growing	→	Forcing	
Method	Into soil	Foliar	Foliar	Foliar	Foliar	
No. of applications	1	3	5	2	2	
Desired crop colour		Pale yellow-green	Darker yellow-green	Dark green	All green	
Fertilizer	Rate (kg ha⁻¹ application⁻¹)					Total (kg ha⁻¹ year⁻¹)
Urea	22	22	33	45	55	450
Potassium nitrate	22	22	33	45	55	450
Iron sulphate	1.5	1.5	2.25	3.0	370	30
Zinc sulphate	0.5	0.5	0.75	1.0	1.25	10
Magnesium sulphate	2.75	2.75	3.75	5.5	6.75	55

highly effective when applied at night or during the early morning hours when the stomates are open (Bartholomew and Kadzimin, 1977). In warm seasons, forcing success may be greater on days when temperatures are less than 30°C. A high N level in the plant at the time of forcing may further reduce forcing success in warm weather, while withholding N fertilizer for 4–6 weeks before forcing can improve induction. Plants approaching the natural period of flowering show greater susceptibility to induction (Fig. 12.8), while smaller plants are not as susceptible to natural flowering and are not as easily forced to flower. The minimum size for forcing is much larger under optimum growing conditions than when growth is restricted by nutrient, water or low-temperature stress. Plant weights are used to estimate suitability for forcing, with a total plant fresh mass in the range of 2–4 kg being used, depending on plant-growth status and weather. Ratoon suckers are more easily induced to flower than the plant crop. The 'Spanish' group is also easier to force in conditions impossible for 'Cayenne'.

Pest Management

Except for small plantings in subsistence farming systems, pineapple is largely a monocultured crop involving relatively large fields under more or less

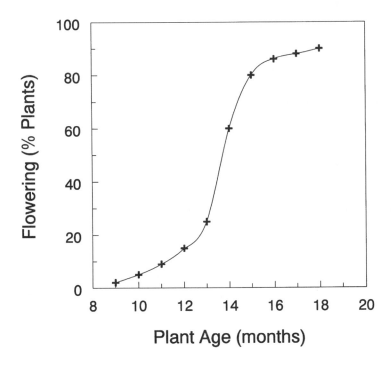

Fig. 12.8. Effect of plant age on susceptibility to natural flowering (Lacoeuilhe, 1975).

standardized cultural practices. Monoculture is conducive to accumulation of diseases and pests, some of which are confined to certain regions, while most have become established worldwide (Lim, 1985; Rohrbach and Schmitt, 1994), as vegetative germ-plasm has been transported around the world since the 1600s. The 'Smooth Cayenne' cultivar is relatively resistant to most pineapple diseases (Rohrbach and Schmitt, 1994). Occasionally, damage from specific pests grows to critical levels.

Diseases and nematodes

The most important disease is mealy-bug wilt (Table 12.4). This wilt is related to the mealy bugs *Dysmicoccus brevipes* and *Dysmicoccus neobrevipes*. The wilt has been thought to be either viral or a phytotoxin transmitted by the mealy bug. Lack of clear-cut evidence to support either theory attests to the complexity of the disease. Closely associated with mealy bugs are the attendant ants, which protect the mealy-bug colonies and transport them from plant to plant. The dominant species of ant on pineapple in Hawaii and South Africa is *Pheidole megacephala,* or the big-headed ant. Ant control is essential for the control of mealy bugs and the wilt. Mealy bugs commonly occur on the roots, on the stem above the roots, on leaf sheaths, on leaves and in fruitlet floral cavities. Wilt-affected plants are stunted, becoming yellowish first and reddish later. These symptoms resemble the effects of severe drought (Rohrbach *et al.*, 1988).

Some clones of 'Smooth Cayenne' possessed partial resistance to the wilt. However, this was not pursued, as it only offered partial resistance and the breeding programme in Hawaii had other higher-priority objectives. The only practical control of mealy bugs and the associated ants is the use of insecticides to control the ants. Mealy-bug concentration begins along the edges of fields closest to uncultivated grassy or weedy borders and gradually moves into the field. This led to a practice of planting several beds of pineapple in a strip parallel to the edge of the field and separated from the main field by a road as a buffer. These buffers were sprayed regularly. Ant control using baits is more common in Hawaii.

Heart and root rot, caused by *Phytophthora cinammomi* Rands and *P. parasitica* Dastur., do not occur uniformly and levels of infection can vary from year to year. Conditions conducive to *Phytophthora* rots are poorly drained soils, high rainfall and a soil pH of about 5.5. Crown plantings, especially during the wet season, are highly susceptible to *Phytophthora* rots, with the loss of 30–100% not being uncommon. Numerous fungicides and various methods of application have been tested in attempts to control these rots. Preplant fungicide dips of planting materials have shown good results. Foliar postplant spray has also given good results when applied immediately after planting. More recent fungicides, such as Aliette® (fosetyl-Al) and Ridomil® (metalaxyl), applied as preplant dips and at planting, followed by a second application 1–2 months later, give good control.

Table 12.4. Some important diseases and nematodes of pineapple.

Common name	Organism(s)	Parts affected and symptoms	Region or country of importance
Mealy-bug wilt	Viral (?)	Plant	Universal
Pink disease	*Gluconobacter oxydens, Acetobacter aceti, Erwinia herbicola*	Dark brown discoloration of infected fruit flesh upon heating	Hawaii, Australia, the Philippines
Interfruitlet corking	*Penicillium funculosum*	Fruitlets	Hawaii
Leathery pocket	*P. funculosum*	Fruitlets	Hawaii, South Africa
Fruitlet core rot	*P. funculosum, Fusarium moniliforme*	Fruitlets	
Pineapple butt rot	*Ceratocystis paradoxa*	Stems of planting material and fresh-fruit rot	Universal
Root/heart rot	*Phytophthora cinammomi, Phytophthora parasitica*	Root, plant, fruit	Universal
Root-knot nematode	*Meloidogyne javanica*	Galls, root destruction, stunts plant	Universal
Reniform	*Rotylenchulus reniformis*	No galls; destroys lateral root system, retards growth	Universal
Root lesion	*Pratylenchus brachyurus*	Lesion on root and destruction	Universal
Root lesion	*Rotylenchulus unisexus*	Root destruction	South Africa
Spiral	*Helicotylenchus dihystera*	Root destruction	Ivory Coast, South Africa, Puerto Rico, Hawaii

Black rot, also called *Thielaviopsis* fruit rot, water blister, soft rot or water rot, is a universal fresh-fruit problem, characterized by a soft, watery rot. Black rot is caused by the fungus *Chalara paradoxa* (De Seynes) Sacc. (syn. *Thielaviopsis paradoxa* (De Seyn.) Hohn (teleomorph *Ceratocystis paradoxa* (Dade) C. Moreau). The severity of the problem is dependent on the degree of bruising or wounding during harvesting and packing, the level of inoculum on the fruit and the storage temperature during transportation and marketing (Rohrbach and Schmitt, 1994). Infection occurs within 8–12 h following wounding. Black rot is commercially controlled in fresh fruit by minimizing bruising of fruit during harvest and handling, by refrigeration and with fungicides. Fruit must be dipped in a fungicide within 6–12 h following harvest prior to packing and shipping (Rohrbach and Schmitt, 1994). Susceptibility varies with the cultivar, 'Red Spanish' types being more resistant than 'Smooth Cayenne'.

Other fruit diseases – fruitlet core rot, interfruitlet corking, pink disease and marbling disease (Table 12.4) – are universally distributed and occasionally can be important (Rohrbach and Apt, 1986). Brown rot or fruitlet core rot occurred in only 7% of the inspected fruit. Other postharvest pineapple diseases that begin prior to harvest may cause sporadic economic problems (Rohrbach, 1989).

Nematodes (Table 12.4) are a serious problem wherever pineapple is grown on the same land over many years. Yield losses of one-third are not uncommon in the plant crop, with total failure in the first ratoon crop without nematode control. The feeding by nematodes on the roots is compounded by the roots' non-regenerative nature. Fumigants do not completely eradicate nematodes. Eggs and dormant larvae are harder to kill, repopulating the soil with time and necessitating periodic postplant application. For maximum efficiency of fumigants, soil must be well prepared, with deep ploughing and the breaking of all clods. Too much soil moisture inhibits good dispersion of the fumigant in the soil, while fumigant escape is rapid when the soil is too dry (Rohrbach and Apt, 1986). Polyethylene mulch over beds helps to retain the gas in the soil. Application of nematocides by drip irrigation has shown promise, particularly in the ratoon crop. Nematocides such as Telone (1,3-dichloropropene), oxamyl (Vydate®), Fenamiphos (Nemacur®) and Mocap® are registered for use in pineapples.

Pests

Insects can cause direct damage of various plant parts or be vector agents of diseases (Table 12.5). Prior to 1953, pineapple fruit were fumigated with methyl bromide before importation into continental USA. Pineapple fruit that are more than 50% 'Smooth Cayenne' are not now regarded as a host for tephritid flies – Mediterranean fruit fly, *Ceratitis capitata* (Wiedermann), the melon fly, *Dacus cucurbitae* (Coquillet), and the Oriental fruit fly, *Dacus dorsalis* Hendel; hence, insect disinfestation is no longer required (Armstrong and Vargas, 1982).

Table 12.5. Some important insect pests of pineapple.

Common name	Organism	Parts affects and symptoms	Region or country
Big-headed ant	*Pheidole megacephala*	Associated with mealy bugs and wilt	Hawaii
Ants	*Solenopsis* sp.	Associated with mealy bugs and wilt	Guyana
	Anaucomyrmex sp.		
Pineapple scale	*Diaspis bromeliae*	Leaves, fruit	Universal
Souring beetle (pineapple beetle)	*Carpophilus humeralis*	Plant, ripe fruit	Universal
Mealy bug	*Dysmicoccus brevipes*	Roots, stems, leaves	Universal
Black maize beetle	*Heteronychus arator*	Lower stems and roots	South Africa
White grub	Several beetle species	Root system	South Africa
Thrips	*Thrips tabaci*	Yellow-spot disease (virus transmitted by thrips from *Emilia sonchifolia*)	Hawaii, other areas
Batrachedra butterfly	*Batrachedra methesoni*	Fruit at flowering stage, causes gummosis, larvae cause damage	Caribbean
Thecla butterfly	*Thecla basiliodes*	Fruit at flowering stage, causes gummosis, larvae cause damage	Mexico, Central America, South America, Caribbean
Symphylids	*Hanseniella unguiculata*	Root tips, severe stunting	Universal

The pineapple scale, *Diaspis bromeliae* (Kerner), occurs wherever pineapple is grown. Normally, in Hawaii, pineapple scale is not a major problem in fields, probably because of scale parasites and predators. However, because of the US quarantine requirement, fruit have to be insect-free, and even low levels of pineapple scale at harvest present quarantine problems. Scale can be controlled relatively easily by preharvest insecticide applications, taking into consideration label requirements relating to last application prior to harvest time.

The pineapple fruit mite, *Steneotarsonemus ananas* Tryon, occurs universally on the growing plant, developing inflorescence, fruit and crown. The fruit mites feed on developing trichomes on the white basal leaf tissue and flower bracts and sepals, causing light brown necrotic areas. The pineapple red mite, *Dolichotetranychus floridanus* Banks, feeds on the white basal leaf tissue, particularly of the crown. Severe damage occurs when the fruit mature under drought conditions and it may cause death of the basal crown leaves, thereby affecting fruit quality (Rohrbach and Schmitt, 1994).

Other insects are mentioned frequently in the literature (Table 12.5). In Latin America and in some Caribbean islands, the larvae of the *Thecla* butterfly (*Thecla brasiliodes*) causes fruit damage, occasionally in serious proportions. Dusting or spraying the fruit at flowering stage with appropriate insecticides has given satisfactory control. White grubs and black maize beetle primarily feed on the root system and stunt the plant. In the Caribbean region, the larvae of the *Batrachedra* butterfly (*Batrachedra* sp.) damages fruit at the flowering stage, causing gummosis. Field observations in Jamaica during 1972 showed a high incidence of damage in the 'Red Spanish' but not in the 'Smooth Cayenne'.

Weed Control

Neglect of weeding at monthly intervals can result in a 20–40% decline in yields. Average fruit weight in Guinea can be 0.60 kg from an unweeded field, as compared with 1.55 kg with good weed control. Weed control is a major cost item in pineapple production and is essential to ensure high yields, as well as removal of sites that harbour nematodes and insects, such as mealy bugs. The weed problem can be lessened by complete mulching of the beds and interbeds and hence delaying or minimizing weed growth. Polyethylene-sheet mulching of beds substantially alleviates weeding within beds.

Weed control in commercial plantations is achieved by chemical means, unless labour is abundant and inexpensive. Chemical weed control is efficient and rapid if done correctly at appropriate times, using mechanical applicators. Pre-emergence application, either before or immediately after planting, is the least costly and most effective and desirable management practice. Diuron is a commonly used effective pre- and postemergence herbicide for broad-leaved weeds. For many herbicides, there are limits as to the quantity that can be used per hectare, per crop cycle or per year.

HARVESTING AND POSTHARVEST HANDLING

Harvesting

Prior to the use of chemicals to induce flowering, many passes were required to complete harvesting of a large field, due to the wide variation in time of flowering during the winter season when most flower induction occurred. Chemical induction concentrated the period of flowering, thus condensing harvesting to two to four passes. Natural or precocious flowering during cool weather with short days can significantly disrupt the harvesting and marketing schedules. In Hawaii, fruit developing from precocious flowering in some years leads to a second peak, followed by a midyear dip in production (Fig. 12.9).

Controlled ripening of fruit with ethephon prior to harvest can further reduce the number of harvesting passes. Ethephon is applied 48 h or more before harvest to accelerate shell degreening (Soler, 1992). This accelerated shell degreening is due to destruction of chlorophyll, giving the shell a more uniform colour. The application should occur when natural colouring has started to ensure good fruit quality (Paull, 1997) and is sometimes less effective in hot weather.

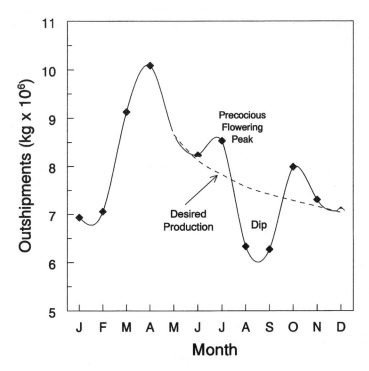

Fig. 12.9. The effect of precocious flowering in the January–February period on outshipment of pineapples from Hawaii, contrasted with desired production.

Fruit destined for the cannery are usually harvested at the one-half to three-quarters yellow stage. Fruit to be transported to distant fresh-fruit markets are picked anywhere from mature green (no yellow colour) to one-quarter coloured stage. Fruit maturity is evaluated on the extent of fruit 'eye' flatness, skin yellowing and acidity to sugar measurements (Plate 30). Consumers similarly judge fruit quality by skin colour and aroma. A minimum reading of 12% TSS (sugar) is required for fresh fruit in Hawaii, while others have suggested 14%. A sugar-to-acid ratio of 0.9–1.3 is recommended (Soler, 1992).

Pineapple for the fresh market is hand-harvested, with pickers being directed as to stage or stages (shell colour) of ripeness required. Fruit is either packed in the field or at a central packing shed. In Hawaii, pickers walk in the interrows and place the fruit on a conveyor belt running on a boom, which transfers the fruit to a truck field bin (Plate 31). Fruit is placed in the bin by hand, upside down on the crown, to avoid injury. Fruit are also harvested by pickers carrying large baskets on their backs. When the baskets are full, the fruit are dumped at the ends of the rows or at the side of field roads, later to be loaded into trucks or trailers. When the fruit arrives at the packing shed, it is unloaded by hand, by submerging the field bin in water or by sliding the fruit out of the field bin into water. Fruit with high translucency ('sinkers') are separated at this step. Full-ripe translucent fruit are unsuitable for transporting to distant markets and less mature fruit are selected. Immature fruit are not shipped, since they do not develop good flavour, have low sugar and are more prone to chilling injury (CI). Care is taken to avoid mechanical damage to the crown leaves and mechanical injury to the fruit shell (Paull, 1997).

Postharvest Handling

Fruit are waxed after washing, frequently with polyethylene and paraffin or carnauba and paraffin waxes. The selected waxes reduce internal browning symptoms of CI, reduce water loss, improve fruit appearance and ensure a more even application of fungicide used to control rots (Paull and Rohrbach, 1985). There is no worldwide uniformity in acceptance of wax components, so importing country restrictions need to be considered. If the wax injures the crown leaves, only the fruit body is waxed. Postharvest fungicide is normally included in the wax.

Fresh-fruit quality standards are based upon recognized appearance characteristics (Table 12.6). The fruit needs to be mature, firm, well formed and free of defects and to have flat eyes and a minimum TSS of 12% in Hawaii. Crown size is a crucial grade component, with a minimum size and a set ratio of crown to fruit length (0.33–1.5) for the higher grades. Crowns that develop during the summer in Hawaii tend to be larger and may require 'gouging' at harvest to meet the standard. This 'gouging' leaves a wound for possible disease entry and detracts from overall appearance. 'Gouging' 2 months before harvest limits crown growth and avoids visible scarring. Fruit are graded by degree of skin coloration, size (weight), absence of defect and disease and other market needs before packing (Soler, 1992).

Temperatures in the range 7.5–12°C are recommended for storage, with relative humidities of 70–95%; the higher humidities significantly reduce water loss. At a temperature of 0–4°C, the fruit may be stored for weeks, but, upon removal, the fruit fails to continue ripening and shows severe CI (Paull, 1993). Half-ripe 'Smooth Cayenne' fruit can be held for about 10 days at 7.5–12.5°C and still have about a week of shelf-life, with no chilling-induced internal browning.

Physiological Disorders

Chilling injury

The maximum storage life of pineapple at 7°C is about 4 weeks; however, when removed CI develops within 2–3 days. The symptoms of CI include: (i) wilting, drying and discoloration of crown leaves; (ii) failure of green-shelled fruit to yellow; (iii) browning and dulling of yellow fruit; and (iv) internal flesh browning (Paull and Rohrbach, 1985). Preharvest shading and pre- and post-harvest low temperature are the major factors increasing CI symptom intensity. Chilling-injury symptoms have been called endogenous brown spot, physiological breakdown, blackheart and internal browning. Postharvest CI symptoms develop after fruit are returned to physiological temperatures (15–30°C). Susceptible fruit are generally lower in ascorbic acid and sugars and are opaque (Teisson, 1979a, b). Partial to complete control of CI-symptom development has been achieved by waxing, polyethylene bagging, heat treatments, controlled atmospheres and ascorbic-acid application.

Flesh translucency

Flesh translucency increases fruit sensitivity to mechanical injury. This condition begins before harvest and continues after harvest. Total soluble solids, flesh pigments and palatability increase to a maximum at about 60% translucency and then decrease in fruit with greater translucency. Translucency is more severe and has a higher incidence when maximum and minimum temperatures 3 months before harvest are both low, less than 23° and 15°C, or, to a lesser extent, high, greater than 29 and 20°C, respectively. Fruit with larger crowns have a lower incidence and severity of translucency (Paull and Reyes, 1996).

Bruising

Fruit bruising is a major problem during harvesting and packing. Bruising can be caused by impact damage, a 30 cm drop causing significant damage. This

Table 12.6. United States fresh-pineapple fruit standards. All grades have 10% limits on defects, 5% on serious damage and 1% on decay.

Standards	Varietal characteristics	Maturity	Free from	Slips knobs	Crown-to-fruit ratio length
US Fancy	Similar	Mature, firm, dry, well formed, well-developed eyes	Decay, sunburn, injury, bruising, well-cured butt	Single crown, no slips	Ratio less than 1.5, not less than 12.7 cm in size
US No. 1	Similar	Mature, firm, dry, well formed, well-developed eyes	Decay, sunburn, injury, bruising, fairly well-cured butt	< 5 slips	Ratio less than 2.0 and greater than 10.2 cm in size
US No. 2	Similar	Mature, firm, dry, fairly well formed	Serious decay, sunburn, injury, bruising	Slips	Any ratio Double crown

injury is normally confined to the impact side of the fruit. The damaged flesh appears slightly straw-coloured. Mechanical injury of translucent fruit can lead to leakage of fruit-cell contents and loss of marketable fruit.

Sunburn

Sunburn is common during hotter periods (> 35°C) of the year, when the fruit is not shaded by leaves and especially in ratoon crops (Plate 29). The condition is more prevalent in the outer rows and when fruit is lodged. Sun-scorched fruit first show a bleached yellow-white skin, which turns pale grey and brown, with damage to the flesh underneath. These damaged areas are more susceptible to disease organisms, particularly yeasts and bacteria.

Malformations

Knobs on the base of fruit occur in off types. Culling of the crowns of these fruit as planting material reduces the subsequent field incidence. These fruit are not marketed, since trimming generally breaks the fruit skin and allows rots to develop. The other genetic off type is multiple crowns (fasciation), two or more on each fruit (Fig. 12.5), with the fruit taking on a flattened appearance. This condition is often related to high-temperature injury after forcing. These fruit should not be marketed and crowns should not be used for planting. Fruit with pronounced 'eyes' or fruitlets normally do not meet most grade standards and the thicker skin means lower flesh recovery. This condition is common in fruit that flower during cool weather. Some 'Spanish' varieties are susceptible to broken core, in which the central fruit core has a transverse break, leading to the upper part of the fruit ripening ahead of the bottom.

UTILIZATION

Pineapples do not sweeten after harvest, although the acid level may decline. Canning is more successful with an acid fruit, while a better balance of sugar to acid is required for the fresh market. The sugar-to-acid ratio varies widely with cultivar, growing condition and stage of harvest, from 80 to 200 g kg^{-1} fresh mass (FM) of flesh. Ascorbic acid also varies widely with cultivar, from 2.5 to 180 g kg^{-1} FM of flesh. Pineapple is a good source of ascorbic acid (vitamin C), some vitamin A, Ca, P, Fe, K and thiamine (Table 12.7). It is low in sodium.

The bulk of the fruit produced goes into the fresh-fruit market. There is an increasing interest in minimally processed pineapples, with the shell and core removed just before purchase. The major processed products are the canned slices or solid pack, with a recovery percentage varying from 20% in 'Singapore Spanish' to *c.* 60% for 'Smooth Cayenne'. Cans of several sizes are

Table 12.7. Proximate analysis of 'Smooth Cayenne' fruit (Wenkam, 1990).

		Amount per 100 g edible portion
Proximate		
Water	g	86
Energy	kJ	218
Protein	g	0.5
Lipid	g	0.2
Carbohydrate	g	13.5
Fibre	g	0.5
Ash	g	0.3
Minerals		
Calcium	mg	18
Iron	mg	0.3
Magnesium	mg	12
Phosphorus	mg	12
Potassium	mg	98
Sodium	mg	1
Vitamins		
Ascorbic acid	mg	10
Thiamine	mg	0.09
Riboflavin	mg	0.04
Niacin	mg	0.24
Vitamin A	IU	53

filled with the best slices. Uncanned and broken slices are packed as chunks. Flesh remaining on the shell, cut ends, core and trimmings are processed into crushes and juice. Much of the juice is now concentrated and frozen to maintain a higher flavour quality than the single strength. Another by-product of the cannery fruit residue is pineapple bran, used as cattle feed. This residue from pressed fruit shells and pulp has been sold to dairy farms, either wet or dried. Bromelain, the proteolytic enzyme found in pineapple plants, particularly in the stem, has a number of industrial and medicinal applications.

FURTHER READING

Bartholomew, D.P. and Malezieux, E.P. (1994) Pineapple. In: Schaffer, B. and Andersen, P.C. (eds) *Handbook of Environmental Physiology of Fruit Crops*, Vol. II, *Subtropical and Tropical Crops*. CRC Press, Boca Raton, Florida, pp. 243–291.

Leal, F. and Coppens d'Eeckenbrugge, G. (1996) Pineapple. In: Janick, J. and Moore, J.N. (eds) *Fruit Breeding*, Vol. I, *Tree and Tropical Fruits*. John Wiley & Sons, New York, pp. 515–557.

Py, C., Lacoeuilhe, J.J. and Teisson, C. (1987) *The Pineapple, Cultivation, and Uses*. G.P. Maisoneuve et Larose, Paris.

Rohrbach, K.G. and Schmitt, D.P. (1994) *Compendium of Tropical Fruit Diseases – Part IV. Pineapple*. American Phytopathological Society, St Paul, Minnesota.

Soler, A. (1992) Pineapple – quality criteria. CIRAD-COLEACP, Montpellier, France, 48 pp.

OTHER ASIAN TROPICAL FRUIT

BREADFRUIT, JACKFRUIT AND CHEMPEDAK

BOTANY

Introduction to Family, Important Genera and Species

The family *Moraceae* includes the fig and mulberry. The genus *Artocarpus* contains about 50 species; most are native to Asia and 15 produce edible starchy fruit, which are frequent staples. The three most important species are the more tropical breadfruit, *Artocarpus altilis* (Parkins) Fosb. (syn. *Artocarpus communis,* Foster; *Artocarpus incisus* L.; *Communis incisa*), the jackfruit, *Artocarpus heterophyllus* Lam. (syn. *Artocarpus integer* (Thumb.) Merrill; *Artocarpus integrifolius*) and its close relative, chempedak, *Artocarpus integrifolia* L. (syn. *Artocarpus polyphema* Persoon; *Artocarpus chempeden* (Lour.) Stokes) (Rajendran, 1992).

Areas of Origin and Distribution

The origin of breadfruit is unknown, with diverse material spread from Indonesia to Papua New Guinea and into the islands of the Pacific (Fosberg, 1960). Since the time of Columbus, it has been spread throughout the humid tropics. In many regions, the seeded and seedless varieties have different common names. Seeded breadfruits are called breadnut (English), kelur or kelor (Indo-Malaya), and kamansi (Philippine), while seedless are sukun and rimas, respectively. Other names are arbre à pain (French), sake (Thai and Vietnamese), panapen (Spanish) and fruta pao (Portuguese).

Jackfruit, jacquier (French), nangka (Javan and Malay), langka (Philippine), khnaor (Cambodia), makmi, khanum, banum (Thailand), or mit (Vietnamese) is indigenous to south-western India and has been introduced throughout South-east Asia and to most of the tropics. Chempedak (Malaysian), sonekadat (Burmese) or champada (Thai) is distributed in

Burma, peninsular Thailand and Malaysia, the Indonesian islands and western New Guinea.

BREADFRUIT

ECOLOGICAL REQUIREMENT

Soil

Breadfruit grows well in a variety of soils with sufficient depth and drainage. High organic matter and fertility are recommended. On Pacific islands, breadfruit does grow on shallow coralline soils, showing considerable varietal adaptability.

Climate

Regular rainfall of 200–300 cm year^{-1}, and high humidity of 70–90% are preferred. Rainfall or irrigation is necessary for vegetative growth, flowering and fruit growth. The tree is chilling-sensitive, with no growth at or below 5°C. The tree is well suited to hot, humid tropical lowlands and temperatures up to 40°C. Full sun is required. No photoperiodic events have been noted, but breadfruit trees are seasonal in their production. Although day length cannot be ruled out, a period of drought followed by rainfall is more likely to be the trigger for vegetative growth and subsequent flowering.

GENERAL CHARACTERISTICS

Tree

The breadfruit tree grows up to 30 m, is evergreen when grown in humid, wet areas and is deciduous under drought or during the dry part of a monsoon climate. The 30–60 cm alternate ovate leaves are dark green, with three to nine lobes towards the tip (Fig. 13.1). The trunk is straight, with thick branches terminating in 10–20 cm-long branches covered with conical, keeled stipules. This fast-growing tree lives for 30–50 years and suckers begin bearing in 5–6 years and seedling plants in 8–10 years. Canopy volume is a good measure of yield.

Flowers

This monoecious species has the male and female inflorescence on auxiliary 4–8 cm peduncles in separate leaf axils (Fig. 13.1). The drooping, spongy,

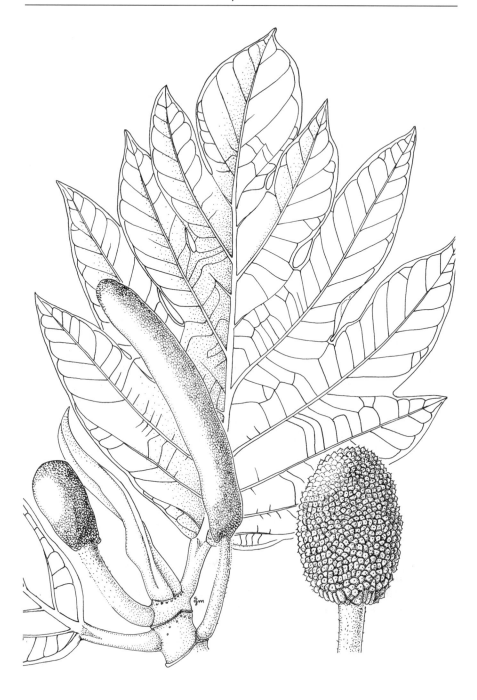

Fig. 13.1. Breadfruit leaves, male and female inflorescence. Jackfruit and chempedak inflorescences are similar in shape, though found in different locations on the tree.

club-shaped male (15–20 cm long, 3–4 cm wide) inflorescence has minute flowers, with a single stamen. The globose female inflorescence (6–10 cm) is covered with numerous flowers on a receptacle. In seedless varieties, each female flower has a tubular calyx, a two-celled ovary and a two-lobed stigma on a short style below the perianth tip.

Pollination and Fruit Set

Rain encourages vegetative growth and flowering. Some cultivars are capable of flowering throughout the year under the right environmental conditions. Cross-pollination is ensured by the male inflorescences maturing before the female. Following wind or insect pollination, fertilization occurs over 3–6 days in seeded varieties. A high percentage (75%) of the florets are set, with the percentage being reduced in rainy weather. This reduction suggests that pollination is necessary to stimulate parthenocarpic growth (Hasan and Razak, 1992). However, pollination is difficult, as the rudimentary perianth acts as a physical barrier to pollination and argues against the fruit being parthenocarpic.

Fruit

The fruit develops from the whole inflorescence and is normally round, sometimes cylindrical and 10–30 cm in diameter (Fig. 13.2). The yellow-green, thin, reticulated skin turns yellow-brown when ripe. The receptacle is surrounded by a pale yellow-white edible pulp. Most varieties are seedless; seeded wild varieties have from 10 to 150, 2.5-cm-long brown seeds (Bennett and Nozzolillo, 1987). The fruit and all other parts of the tree have a white latex. The fruit matures 60–110 days after inflorescence emergence. The difference in time is due to the variety and fruit growth temperature.

CULTIVAR DEVELOPMENT

Numerous varieties are described (2n = 27, 28, 56), although few have been compared at the same location. Triploidy is common in seedless varieties and some seedless varieties may be a hybrid of *A. altilis* and *Artocarpus mariannensis*. Seedy varieties are more common in the western South Pacific. No breeding or selection work has been reported. Variability has been observed in growth form, leaf shape, fruit quality, time to bearing, seasonality, keeping quality of fruit and salt tolerance (Ragone, 1988).

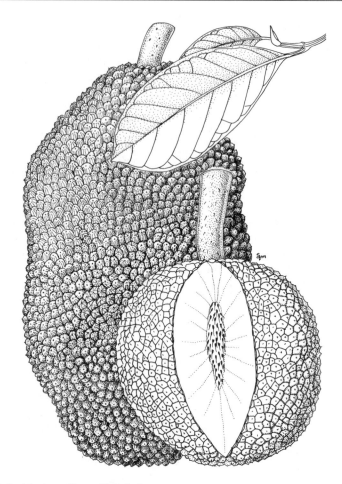

Fig. 13.2. Jackfruit and breadfruit fruit.

CULTURAL PRACTICES

Propagation and Nursery Management

Seeds have 90–95% viability as soon as removed from the fruit. Viability is lost in 2 weeks. Root suckers and cuttings (*c*. 2.5 cm diameter, 20 cm long) are the traditional propagation method, although plant losses can be high with root suckers (Rowe-Dutton, 1976). Rooting of the cuttings may take 2–5 months, if kept well watered (Hamilton *et al.*, 1982). Some success has been achieved with air layering, budding and grafting.

Field Preparation

This is the same as for most other fruit trees.

Transplanting and Spacing

Plants are set out at 7–15 m, depending on variety and growing conditions. Shade is provided for the first year as the young tree's growth resumes.

Irrigation Practices

Continued vegetative growth requires irrigation, especially during periods of drought.

Pruning

The tree may be pruned to improve shape. Regular pruning is not normally carried out. Some growers suggest that pruning of branches that have borne fruit stimulates new shoots and limits tree height.

Fertilization

Requirements have not been determined. Mulching and the application of organic manure two or three times a year, sometimes mixed with fertilizer, are used to increase and maintain growth rate.

Pest Management

'Pingelap', causing dieback from the top branches and tree death, is caused by an unknown organism and there is no method of control. This is a major problem in Micronesia. Other diseases are dieback (*Fusarium*, *Pythium* and *Rosellinia*), leaf spot (*Cercospora*), leaf rust (*Uredo artocarpi*), root rot (*Phillinus noxius*), fruit rot (*Phytophthora*, *Phyllosticta* and *Rhizopus*) and stem-end rot (*Phomopsis*, *Dothiorella*) (Trujillo, 1971). Mealy bugs, scales and twig borers are the major pests, with no control usually practised. The fruit is a fruit-fly host.

Weed Management

Mulching around the base of the trunk is practised to control weeds, conserve moisture and provide nutrients.

Orchard Protection

A windbreak is not usually necessary. Breadfruit trees are occasionally used as windbreaks and shade for other trees.

HARVESTING AND POSTHARVEST HANDLING

A tree normally produces 10–600 fruit year^{-1} ranging in weight from 0.5 to 3 kg. Fruit at different growth stages are harvested for different uses. Mature green fruit are harvested as a starch vegetable, while some people prefer to eat the ripe sweet fruit. Harvested green fruit produce copious latex, especially from the cut peduncle and injuries on the fruit. Maturity is indicated by larger size, a slight change to yellowish and small drops of latex on the rind.

The latex from the cut peduncle is allowed to drain in air or the fruit is submerged in water. Collected fruit is packed in baskets or cartons. Fruit wrapped in polyethylene and stored at 12°C last about 20 days. Lower temperature causes chilling injury. The fruit ripens in 3–7 days. During ripening, the flesh softens, starch is converted to sugars and flesh colour changes from creamy white to fawn. Fruit is also stored cooked. In the South Pacific, fruit are preserved by pit fermentation, requiring 2–3 months to produce a palatable, strong-smelling, cheesy product (Atchley and Cox, 1985).

UTILIZATION

The sweet ripe fruit is eaten as a dessert. Fruit can be boiled, dried, used in bread-making or fermented, while slices can be fried or stored in brine (Whitney, 1988; Bates *et al.*, 1991). The edible flesh (70%), is 60–85% water, 1.2–2.4% protein, 22–37% carbohydrate and 0.2–0.5% fat (Table 13.1). The carbohydrate in mature fruit is mainly starch (Graham and Negron, 1981). The cooked seeds are also eaten. The leaves and fallen fruit are fed to animals. The collected latex is used as a caulk, glue and chewing-gum (Wooten and Tumoalii, 1984).

JACKFRUIT AND CHEMPEDAK

The chempedak can be distinguished from jackfruit by having smaller fruit with thinner rind and more juicy flesh, which is a darker yellow when ripe (Jarrett, 1959).

ECOLOGICAL REQUIREMENT

A variety of well-drained soils can be used, with pH 6–7.5. Sandy and clay loams are preferred. Cold, drought and flood tolerance limits the distribution of jackfruit

Table 13.1. Proximate analysis of ripe breadfruit, jackfruit, chempedak, durian, lanson, mangosteen and wax apple per 100 g edible portion (Leung and Flores, 1961; Siong *et al.*, 1988; Wenkam, 1990; Dignan *et al.*, 1994).

		Breadfruit	Jackfruit	Chempedak	Durian	Lanson	Mangosteen	Wax apple
Edible portion	%	70	28	22	22	41	24	95
Proximate								
Water	g	62	83	67	64	86	88	91
Energy	kcal	561	301	490	640	230	34	80
Protein	g	1.3	1.6	2.5	2.7	0.9	0.6	0.6
Fat	g	0.18	0.2	0.4	3.4	0.3	1.0	0.1
Carbohydrate	g	37.0	25.4	25.8	27.9	12.1	5.6	8.0
Fibre	g	1.45	5.6	3.4	0.9	0.4	5.1	0.7
Ash	g	1.2	2.2	1.2	1.0	0.5	0.1	0.4
Minerals								
Calcium	mg	21	37	40	40	12	7	6
Phosphorus	mg	48	26	5	44	30	13	16
Potassium	mg	551	292	246	70	142	45	–
Iron	mg	0.26	1.7	1.1	1.9	0.3	4	0.4
Sodium	mg	13	48	25	40	2	7	–
Vitamins								
Vitamin A	IU	41	66	48	90	0	0	Trace
Thiamine	mg	0.12	–	–	0.35	–	0.06	0.03
Niacin	mg	1.54	0.4	0.5	0.7	1.4	0.3	0.3
Riboflavin	mg	0.06	0.06	0.15	0.2	0.1	–	0.03
Vitamin C	mg	20.5	7.9	17.7	23	1.7	4.2	13

to areas having more than 150 cm rainfall, without a prominent dry season. Warm and humid frost-free climates below 1000 m and 25°N and S are desirable for good bearing. Temperature below 5°C severely damages trees. The trees have some salt tolerance and poor drought and flood tolerance. No photoperiod response has been reported. Chempedak is restricted to South-east Asia, with some trees in Australia and Hawaii, while jackfruit is spread throughout the tropics.

GENERAL CHARACTERISTICS

Tree

These monoecious evergreen latex-producing trees are up to 20 m in height, with dark green entire leaves. The chempedak has long, wiry, brown hairs on the leaves (< 3 mm long), stipules and twigs. All parts have a milky white latex (Soepadmo, 1992).

Flowers

The jackfruit flowers are borne on the trunk and older branches on short shoots. Chempedak male inflorescences are cylindrical, 3–5.5 cm long and 1 cm in diameter. Jackfruit male inflorescences are 5–15 cm long and 2–4.5 cm wide. Anthesis commences 2–3 weeks after emergence and lasts for about 2 weeks, the flower normally rotting before abscission and attracting numerous insects by the smell. The female flower clusters are elliptical or round. The stigmatic surface is composed of papillae, which become sticky 1–2 weeks after exertion and remain so for a further 2 weeks (Moncur, 1985). The inflorescences are similar to those of breadfruit (Fig. 13.1).

Pollination and Fruit Set

The male inflorescence matures 3–5 days before the female. The sticky yellow pollen is released with a sweet scent that attracts small insects; however, the flowers may be wind-pollinated (Sambamurty and Ramalingam, 1954). In suitable environments, jackfruit bear flowers and fruit throughout the year. A load of fruit, however, may suppress further flowering. Chempedak is more seasonal than jackfruit, with blooms being more common in February–April and August–October in peninsular Malaysia.

Fruit Morphology

The fruits are pear- or barrel-shaped syncarps, with jackfruit (Fig. 13.2) having larger fruit than chempedak. Both are borne on a 5–10-cm stalk. The thick skin (1 cm) in jackfruit has short protuberances. The jackfruit receptacle is not separable from the waxy, firm to soft, golden yellow, fleshy, edible perianth (25–40% of the total fruit) that surrounds the seed (5% of the total weight). The fruit can have up to 500 seeds, 2–4 cm long by *c.* 2 cm, each surrounded by a horny endocarp and subgelatinous exocarp. The fruit matures in 3 months (Fig. 13.3) and may take up to 6 months at higher altitudes and in cooler areas. Unfertilized flowers develop as strap-like tissue between fertilized developing fruitlets. Total starch content is high at harvest and rapidly declines during ripening, leading to a significant increase in soluble sugars, while acidity shows little change (Fig. 13.4). Fruit quality after harvest is very dependent on maturity at harvest. In peninsular Malaysia, the nearly mature fruit are often wrapped in palm leaves or bagged, ostensibly to protect against bats, rats and fruit flies and to attract ants, which keep other insects away.

CULTIVAR DEVELOPMENT

There is high variability in both jackfruit (2n = 56 (tetraploid)) and chempedak (2n = 56) characters: length of juvenile stage, seed germination,

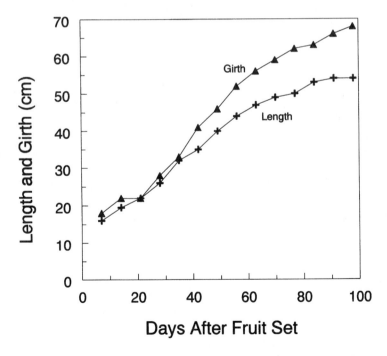

Fig. 13.3. Change in jackfruit cv. NS1 girth and length after fruit set (redrawn from Muda *et al.*, 1996).

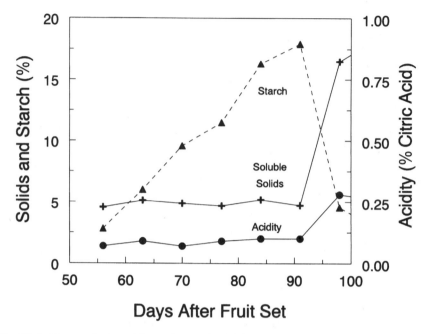

Fig. 13.4. Changes in titratable acidity, total soluble solids and starch during jackfruit growth and development (redrawn from Muda *et al.*, 1996).

fruit shape and size, flesh appearance, amount of latex, flavour, aroma and fruit maturation time. No breeding or selection programmes have been undertaken, but many local varieties exist. Jackfruit and chempedak occasionally hybridize and a clone has been selected in Malaysia called 'Nangka-chempedak CH/NA'.

Ripe jackfruits are divided into two types, based upon edible pulp. The first type has thin, fibrous, soft edible flesh, acid to very sweet with a strong aroma. The other type has thick, firm to crisp flesh with less aroma. There are many varieties of both types. Some are more suited to canning than other fresh fruit varieties. A number of chempedak clones have been selected in Malaysia. Some of the preferred clones have an attractive orange flesh and higher yields.

CULTURAL PRACTICES

Propagation and Nursery Management

Seed is the major means of propagation. The seed loses viability within 3 months of removal from the fruit and should be planted immediately (Chatterjee and Mukherjee, 1980). Germination is improved by soaking in napthalene acetic acid (NAA) or GA. Seedlings are best grown under shade.

Root cuttings can be used, with stem cuttings and air layers also being successful (Mukherjee and Chatterjee, 1979). Budding, grafting and inarching are carried out on 12-month-old rootstocks of *A. integer, A. heterophyllus,* other *Artocarpus* species, and the same species. Jackfruit can be propagated *in vitro* (Roy *et al.,* 1990).

Field Preparation

Orchards are prepared as for other tree crops.

Transplanting and Spacing

A spacing of 6–12 m is recommended for these slow-growing trees. Transplanting needs to be carried out with care to avoid damage to the tap root and is best done before the trees are 1 year old.

Irrigation Practices

Due to poor drought tolerance, irrigation is required.

Pruning

Thinning of shoots and clearing of branches to allow harvesting access are sometimes practised.

Fertilization

The Malaysian recommendation is nitrogen (N) : phosphorus (P) : potassium (K) : magnesium (Mg) (8 : 4 : 2 : 1) 30 g tree^{-1} at 6 month, doubled every 6 months to 2 years. Older trees receive 1 kg tree^{-1} of 4 : 2 : 4 : 1 every 6 months. Higher rates of 2–3 kg are recommended in the Philippines. Application occurs before and at the end of the wet season.

Pest Management

Seed and blossom rots, leaf spots, pink disease and fruit rot occur on jackfruit (Bhutani, 1978). The blossom and fruit rot are caused by *Rhizopus artocarpi* (McMillan, 1974). Bacterial dieback (*Erwinia canetorora*) can be a problem. *Corticum salmonicolor* causes pink disease of jackfruit. Jackfruit is reported to be attacked by shoot borers, bark borers, bud weevils, spittle bugs, mealy bugs, scale insects and aphids. Borers are the major pests.

Weed Management

Once the tree is established, weeds are not a problem, due to dense shade.

Orchard Protection – Windbreaks

The tree can withstand moderate wind. Depending upon wind strength and duration, a windbreak may be required.

HARVESTING AND POSTHARVEST HANDLING

Jackfruits can produce 20–250 fruit tree^{-1} year^{-1}, sometimes up to 500, each fruit weighing 10–30 kg. There is a tendency for both species to be somewhat biennial in bearing. Commercial yields average 70–100 kg tree^{-1} year^{-1}. Chempedak yields are similar to those of jackfruit. Seedling trees start bearing in 4–14 years. India is probably the largest producer (Thomas, 1980).

The fruit is sometimes allowed to fall and must be collected daily for it has only 2–3 days' shelf-life. Letting the fruit fall can cause damage, loss of shelf-life and premature ripening. Fruit maturity can be judged by a dull hollow sound when tapped, by skin colour changing from pale green to yellowish brown and by a characteristic odour and flattening of the surface spines. Fruit harvested after 12 weeks in tropical areas should be organoleptically acceptable after ripening. The peduncle is cut and the latex is allowed to drain. No grading is normally practised and fruit are rapidly transported to the markets. The fruit is frequently cut open and sliced into pieces for sale. Mature undamaged fruit can be stored at 12°C for about 3 weeks and ripen in 3–7 days, depending upon the stage of maturity.

UTILIZATION

Fruit are eaten unripe at 25–50% full size as a vegetable or ripe as a fruit. Its major use is as a vegetable, where it is peeled, sliced and boiled and then seasoned or mixed with other food. The fruit is a good source of carbohydrates and vitamin A and a fair protein source (Table 13.1). Chempedak is very similar to jackfruit in its composition. The seeds of both species, boiled or roasted, are also eaten. The jackfruit pulp is canned in syrup or used to make jams, jellies and ice cream. Young fruit are cooked in coconut milk and eaten curried. Many Westerners find the smell and taste of chempedak only slightly less objectionable than those of durian.

DURIAN

BOTANY

The tropical family *Bombacaceae* is known for its showy flowers and pods with seeds covered with cotton-like fibres. The family is found in both South America and Asia. The fruit of the genus *Durio* differ in having large seeds with fleshy arils (Malo and Martin, 1979). The genus *Durio* has 27 species, six producing edible fruit with *Durio zibethinus* Murray (syn. *Durio acuminatissina* Merr) durian being the most widely cultivated (Soegeng-Reksodihardjo, 1962). The genus specific *zibethinus* is derived from the Italian 'zibetto' for the civet, a cat-like animal with a musky smell. The ornamental *Durio kutjensis* has highly regarded fruit, with no offensive smell and soft pliable spines, but has proved less adaptable outside its native Borneo.

The genus is native to South-east Asia, with wild durian found in Borneo and Sumatra (Nanthachai, 1994). The tree is mainly cultivated in Sri Lanka, southern India, southern Burma, Thailand, Cambodia, Vietnam, Malaysia, Indonesia, Borneo, Mindanao (the Philippines) and New Guinea. It has been spread throughout the tropical world, with the general name of durian (Indo-Malay) or variants – duren (Indonesian), duyin (Burmese), thureen (Cambodian), thurian (Thai) and saurieng (Vietnamese).

ECOLOGICAL REQUIREMENT

Soil

Deep, well-drained sandy clay or clay loam is best. Poor drainage in heavy soils is conducive to *Phytophthora* root rot.

Climate

The tree requires abundant rainfall; most production occurs in areas with a mean yearly rainfall of 200–300 cm, well distributed throughout the year. Higher rainfall is recommended. Drought for more than 3 months leads to irreversible damage, but irrigation overcomes this limitation.

Growth is limited below a 22°C mean monthly temperature, although flowering and fruiting appear to be more prolific. Temperatures below 10°C causes premature leaf abscission, but temperatures up to 46°C can be tolerated. It is grown successfully up to 800 m near the equator and 18°N and

S latitude. Shade (30–50%) is necessary for young trees before they reach 0.8 m in height following field planting. The shade is reduced slowly over 12 months to full sun. Durian is not photoperiodic for flower initiation.

GENERAL CHARACTERISTICS

Tree

Durian trees 20–40 m tall are not uncommon, although grafted trees are normally 8–12 m. The tree has a straight low-branched trunk with an irregular and dense crown, tending to conical. The grafted tree is more irregular and spreading. Leaves are simple, alternate, leathery, drooping, oblong or elliptic, 8–20 cm long and 4–6 cm wide, with a light to dark green upper surface and silvery or brown rusty underneath (Fig. 13.5). The wood is brittle, grainy and dark brown to reddish, and shrinks on drying.

Flowers

Flower clusters arise on laterals, main branches and, occasionally, the tree trunk. Clusters of 1–45 long-stalked (5–8 cm), pendulous, white-petal flowers open over a period of 2–3 weeks. A durian tree can produce 20,000–40,000 flowers per tree. The fleshy outer covering of the flower, the epicalyx, splits into two to four on the day when full anthesis occurs between 3 p.m. and early evening, when flowers emit nectar and a sour-milk odour. The five distinct bundles of 'shaving-brush' clusters of stamens emerge and pollen is released between 8 p.m. and midnight. Hence, the stigma is receptive before and not during peak pollen release. The epicalyx, sepals, petals and stamens then all drop. The pistils fall 3–7 days later if not pollinated.

Temperature does not initiate flowering. Flowering occurs normally during or near the end of the dry period, which has reduced shoot growth (Fig. 13.6). This period of reduced shoot growth should follow a period of vegetative growth, when the tree accumulates photosynthate. A 7–14 day continuous dry period is necessary for cv. 'Chanee'. In monsoon climates, flowering takes place in the dry season. Near the Equator, two crops a year can occur, e.g. Malaysia, early March or April and September or October. In Thailand, flowering occurs from March, after the cool dry season. Low doses of growth retardants (paclobutrazol) stimulate flowering especially as a growth flush finishes, supporting the idea that a period of slow vegetative growth is needed for flowering (Chandraparnik *et al.*, 1992a). Flower buds emerge in 6–8 weeks, while others can remain quiescent as small clusters for up to 32 weeks before full development (Chandraparnik *et al.*, 1992b). Dormancy at this stage is observed when young flower buds at the 'dot' stage are exposed to a small amount of rainfall and continuous growth is

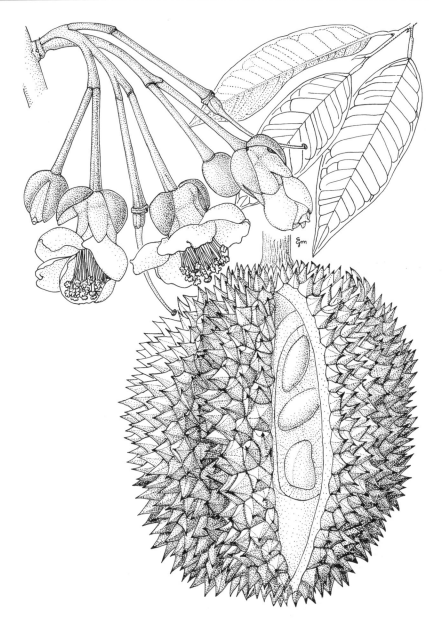

Fig. 13.5. Durian leaf, flower and fruit.

interrupted. Some flowers do develop, leading to more than one fruit set on a tree. Ethephon, diaminozide and NAA do not induce early flowering and GA$_3$ delays flowering (Punnachit *et al.*, 1992).

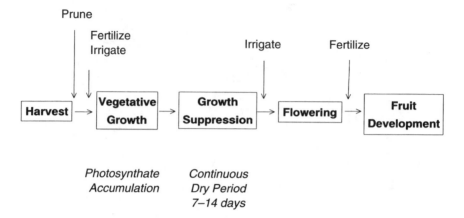

Fig. 13.6. The harvesting-to-fruiting cycle of durian. Flowering is triggered by a continuous dry period of 7–14 days on trees that have sufficient carbohydrate reserves and are competent to flower (redrawn from Salakpetch, 1996).

Pollination and Fruit Set

Pollen from undehisced anthers germinates poorly (3–5%), and germination declines rapidly, after shedding, from 80% to less than 20%. Germination requires a 20–35% sucrose solution, similar to the concentration on the stigmatic surface. Rain or dew can therefore dilute the stigmatic exudate, leading to low pollen germination, with only 10% germination in a 10% sugar solution. The pollen grain also requires 50–90 p.p.m. calcium (Ca), 30–60 p.p.m. boron (B), 15–30 p.p.m. Mg and 15–30 p.p.m. K to germinate (Salakpetch *et al.*, 1992a). Self-incompatibility is common and cross-pollination between different cultivars or trees is frequently necessary. Cross-pollination in 'Chanee' can increase fruit set from 0–6% to 30–60%. However, single-variety plantings in Thailand show no apparent reduced fruit set. Poor pollination gives poor seed and aril development and irregular-shaped fruit.

Pollination is by nectarivorous bats and moths at night, when bees are not active. Due to the size of the sticky pollen, which tends to clump, wind pollination is probably not important. Application of GA$_4$ paste to durian pedicels at 6 weeks after anthesis reduces fruit drop and increases fruit size and mass by up to 30% (Mamat and Wahab, 1992).

Fruit

The fruit is an ovoid or ellipsoid capsule (Fig. 13.5) derived from a single ovary, which dehisces when ripe into three to five segments. The green to brownish fruit, weighing up to 8 kg, can be up to 30 cm long and 20 cm in diameter and is densely covered with stout, sharp, pyramidal spines (1 cm) on a thick fibrous rind. The fruit is divided into three to five smooth-walled compartments, each containing one to six glossy, creamy to red-brown seeds, 2–6 cm long, covered by a white to yellowish soft sweet aril (Fig. 13.5). The aril can be odourless or have a powerful odour, suggesting garlic, onion and strong cheese with a fruity background smell. The smell is due to thiols, esters, hydrogen sulphide and diethyl sulphide (Baldry *et al.*, 1972). The edible aril (20–35% total mass) has a smooth, custard-like to firmer texture. In selected cultivars, the seeds are rudimentary or small compared with the wild types.

Excessive flowering and fruit set, such as in Thailand, may require flower or young fruit (4–6 weeks after anthesis) thinning, the degree of thinning depending upon variety. One to two fruit are left per panicle with 50–150 fruit per tree. There is a positive relationship between aril weight and seed number. The fruit matures in 90–150 days, with a sigmoid growth pattern, depending upon variety and temperature (Fig. 13.7). During the maximum growth phase, a fruit can accumulate up to 16 g dry matter (DM) fruit^{-1} day^{-1}, 50–80 days after anthesis. At maturity, the fruit abscises at the articulation of the pedicel with the inflorescence, ripening in 2–4 days, with

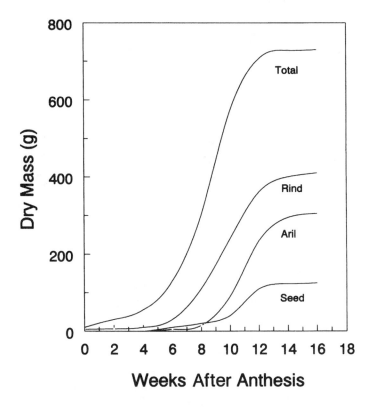

Fig. 13.7. Growth of durian fruit, rind, flesh and seed (redrawn from Salakpetch, 1996).

the fruit normally splitting into segments of irregular width at the stylar end. Ripening, which results in an increase in soluble sugars and a decrease in starch and aril firmness, is still occurring when dehiscence starts (Fig. 13.8).

Leaf flushing during early fruit development, 20–55 days from anthesis, may reduce the number of fruitlets and increase the percentage of deformed fruit. Later flushes can also reduce flesh quality, indicated by hard flesh, brown flesh discoloration and uneven flesh colour development. Use of high K and P, and no foliar fertilizer and growth inhibitor may therefore have a role in modifying young leaf flushing and fruit development needs. The balance in carbohydrate demand from fruit with a rapid growth rate and vegetative leaf flushing is crucial in developing high quality fruit. Vegetative sinks may be stronger than developing fruit, hence vegetative growth should be minimized. A single flowering peak is desirable to avoid interfruit competition and fruit thinning is necessary to reduce competition within a single flowering flush. A heavy fruit load greatly reduces root growth (Salakpetch, 1996).

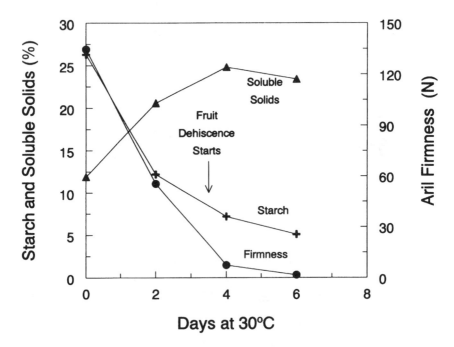

Fig. 13.8. Changes in starch, soluble solids and firmness of the harvested 'Chanee' durian aril during ripening (redrawn from Ketsa and Pangkool, 1994) and fruit dehiscence (redrawn from Sriyook et al., 1994).

CULTIVAR DEVELOPMENT

Breeding

Durian (2n = 56) show considerable diversity, with only a few of the 300 named varieties in Thailand being commercially grown. Hybridization within the genus is reported to be relatively easy and wild species may be able to contribute to disease resistance.

Pollen is released over 3–4 h and collected pollen remains viable for 24 h in a refrigerator, available for hand-pollination the next day. Hand-pollination is carried out on emasculated flowers a few hours before dusk when the stigma becomes sticky; receptivity is considerably reduced the following day. The long period (7–10 years) before bearing, however, limits breeding work.

Selection criteria include an aril with a sweet flavour and good texture, few or small seeds, large aril percentage, tree retention before fruit abscission, high yield of fruit with marketable weight (1.5 kg), elongated to round shape, good shelf-life, good rind colour and thickness, reduced rind dehiscence and increased length of harvest season. Superior varieties have thick, yellow, fibreless, crisp pulp and small, aborted or rudimentary seeds (Table 13.2).

Table 13.2. Comparison of three commercial Thai durian varieties (Hiranpradit *et al.*, 1992b).

Characters	Mon Thong	Chanee	Kan Yao
Fruit set	Very good	Poor	Good
Effect of high fruit retention	–	–	Poor fruit quality Branch dieback
Phytophthora resistance	Poor	Good	Poor
Fruit			
Seed number			High
Seed size			Large
Aborted seed	High		
Edible flesh	High		
Flesh texture	Coarse, sticky	Delicate	Delicate, sticky
Flesh fibre	High	High	Low
Flesh colour	Mild	Attractive	Uniform
Odour	Mild		
Physiological disorders	Few	Many	Few
Ripening	Not uniform	Uniform	
Full ripe	Firm flesh	Wet, pungent	Firm, fleshy
Processing			
Preserved	Very good	Medium	Poor
Frozen	Very good	Acceptable	

Major Cultivars

Selection programmes in Thailand and Malaysia have involved hundreds of cultivars, with only a few being recommended (Hasan and Yaacob, 1986; Hiranpradit *et al.*, 1992a). Malaysia has developed three F_1 hybrid clones. The variety 'Chanee' (Table 13.2) has less garlic flavour and is in high demand in Thailand. Others in Thailand include 'Monthong' (Golden Pillow), 'Kan Yao' (Long Stalk), 'Luang' and 'Kradum Thong'. Malaysian varieties have a 'D' and a number, indicating characterization by the government department and later a research institute; D-2 'Dato Nina' D-7 'Repok B-2', D-10 'Durian Hyan', D-24, D-98 'Katoi', D-99 (a Thai introduction), D-114 and D-117 'Gombak', 'Sitokong', 'Sukun', 'Mas', 'Parang' and 'Bakul'. The clone D-24 is most popular in Malaysia. Old Singapore varieties are 'Tan Chye Siam' and 'Jurong'. In Malaysia, a variety mix of 60% D-24, 30% D-99 and 10% D-98 or D-10 is recommended to ensure pollination. Similar mixed stands extend the harvest period.

CULTURAL PRACTICES

Propagation and Nursery Management

The seed is sexually produced and monoembryonic, giving viable progeny. The seed is short-lived (1 week), especially if exposed to sunlight, and can germinate (3–8 days) after the aril is removed and seed planted. Uniform trees are always obtained by grafting and rooting of cuttings, with air layering not being successful. Inarching has a 50% success rate but is not popular because of the many months necessary (Chua and Teoh, 1973). Patch or cleft budding from flushing shoots on to pencil-thin 2-month-old rootstocks is used (Chua and Young, 1978). 'Chanee' rootstock is frequently used in Thailand, with some related species being used to achieve some root-rot resistance.

Field Preparation

Deep ripping and discing, followed by incorporation of manure at the tree-planting site 6–12 months before harvest, is recommended. Frequently, preparation is performed in the wet season.

Transplanting and Spacing

Planting distances are commonly 8–16 m (40–156 trees ha^{-1}) in either a square or a triangle. The wider spacing being required to prevent canopy competition in mature trees. The narrow spacing can be used initially for some production before the canopy closes and then trees are thinned in the eighth to tenth year. In

Thailand, interplanting is sometimes practised with rambutan and langsat. Transplanting is performed just before the wet season, using 1-year-old trees.

Irrigation Practices

Total rainfall and its distribution are crucial for regular production and tree growth. Irrigation is essential if the dry period is longer than 3 months or during flowering and fruit development. Microsprinklers and drip tubes are used. 'Chanee' is reported to require lower amounts of irrigation than other varieties. Mild stress is needed to initiate flowering.

Pruning

Trees are trimmed to remove laterals and water shoots, leaving only a central leader. In Thailand and Malaysia, trees may be left to grow unrestricted for the first 2–3 years before the interior of the tree is thinned. The object is to have the tree produce only one flower flush per year and pruning is carried out immediately after harvest to improve light penetration and tree structure. Irrigation and fertilization are then carried out to induce vegetative growth (Fig. 13.6). Flower and fruit thinning is carried out to limit interfruit competition and to prevent fruit bearing on upper limbs, except near their bases, because of the load. Fruit are also removed from the main trunk to prevent excessive competition. Heavy crops may require propping or limb tying. Sometimes trees are topped at 10 m to aid in management.

Fertilization

Animal manures and mulch are still used in Malaysia and Indonesia but are not favoured in Thailand (Watson, 1983). Up to 15 g P is added to the planting hole. Fertilizer can be applied in the irrigation water. During the first 5 years, a 14 : 4 : 3.5 mix is recommended for non-bearing trees and then 12 : 4 : 7 twice yearly, increasing from 100 g to 4 kg tree^{-1} per annum to 12 years of age. Compound fertilizer is recommended at flower-bud emergence, supplemented by a side-dressing if there is a good fruit set, with a third application after harvest. The nutrients removed by 1 Mt fruit – 2.4 kg N, 0.4 kg P, 4.2 kg K, 0.3 kg Ca and 0.4 kg Mg – indicate a need for adequate N and K fertilization (Ng and Thamboo, 1967). Calcium fertilizer may be needed.

Pest Management

The major problem is patch canker and root rot, both caused by *Phytophthora palmivora* (Lim, 1990). Patch canker occurs at or just above ground level, eventually girdling and weakening the tree (Tai, 1973). Good drainage, removal of vegetation and use of resistant root stocks are the most common control measures. The disease cannot be controlled by cultural practices once trees are infected, although newer fungicides show promise. Inarching of seedling to mature trees is practised to control the disease. Root rot can also be caused by the same organism, with 'Mon Thong' being a very susceptible variety. The same organism, *P. palmivora*, causes rotting of immature and mature fruit, leading to high losses during rainy weather (Tongdee *et al.*, 1987b). Fruit on the ground can also be attacked by *Sclerotum rolfsii*.

Failure of the aril to soften or to soften evenly is a frequently observed disorder. Another disorder leads to a watery aril, with a flat and dull taste, and occurs especially during the rainy season. The cause of both disorders is unknown (Nanthachai, 1994).

Various stem borers and leaf-eating larvae have been reported on durian; however, damage is minimal. The fruit is not regarded as a fruit fly host if the rind is unbroken. Rats, monkeys, orang-utans, pigs and elephants are attracted to ripening fruit, leading sometimes to significant losses. Four-metre electric fences are used to deter elephants.

Weed Management

The soil around the tree should be kept free of weeds, especially in the first year. Control is by hand or approved herbicide, with mulches that do not touch the trunk also being used.

Orchard Protection — Windbreaks

Shade is needed during the first year and then gradually reduced. Windbreaks are beneficial to prevent damage to flowers, young fruit drop and breakage of branches laden with fruit.

HARVESTING AND POSTHARVEST HANDLING

Harvesting

To prevent natural fruit abscising, fruit may be tied to the limb or harvested at maturity (Plate 32). Maturity is judged by appearance (fruit-stalk thickness and flexibility, abscission zone, carpel sutures, etc.) and a hollow sound when tapped with the finger. Fruit at 85% maturity, based upon days from anthesis and rind characteristics, ripen to excellent quality in less than a week at 22°C. Fruit that is 95% mature has already commenced ripening, while 75% mature fruit may ripen with an inferior quality. Fruit collected from the ground after falling are more subject to disease and fracture and have a short shelf-life – 2–3 days, instead of the 7–8 days if picked from the tree.

Postharvest Treatments

In Thailand, the fruit is harvested with the peduncle attached and the peduncle is wrapped in a leaf or paper, reputedly to extend shelf-life. Fruit are cleaned and, if they are to be exported, brushed to remove insects, especially scales. This climacteric fruit can be stored at 15°C with extended shelf-life; lower temperatures lead to loss of aroma and disrupt aril softening (Brooncherm and Siriphanich, 1991). Fruit are graded on weight, shape, size and defects (Nanthachai, 1994). Defects include disease, insects, mechanical injury and flesh disorders. The standards vary with variety. The fruit is packed into a truck bed, bins or bamboo baskets for the local markets (Plate 33) and into cardboard cartons (four to six fruit per carton) for export.

For 'Chanee' and 'Kan Yao', the respiratory and ethylene peaks plateau or decline when the fruit is overripe, while in 'Mon Tong' the peak occurs when overripe (Tongdee *et al.*, 1987a). 'Chanee' is at optimum eating stage for only a few days, while 'Mon Tong' is at this stage for a longer period.

Marketing

The fruit is marketed fresh, with the short season usually glutting the markets. The main season in Thailand lasts for 2–3 months. Choice varieties demand and receive higher prices than other varieties. Stalls may cut open the fruit and package the soft aril and seed in a shrink- or stretch-wrapped tray. Ripe fruit and soft arils are also frozen for export. Partially ripe fruit, however, are difficult to open without damaging the aril. Fruit maturing out of the peak season commands high prices.

Thailand is the largest producer, followed by Indonesia and peninsular Malaysia. The fruit is highly prized in the markets of South-east Asia and probably has limited appeal elsewhere. Singapore and Hong Kong are major markets, consumers in Singapore preferring fully ripe fruit with no splitting, while Thais prefer the firmer pulp of the less ripe fruit with fewer volatiles. Many prefer the strong-flavoured durian over the milder cultivars. There is a demand among ethnic groups familiar with the fruit in large temperate cities.

UTILIZATION

Peak tree production occurs at 12–15 years and averages 50–100 fruit tree^{-1} year^{-1} (4–18 Mt ha^{-1}). The fruit is most frequently eaten fresh. The aril contains 64% water, 2.7% protein, 3.4% fat, 27.9% carbohydrate and 23 mg kg^{-1} Vitamin C (Table 13.1). The aril pulp is dehydrated and sold as 'durian cake', boiled with sugar, fermented or salted. The dried aril is used as a flavouring in ice-cream, confectionery, pastry and soft drinks. Boiled or roasted seeds are eaten as snacks.

LANGSAT, DUKU AND SANTOL

BOTANY

The family *Meliaceae* has six or seven species native to India, Malaysia, Indonesia, Borneo and the Philippines. The best-known edible fruit species are langsat and duku (*Lansium domesticum* Jack) and santol (*Sandorium koetjape* (Burm.) Merrill. (syn. *Sandorium indicum, Sandorium nervosum*)). There are two major types of *L. domesticum*, langsat (Malay, Thai) and duku (Malay, Thai). Other names include duku-langsat (Malay, Thai), long kong (Thai) and lanson (Philippine). Santol's other names include kechapi, sentol (English), faux mangostan (French), sentul, kecapi and ketuat (Indonesian), thitto (Burmese), toongz (Laotian), sau (Vietnamese) and sathon, matong and krathon (Thai).

   Lansium domesticum originated in the area from peninsular Thailand to Borneo where wild species are still found. The same area is where major cultivation occurs. Along with the Philippines, it is also cultivated in Vietnam, Burma, India, Sri Lanka, Australia, Surinam and Puerto Rico. Santol is native to Indonesia, Malaysia, Borneo and the Philippines and is cultivated in a narrower area than *L. domesticum*, including Indonesia, the Philippines, Thailand and Vietnam.

ECOLOGICAL REQUIREMENT

Soil

Langsat, duku and santol can be successfully grown on a number of soil types (Yaacob and Bamroongrigsa, 1992). Well-drained soils are preferred.

Climate

Langsat and especially duku require high soil moisture, with adequate rainfall being essential to prevent flower and fruit drop. The more drought-tolerant santol can survive and bear fruit down to 800 mm per annum. Damage occurs at temperature less than 6°C in both langsat and duku. The trees can withstand 40°C and grow better at a mean temperature of 22°C. Santol is more cold-tolerant and is grown up to 1200 m in Java. Santol trees are reported to recover from the severe damage caused by a −3°C frost. Santol does best in a wet monsoonal climate.

GENERAL CHARACTERISTICS

Tree

The langsat and duku can grow to 30 m, although in cultivation it is only 5–10 m tall. The bark is irregularly fluted, mottled grey and orange with a sticky milky sap. The glossy leaves are alternate, 30–50 cm long, on petioles up to 7 cm long (Fig. 13.9). Langsat leaves are faintly hairy underneath, while duku are hairless. Santol is a semideciduous, can also grow to the same height as langsat and also has milky sap. The leaves are also alternate trifoliate on long petioles. The leaves are glossy green above and light green below.

Flowers

Langsat and duku bear many-flowered racemes (100–300 mm long) sometimes in groupings of two to five on the trunk and large branches. The individual perfect sessile flowers (20–30 per raceme) are 12 mm wide and 5 mm long, with five sepals and five petals, with the stamens united to form a fleshy tube of ten anthers (Fig. 13.9). Santol flowers are similar in size to langsat and occur on loose-hanging bunches arising on the branches.

Pollination and Fruit Set

Langsat pollen is thought to be sterile. Santol is pollinated by insects, with self-incompatibility being the rule and few flowers setting fruit (Sotto, 1992). Langsat and duku flower-bud differentiation occurs early in the dry season and flowering occurs after the commencement of the monsoon rains, in about 7 weeks. The fruit develops parthenocarpically and the seed apomictically (Salma and Razali, 1987). There is substantial early fruit drop. Santol flowering lasts about 3 months, following a brief deciduous or partial leaf shed on the new flush.

Fruit

Duku fruit are round (40–50 mm), while langsat are slightly ovoid (30–50 mm). There are 15 to 25 fruits per langsat raceme and four to 12 in duku (Table 13.3). The pale green immature fruit with white latex ripen to a pale

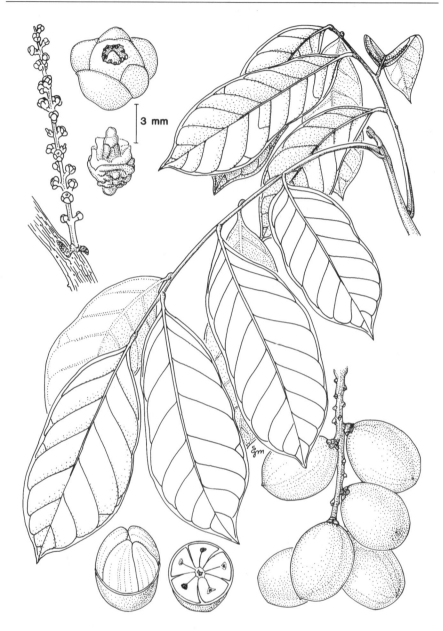

3 mm

Fig. 13.9. Langsat leaves, flower and fruit.

yellow, frequently with brown blemishes. Langsat pericarp is thin with a sticky sap, while duku has a thicker pericarp and no sap. The pericarp peels easily to reveal a clear, white, translucent and juicy adhering aril. Langsat tends to vary

Table 13.3. Criteria used to separate langsat from duku.

	Langsat	Duku
Tree	Slender	Spreading
	Upright branches	Dome-shaped
	Sparse dark green foliage	Bright green leaves
	More adaptable	Humid tropics
Fruit	Spikes long (10–30 cm)	Short spikes
	15–25 fruit per spike	4–12 fruit per spike
	Ovoid (30–50 mm)	Round (60–100 mm)
	Thin-skinned	Thick skin up to 6 mm
	Latex until fully ripe	Low latex
	Subacid taste	Sweet and aromatic
Varieties	Uttarodit (Thailand)	DU-1 (Malaysia)
	Paete (Thailand)	

from sweet to sour, with duku being sweet. Both fruit have five separate segments, with one to five seeds in langsat and one or two in duku.

The santol is a golden-yellow-skinned berry (50–100 mm across), which is firm and downy. The aril is thin (15 mm), white, juicy and translucent, surrounding three to five seeds (15–20 mm long). The skin is thick, soft and hairy.

Langsat and duku parthenocarpic fruit take from 3 to 4.5 months to develop. Fruit development is slow for the first 90–100 days and growth then increases for 120 days from anthesis to maturity (Fig. 13.10). During this final stage, aril sugar increases while acids and phenols decrease between 110 and 130 days from anthesis (Fig. 13.11). Santol fruit takes 5–6 months to develop.

CULTIVAR DEVELOPMENT

The langsat and duku (2n = 144) and santol (2n = 22) have received little genetic attention. *L. domesticum* has appreciable diversity, with wild and cultivated forms. The cultivated forms range from langsat to duku, with many forms in between. Since the cultivated form apparently does not form pollen, due to androecium degeneration, the wild type must produce viable pollen to explain the heterozygosity. The intermediate types (duku-langsat (Malaysia), long-kong (Thailand), kokosan (Indonesia)) are also variable, with long-kong being nearly seedless, having a brittle skin and a soft aril. Kokosan has hairy leaves and very compact fruit branches and is dark yellow, with a sour aril. The *L. domesticum* heterozygosity has led to considerable problems in describing varieties. The langsat and duku have

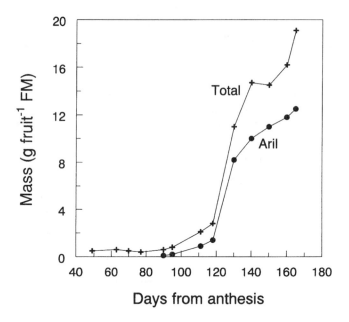

Fig. 13.10. Growth of langsat fruit and aril (redrawn from Paull *et al.*, 1987). FM, fresh mass.

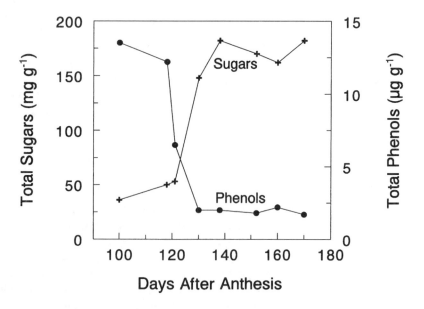

Fig. 13.11. Changes in aril total sugars and phenol during the final stages of langsat fruit development. Titratable acidity declines from *c.* 38 mequiv 100 g^{-1} to 22 mequiv 100 g^{-1} over the same period. (Redrawn from Paull *et al.*, 1987.)

general characteristics allowing for classification (Table 13.3); the intermediate forms are not as readily described.

Santol was previously divided into yellow and red species (*S. indicum*,
yellow; *S. nervosum*, red), with characteristics used not being consistent. The
yellow santol's leaves turn yellow before abscission, while the others turn red.
A number of santol varieties have been described: in Thailand, 'Barngklarng',
'Eilar', 'Tuptin' and 'Teparod'. The last variety is a tetraploid and is called
'Bangkok' in the Philippines and referred to as 'Manila' in Florida.

CULTURAL PRACTICES

Propagation and Nursery Management

Langsat and duku are commonly grown from the larger seeds or seedlings
found under the trees. Early seedling growth is slow; planting out occurs when
1 m tall. Santol seeds germinate in *c.* 20 days and have very rapid early
growth.

Approach, cleft, whip or side-veneer grafts have high success rates and,
along with air layers and cutting, with care, have been used for langsat and
duku (Blackler, 1976). Grafting is preferably done near the end of the wet
season. Santol can be propagated by air layers, budding, grafting and
inarching.

Field Preparation

Field preparation is similar to that used for other tropical trees.

Transplanting and Spacing

Trees are frequently planted out at the start of the rainy season, if no
supplementary irrigation is available. Spacing varies widely, from 8m × 8 m to
12 m × 12 m (70–156 trees ha^{-1}), with companion trees. The wider spacings
are used for the 'long-kong' types in Thailand. Similar spacings are used for
santol.

Irrigation

Irrigation is probably essential for both species in drier areas. Santol is more
tolerant of drought than langsat and duku. Water stress should be avoided
during flowering and fruit development.

Pruning

Trees are rarely pruned except for the removal of dead and crossed branches. Langsat, with its upright branches, can be headed to induce a more branched tree.

Fertilization

General recommendations for NPK rates are not available. Langsat and duku are slow-growing trees and may only require limited fertilizing. Mulching is suggested for langsat, duku and santol.

Pest Management

Collar rot (*Stibella cinnabarina*) of langsat and duku can be controlled by copper treatments. Root rot and preharvest and postharvest fruit anthracnose (*Colletotrichum gloesporioides*) can be serious. Collar rot (*S. cinnabarina*) can also occur on santol, though pink disease (*C. salmonicolor*) is more common.

Langsat and duku are fruit fly hosts and subject to quarantine control in some countries. Santol, with its thick rind, may have some resistance to the Caribbean fruit fly. Larvae of various species can lead to defoliation and damage to flower buds. A gall-forming mite (*Eriophyes sandorici*) is a serious pest of santol.

Weed Management

Control around the trunk is desired and any mulch used should not touch the trunk.

Orchard Protection — Windbreaks

Santol branches are easily broken by wind, and windbreaks are necessary in exposed areas. Trees of both species should be protected from wind to avoid fruit bruising.

HARVESTING AND POSTHARVEST HANDLING

Maturity is judged by skin colour (green to yellow) and lack of latex. Not all fruit on the tree ripen together, although all fruit on one raceme ripen within a short time. The fruit raceme is cut and fruit are either individually removed or sold as a raceme. Santol maturity is judged by their golden-yellow colour and softness.

Langsat and duku require careful handling to avoid bruising and skin discoloration (Anon., 1984). Fruit are graded for size and bruised and damaged fruit removed. Skin browning occurs rapidly if stored at temperatures less than 12°C, maximum storage life is about 2 weeks at 15°C, with 85–90% relative humidity (RH). Benomyl dip can assist in disease control. No in-depth storage studies have been reported for santol.

UTILIZATION

Langsat and duku yield is variable. 'Long-kong' trees (10 years old) can produce 40–50 kg tree^{-1} increasing to 80–150 kg at 30 years. Other data suggest 25 Mt ha^{-1} year^{-1} in the Philippines to 5.6 Mt ha^{-1} year^{-1} for the duku type in Thailand. Santol yield data are very scarce, although experimental observation suggests up to 14 Mt ha^{-1} year^{-1}.

The langsat and duku are eaten fresh and are low in vitamins A and C (Table 13.1). Seedless fruit can be bottled in syrup. Santol taste can range from sweet to sour and has about 54% edible pulp. The fruit is usually eaten fresh, but can be processed into candy, chutney, jams and jellies. The rind and the pulp that does not adhere to the seed are edible.

MANGOSTEEN

BOTANY

The family *Guttiferae* has 35 genera, with eight genera producing edible fruit (Alexander, 1983). The large Old-World tropical genus *Garcinia* consists of dioecious evergreen trees or shrubs. The many edible fruits of this genera range from 1 to 12 cm in diameter and have a woody skin, which varies from thin to thick (Almeyda and Martin, 1976). *Garcinia mangostana* L. (syn. *Mangostana garcinia* Gaertner (1790)), one of the most praised of tropical fruit, is known as mangosteen (English), mangostanier (French), mangostan (Spanish), manggis (Malaysian), manggustan (Philippine), mongkhut (Cambodian), mangkhut (Thai) and cay mang cut (Vietnamese). *Garcinia atrovirid* is a native of peninsular Malaysia, Thailand, Burma and Assam, *Garcinia dulcis* is a native of the Philippines and Indonesia and *Garcinia xanthochymus*, from India, is also cultivated. The other well-known fruit in the tropics is another genus: *Mammea americana* L. Mammy.

Mangosteen is only known as a cultivated plant (2n = 88–90?) and may be an allopolyploid hybrid, as a female, between *Garcinia hombroniana* Pierre (2n = 48) and *Garcinia malaccensis* (2n = 42?) based on morphological characters (Richards, 1990). Production occurs in South-east Asia (Thailand, Malaysia, Indonesia), where it is thought to have originated. In the last two centuries, it has been spread to Sri Lanka, southern India, Central America, Brazil, northern Australia, Hawaii and other countries in the tropics.

ECOLOGICAL REQUIREMENT

Soil

The tree grows on a wide range of soils (Campbell, 1967). Heavy soils are tolerated, even with a reported weak root system. The tree is not adapted to limestone soils, sandy alluvial soils or sandy soil low in humus. The tree can withstand some waterlogging.

Climate

The crop is found in the humid tropics that have a short dry season (15–30 days) to stimulate flowering and then an uninterrupted water-supply. Stress is to be avoided and irrigation may be needed during the dry season if the annual rainfall is less than 127 cm. Growth is slow below 20°C and the trees are killed at 3–5°C. The photosynthetic rate is steady over a 27–36°C temperature range, in 20–50% shade (Weibel *et al.*, 1993). The upper temperature limit is 38–40°C, with both leaves and fruit being susceptible to sunburn. In the tropics, it is grown up to 1000 m, though at a slower growth rate. Shade is essential during the first 2–4 years in the nursery and field. The tree is regarded as a shade-tolerant, lower-canopy tree, adapted to humid tropical lowlands. There are no reports of photoperiod responses.

GENERAL CHARACTERISTICS

Tree

The 6–25 m tall female tree has a straight trunk, symmetrical branching and pyramidal crown. Leaves are opposite on short petioles (Fig. 13.12), oblong or elliptical blades 15–25 cm long by 7–13 cm wide, thick, leathery, glabrous and olive-green above and yellow- green below. Early seedling growth is dominated by shoot development and slow root development. Secondary and tertiary root development does not occur until 3–4 months after germination. Root hairs are very sparse on all roots (Rukayah and Zabedah, 1992).

Flowers

Flowers are terminal, solitary or paired, with four sepals in two pairs, four

Fig. 13.12. Mangosteen leaves, flower and fruit.

thick, fleshy, yellow-green petals with reddish edges and a globose ovary, with a four- to eight-lobed stigma and staminodes (Fig. 13.12). Flowers are borne at the tip of older mature shoots and more hidden branches. Flower initiation is noticed as tip swelling, and the bud stage to anthesis takes 25 days. Male trees are believed to be non-existent.

Pollination and Fruit Set

Trees tend to bear in alternate years and bearing varies from tree to tree. Trees tend to flower after vegetative growth flushes and especially after dry weather (Fig. 13.13). Flowering can frequently occur twice a year, depending upon growing conditions and number of growth flushes. In Sri Lanka, low-elevation trees fruit in May and July and higher-elevation trees in September and October, while, in Puerto Rico, unshaded trees fruit in July and August and shaded trees in November and December. India has two flowering periods, during the monsoon (July–October) and another from April to June (Krishnamurthi *et al.*, 1964).

Fruit set is not a problem, although, when 10–15% of the flowers are set, they may inhibit the remaining buds, leading those buds to vegetative development. Irrigation should be scheduled to about 75% of the pan-evaporation rate so as to stimulate about 30–35% of the apical buds to produce flowers.

Fruit

The globose, smooth berry develops parthenocarpically. The fruit is 4–7 cm across, with the persistent calyx at the stem end (Fig. 13.12). The pericarp is 6–10 mm thick and turns purple at ripening. It contains a bitter yellowish latex and purple-staining juice. The edible white aril is four- to eight-segmented, with one or two larger segments containing apomictic seeds. There is no true seed, as the seed develops from the inner carpel wall; it is sometimes

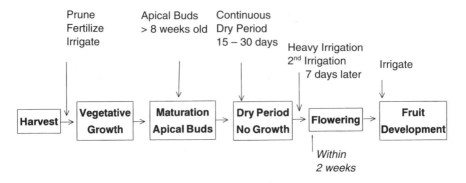

Fig. 13.13. Phenology of steps required to induce flowering in mangosteen (after Salakpetch, 1996). The tree must be vigorous, with bright green leaves and wide, thick leaf blades at flowering.

polyembryonic, with an underdeveloped embryo. At anthesis, it is obvious which fruit segment will be seeded. Fruit development takes 100–120 days from anthesis and up to 180 days in cooler areas or at higher elevations. Initially, growth is dominated by the pericarp, with aril DM not increasing until 20 days from anthesis and continuing throughout fruit growth (Fig. 13.14). Full-grown trees can produce from 200 to 2000 fruit tree^{-1}. The average yield in Thailand is reported as 4.5 Mt ha^{-1}. Older trees (45 years) can yield 3000 fruit tree^{-1} and then decline in yield.

CULTIVAR DEVELOPMENT

The 2n number has been variously reported: 56–76, 88–90, 96 and 120–130; the most commonly accepted is 88–90. The difficulty is to count the numerous small chromosomes in shoot tips, as there is no male meiosis. Apomictic seeds and the absence of male trees or flowers suggests that all trees belong to a single clone. Some differences in field leaf and fruit size have been reported, but they may be due to differences in growing conditions. The nucellar nature of the seeds, absence of diversity and long juvenile period (6–15 years) limit selection work. Collections are maintained in South-east

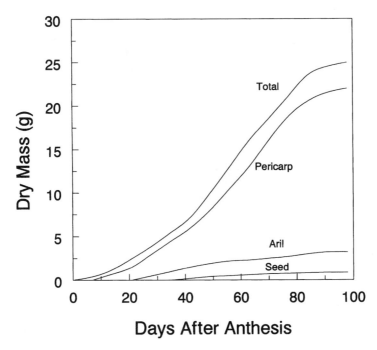

Fig. 13.14. Mangosteen fruit total dry mass, rind, aril and seed increase during growth (after Poonnachit *et al.*, 1992).

Asia, principally in Thailand and Malaysia. No described cultivars are reported.

CULTURAL PRACTICES

Propagation and Nursery Management

The apomictic seeds are viable for a short period (3 days, if dried) and are best kept in moist peat moss or left in the fruit. The heavier seeds (> 1.3 g) give the best seedling survival, as initial growth is slow (Hume and Cobin, 1948). Seeds should be planted in freely draining growing medium under high humidity and shade. Upon germination (2–3 weeks), a radicle and plumule emerge from opposite ends. As soon as an adventitious root develops at the base of the young shoot, the radicle dies. Slow growth is attributed to poor root development, for the roots have no root hairs and few laterals. Growth increases when side-shoots emerge, generally from every node.

Trees can reach 60 cm in 2 years, with one or two pairs of laterals. There is a very large shoot : root ratio (6.24) at the seedling stage (Table 13.4) which declines with age to 4.94 at 24 months. The juvenile phase lasts for about 16 pairs of laterals and the first crop appears after 5–7 years. If growing conditions limit growth, this phase may last 10–15 years.

Grafting on to mangosteen seedlings is not difficult, with the grafted plant being slower-growing and having smaller fruit and a shorter juvenile phase. Various rootstocks have been tested, with variable results (Gonzalez and Anoos, 1951). Positive results have been obtained with top-wedge grafting. Rooting of cuttings and air layers have failed.

Field Preparation

No specific information is available on land preparation, although high rates of organic matter are recommended. Practices normally follow recommendations for other tree crops in the area (Marshall and Marshall, 1983).

Table 13.4. The shoot-to-root ratio of mangosteen seedlings to 24 months.

Seedling age (months)	Shoot : root ratio
6	6.24
12	5.75
18	5.5
24	4.94

Transplanting and Spacing

Due to the long delicate tap root and poor lateral-root development, transplanting must be performed with care. Planting holes are prepared (1.2 m × 1.3 m) in advance and organic matter added a month before transplanting. Plants should have reached 60 cm before transplanting and a large ball of earth is set out and then watered heavily. Because of the need for shade and humidity, trees are frequently not planted in pure stands. Mixed stands are used with durian, rambutan and coconut as the dominant trees. An area of 40–80 m² is allowed per mangosteen tree and trees 0.6 m high are planted at 8–10 m (110–140 trees ha⁻¹) or 11–12 m if equipment access is needed. Shade is maintained for 2–4 years and then gradually reduced to full sunlight.

Irrigation Practices

The trees can withstand some waterlogging but not drought; hence, a constant supply of water is required. A continuous dry period of 15–30 days should be imposed to limit apical bud growth (Fig. 13.13). This dry period is followed by two heavy waterings, spaced 7 days apart, to induce flowering. Regular watering should then occur during fruit growth and development, at least at 80–85% of pan evaporation (Fig. 13.13). Trickle irrigation or microsprinklers are ideal for this crop.

Pruning

The regular pyramidal crown and slow overall growth limit pruning. However, the tall nature of the tree (25 m) and the fact that fruit are borne singly, making harvesting difficult, suggest that dwarfing rootstocks and pruning may be useful. Inside shoots and dead branches are removed, along with suckers at the base of the main trunk. Water sprouts should also be removed. Severe pruning is never desirable. The limited pruning is carried out when there are no flowers, fruit or flushes of new leaves.

Fertilization

Manure is recommended for young trees, along with mulching around the tree base. About 2–7 kg of complete fertilizer (10 : 10 : 19) per year is required for mature trees over 15 years of age. Younger trees receive 70 g N, 6 g P and 50 g K tree⁻¹ year⁻¹ up to 15 years of age. Half is applied when vegetative growth is being stimulated after fruit harvest (Fig. 13.13) and the remainder is applied 2–5 weeks after anthesis. Dolomite at 0.2 kg tree⁻¹ year⁻¹ to 15 years and then a constant application rate for older trees is recommended. Organic manure is also used.

Pest Management

Cankers on stems, young and older branches are caused on the Malay Peninsula by *Zignoella garcineae* P. Henn. The foliage on infested branches withers and eventually the whole tree dies. Trees should be cut and destroyed to arrest the spread. Thread blight, caused by *Pellicularia koleroga* Cooke, has been reported in Puerto Rico under conditions of excess shade and humidity. The smaller stems are first attacked and the disease becomes severe when it attacks the leaves, forming a whitish film over the blade. The leaves turn a clear brown and then darken before abscising. Removal of some shade and the application of Bordeaux mixture or other copper fungicides give control. Postharvest decay can occur, due to *Botryodiplodia theobromae*.

Only a few insect pests have been reported, possibly due to the bitter sap. Ants nesting in the tree can damage the growing tips. Mites can attack the fruit surface and make it unattractive for market. Caterpillars and grasshoppers can cause some leaf damage. Fully ripe fruit are eaten by monkeys, rats and bats.

Weed Management

Due to slow growth, young trees can be quickly overtaken by weed growth. Organic mulch assists in weed control and reduces evaporation from the soil around the trees. Plastic mulch can also be used.

Orchard Protection – Windbreaks

Shade during the first 2–4 years is essential. Shading can be achieved with mixed stands or crops such as pigeon peas, bananas, plantains, rambutan, durian and coconut, at least 1.5 m from the mangosteen. Cover crops (*Crotalaria*, cowpeas, tropical kudzu) also help and have been recommended as long as the area around the tree is clear. Mangosteen must be protected from strong winds and salt spray.

HARVESTING AND POSTHARVEST HANDLING

Harvesting

The fruit are picked when soft and dark purple, with the peduncle attached, by hand or with a pole and basket every 2–3 days. The mangosteen harvest index has been based on the extent and intensity of purple pericarp development (Plate 34; Table 13.5). The stage at harvest depends upon whether the fruit is to be exported or sent to local market, export fruit being picked at an earlier stage (Table 13.5). Fruit should not be harvested before the pericarp is a light

Table 13.5. Mangosteen harvest-index stages (fruit are normally harvested between stages 1 and 4 and eaten from stages 4 to 7), final total soluble solids obtained after the fruit was allowed to ripen at 24°C to stage 5 (after Tongdee, 1985) and fruit detachment force from the tree (after Tongdee and Suwanagul, 1989).

Stage	Pericarp colour	Detachment force (kg)	Latex	Seed/aril	Final eating flavour	Final total soluble solids
0	Yellowish white	2.2	Severe	Not separable	Inferior	15.2
1	Light greenish yellow with scattered pink spots	2.09	Severe	Not separable	Inferior	–
2	Irregular pink-red	1.19	Moderate	Difficult	Minimum stage	16.0
3	Uniform pink background	1.24	Slight	Moderate	Export	–
4	Red–reddish brown	1.32	None	Readily	Export	17.7
5	Reddish purple	1.32	None	Easy	Eating stage	18.3
6	Dark purple, slight red coloration	1.32	None	Easy	Eating	–
7	Dark purple-black	1.32	None	Easy	Eating	–

greenish yellow, with distinct irregular pink red spots over the entire fruit (stage 1). Fruit with less colour development have excessive latex exudation at the peduncle and have inferior flavour when they do darken to full purple stage in about 5 days (Table 13.5). Care is essential to avoid mechanical injury, as a 20 cm fall causes significant damage to the aril. Fruit ripening on the tree takes place over 6–12 weeks. The amount of latex declines with maturity, while soluble solids increase after stage 1 and acidity remains constant (Fig. 13.15).

Postharvest Treatments

Burst latex vessels on the fruit skin leave yellow dried latex (gamboge), which may be scraped off. Fruit are graded to remove damaged fruit and for size and either sold in baskets, strung in long bundles or in cartons (5 kg). The thick fruit wall hardens as the fruit ripens and during storage at low temperatures (< 10°C). Cool storage at 8–10°C and fungicide dips are recommended for long-term (up to 8 weeks) storage. Precooling and tray packs are useful (Augustin and Azudin, 1986).

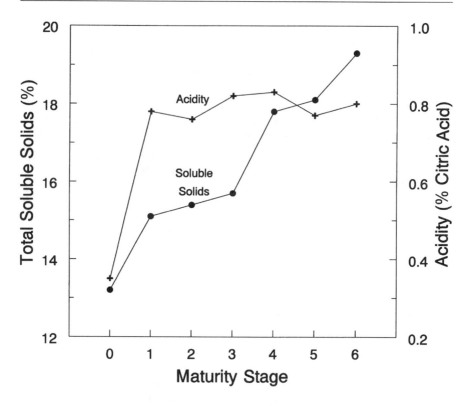

Fig. 13.15. Mangosteen aril acidity and total soluble solids at different maturity stages, described in Table 13.5 (after Tongdee, 1985).

Marketing

The fruit is a fruit-fly host; hence its movement in international trade is limited. Shipments are made by air to Europe, which has no fruit-fly disinfestation requirements. Harvested fruit are carefully cut open and the aril inspected and then the fruit is frozen whole and shipped to Japan.

Fruit damage during harvesting and marketing can affect more than 20% of fruit. The 'gamboge' disorder occurs where the latex seeps into the aril, giving a bitter taste, and is of unknown cause. The disorder makes it difficult to separate the aril from the surrounding tissue, even in ripe fruit, and causes hardening of the pericarp. This should not be confused with impact injury. Hardening of the pericarp at the point of impact and aril collapse, dehydration and pink or brown coloration are common symptoms of mechanical injury. A drop of 10 cm can cause slight pericarp damage, indicated as hardening at the point of impact, higher drops causing significantly greater damage, leading to downgrading of the fruit.

UTILIZATION

There are reports of declining production, associated with the costs of the long juvenile stage, labour cost for harvesting and irregular bearing habit (Achmad *et al.*, 1983). The fruit is one of the most highly regarded fresh fruit in South-east Asia. It is also consumed in a partially frozen state. The fruit aril makes up to 20–30% of the fruit, is 20% DM and has about 20% total soluble solids and 0.8% acid. The fruit is devoid of vitamin A and has a low mineral content (Table 13.1). The delicate aril flavour is due to hexyl acetate and *cis*-hex-3-enyl-acetate. Hex-3-en-1-ol and six sesquiterpenes have also been identified (MacLeod and Pieris, 1982). Attempts to use it as a juice have been unsuccessful due to the delicate flavour. Preserves are made; however, the product darkens and does not possess a unique flavour. Immature fruit can be canned, but mature fruit have little flavour when canned. The pericarp is used to tan leather and to dye fabric black. The rind and bract are used in traditional medicine. The wood is dark red, coarse and strong and can be used in carpentry.

WAX APPLE

INTRODUCTION

The wax apple belongs to the family *Myrtaceae* and includes guava. The genus *Syzygium* Gaertn. (*Jambosa* DC.) is the Old World genus, which was removed from the New World genus, *Eugenia*. *Syzygium* differs from *Eugenia* in having the cotyledons of the embryo free and the inflorescence being mostly panicles. *Syzygium* spp. are mainly found in South-east Asia (Panggabean, 1992).

The genus *Syzygium* includes a number of fleshy fruit-bearing species: *Syzygium aqueum*, rose water apple, bell fruit, jambu air; *Syzygium aimini*, Java plum, jambolan; *Syzygium jambos*, rose apple, malaba plum; *Syzygium malaccense*, Malay apple, rose apple, pomerac, jambu bol, yanbu; and a spice, *Syzygium aromaticum*, clove. More popular is *Syzygium samarangense* (Blume) Merrill & L.M. Perry (*Eugenia javanica* Lam), most commonly called wax apple; other names include Java apple, wax jambu, lien wu (Chinese), jambu air manar (Malaysian) and akopa (Philippine). The sweeter rose water apple and wax apple are both eaten fresh and are not easily distinguished in the market. The sweet water apple tends to be slightly astringent when ripe.

Wax apple and Malay apple are widely grown throughout the tropics. These often occur in home gardens, with limited commercial production in Malaysia, Australia, and Central and South America. Taiwan has a more extensive commercial industry around Pingtung at the southern end of the island.

ECOLOGICAL REQUIREMENT

These tropical trees grow in the warm humid lowlands to 1200 m, with no frost, and they apparently need a dry season. The trees can be grown in a range of fertile soils, with a preference for heavy soils.

GENERAL CHARACTERISTICS

Tree

The 5–10 m evergreen tree has an open, spreading crown. The trunk is 25–50 cm in diameter and has branches near the base. The leaves are opposite, elliptic to oblong, 10–25 cm long by 5–12 cm wide attached via a thick 3–5 mm petiole (Fig. 13.16). The leaves are aromatic when crushed.

Flowers

The flowers occur in drooping single branches or panicles of three to 25 flowers at the shoot tip and in fallen leaf axils (Fig. 13.16). The calyx tube is *c.* 1.5 cm long and lobed, with four petals 10–15 mm long and yellow-white. The stamens are numerous and up to 2.5 mm long. The style also can be up to 3 cm long.

Pollination and Fruit Set

The formation of flower buds does not mean early flowering (Yang *et al.*, 1991). Wax apple commonly flowers early or late in the dry season; the flowering can be triggered by a number of protocols, which have been developed in Taiwan. These protocols (Fig. 13.17) take into account the fact that flower panicles are borne at the tip of shoots and in the axils of fallen leaves, and new shoot flushes are necessary for additional flowers to form (Shu *et al.*, 1996).

Fruit

The broad, bell-shaped, sometimes oval, berry has incurved calyx lobes at the stigma end (Fig. 13.16). The fruit is 5–6 cm long and 4–5 cm wide and usually has 1–4 rounded seeds but sometimes none. The green to light red to dark red skin is very thin and has a wax-like glossy sheen. The white, juicy, aromatic flesh is low-acid, sweet and crisp when ripe, about 30–40 days after anthesis.

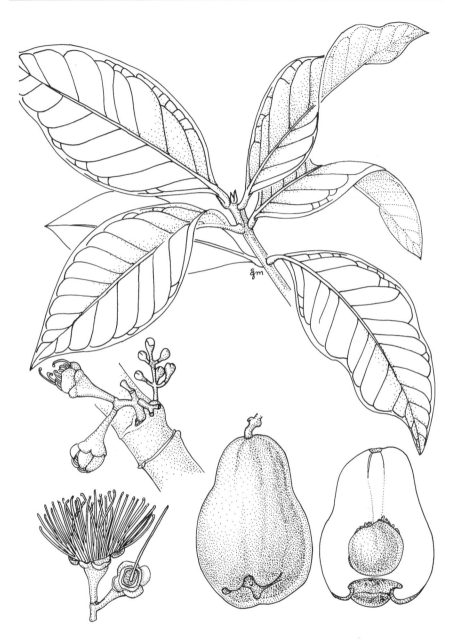

Fig. 13.16. Wax-apple leaves, flowers and fruit.

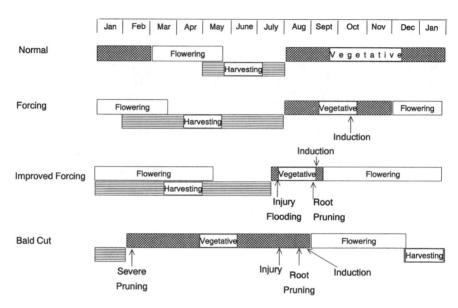

Fig. 13.17. Patterns of flowering and fruiting for wax apple subjected to three different induction protocols (after Shu *et al.*, 1990). Induction can be a single treatment or a combination of cultural and chemical treatments.

CULTIVAR DEVELOPMENT

Varying figures are given for the 2n value 33, 42, 44, 66 and 88, while *S. aqueum* has 2n = 44, and *S. malaccense* 2n = 22.

There has been some selection for superior clones. Breeding has apparently not been carried out. In Thailand, there is a green-fruited cultivar 'Khiew Savoey'. Taiwan has a number of clones based on fruit-skin colour: 'Pink', 'Deep Red', 'Light Red', 'White' and 'Green', with the 'Pink' occupying 99% of the planted acreage.

CULTURAL PRACTICES

Propagation

Seeds germinate rapidly if planted and lose their viability soon after removal from the fruit. Preferred trees are readily obtained by air layering. Trees can also be obtained by budding. Budding on to *Syzygium pycnanthum* Merrill & LM Perry, wild rose apple, provides some protection, as it is resistant to termites.

Culture

The orchard preparation, field planting and earlier management are similar to those for other tropical-tree fruits. Tree spacings of 5–10 m apart are used. The trees require little attention during the juvenile period of from 3 to 7 years, with air-layered trees fruiting in 3–5 years. The trees require a regular water-supply, with a short period of drought. Fertilization after harvest assists in flushing and is recommended after the inflorescences have formed.

There are no easily accessible reports on pruning or thinning to develop tree shape. In Taiwan, a labour-intensive production system has been developed (Wang *et al.*, 1994). Three different protocols involve cultural practices such as pruning and thinning, root pruning, girdling, flooding and fertilization, coupled with the less effective chemical sprays (Cycocel, ethephon, sodium naphthalene acetate (SNA), etc.) to control tree vigour and induce preformed flower buds to flower (Fig. 13.17). There is, however, no set formula for forcing treatment, as it has to be related to tree vigour and field and weather conditions (Plate 35). Individual cultural practices do not induce more flowering than that of the untreated tree (Fig. 13.18), although they induce earlier and more intense flowers, and combinations are more effective. 'Bald cut' pruning is the most severe and involves removing all flowers, fruit, leaves and small branches, leaving only large limbs. This is more appropriate for low-fertility sandy soils. After this pruning, the tree is heavily fertilized with N and then heavily irrigated. The next stage is when the second vegetative flush shows and involves girdling and application of high-P and K fertilizer, without N in late June and July. The flowers are initiated and forced to open in late August–September. However, the success of the flowering is 30% of the trees in an orchard, while failure leads to a 2-month delay in flowering (Shu *et al.*, 1994). The improved forcing protocol is used on fertile soils and is more common in Taiwan than the 'bald-cut' protocol (Shu *et al.*, 1996). The improved forcing protocol can increase fruit yield by 153%. A common chemical forcing treatment is three ethrel sprays (500 μl ml^{-1}) at 3-weekly intervals. Girdling is also used to alter the flowering period (Shu *et al.*, 1996). Leaf analysis is used to control fertilizer application (Wang, 1989).

Pest Management

There are few reports on disease, pest problems or control. The fruit is regarded as a fruit-fly host.

Windbreaks

The large leaf size and ease of fruit bruising necessitate some protection from wind.

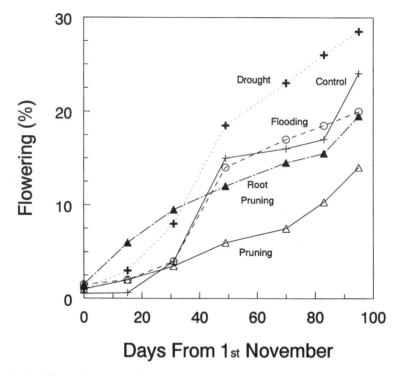

Fig. 13.18. Effect of various cultural practices (drought, flooding, root and tree pruning) prior to 1 November on flowering. Flooding was carried out from 20 July for 40 days, with 7 days' drainage after the first 20 days; drought involved no irrigation; root pruning, circular ditch 60 cm from the trunk, 30 cm wide and deep; pruning erect, thick, dead branches and water sprouts removed (after Shu *et al.*, 1990).

HARVESTING AND POSTHARVEST HANDLING

A fully mature tree can yield from 700 to over 1000 fruit weighing *c.* 65 g. Normal production in Taiwan and Hawaii is from May to July, in Sri Lanka from March to May and in Java from June to August. However, fruit can be harvested from one to three times a year, depending upon when conditions lead to flowering.

The thin, delicate skin is easily damaged and extra care is required in harvesting and handling. Fruit are harvested twice to three times a week when the skin has nearly full colour and the fruit are firm. The fruit are sorted for size and shape and blemished fruit are removed before packing a single layer in tray packs, with padding to limit injury, sometimes with a paper wrap on each fruit. Cracking of mature fruit can be a serious problem, which has no solution at this time.

The fruit is probably non-climacteric, as are other fruit in this genus, and is chill-sensitive. The waxy skin does not readily lose water, but a loss of 2% makes the fruit look slightly shrivelled; at 4% it is shrunken and soft. The fruit last 4–6 days at ambient temperature and become pitted and readily decay if stored at 0–10°C. Sugar content declines about 1% within 4 days after harvest. Water loss and sugar decline are reduced by holding the fruit in a polyethylene-wrapped pack at 12°C.

UTILIZATION

Green fruit are eaten raw with salt or cooked in Malaya. The ripe fruit are eaten fresh, the edible portion being about 80% and water 90%. The proximate analysis indicates that it has very low nutritional value, as do all fleshy-fruited *Syzygium* spp. (Table 13.1), having little vitamin A or C.

FURTHER READING

BREADFRUIT

Fownes, J.H. and Raynor, W.C. (1993) Seasonality and yield of breadfruit cultivars in the indigenous agroforestry system of Pohnpei, Federated States of Micronesia. *Tropical Agriculture (Trinidad)* 70, 103–109.

Marriot, J., Perkins, C. and Been, B.D. (1979) Some factors affecting the storage of fresh breadfruit. *Scientia Horticulturae* 10, 177–181.

Ragone, D. (1991) Ethnobotany of breadfruit in Polynesia. In: Cox, P.A. and Banack, S.A. (eds) *Islands, Plants, and Polynesians – An Introduction to Polynesian Ethnobotany*. Dioscorides Press, Portland, Oregon, pp. 203–220.

Sedgley, M. (1984) Moraceae – breadfruit. In: Page, P.E. (compiler) *Tropical Tree Fruits for Australia*. Queensland Department of Primary Industries, Brisbane, Queensland, Australia, pp. 100–103.

JACKFRUIT AND CHEMPEDAK

Jansen, P.C.M. (1992) *Artocarpus integer* (Thumb.) Merr. In: Verheij, E.W.M. and Coronel, R.E. (eds) *Plant Resources of South East Asia*. No. 2. *Edible Fruits and Nuts*. Prosea, Bogor, Indonesia, pp. 91–96.

Morton, J.F. (1965) The jackfruit (*Artocarpus heterophyllus* Lam.): its culture, varieties, and utilization. *Proceedings of the Florida State Horticulture Society* 78, 336–344.

Primack, R.B. (1985) Composite studies of fruits in wild and cultivated trees of chempedak (*Artocarpus integer*) and terap (*Artocarpus odoratissimus*) in Sarawak, East Malaysia, with additional information on the reproductive biology of the Moraceae in Southeast Asia. *Malayan Nature Journal* 39, 1–39.

DURIAN

Nanthachai, S. (ed.) (1994) *Durian: Fruit Development, Postharvest Physiology, Handling and Marketing in ASEAN.* ASEAN Food Handling Bureau, Kuala Lumpur, Malaysia.

Salakpetch, S. (1996) *Physiology of Durian Production.* 'Tropical Fruit Production' Course Notes, 10–12 December 1996, Chanthaburi Horticultural Research Centre, Chanthaburi, Thailand, 12 pp.

Salakpetch, S., Chandraparnik, S., Hiranpradit, H. and Punnachit, U. (1992b) Source–sink relationship affecting fruit development and fruit quality in durian, *Durio zibethinus* Murr. *Acta Horticulturae* 321, 691–694.

LANGSAT, DUKU AND SANTOL

Almeyda, N. and Martin, F.W. (1977) *Cultivation of Neglected Tropical Fruits with Promise,* Part 4. *The Lanson.* ARS-S-171, US Department of Agriculture Research Service, New Orleans.

Paull, R.E., Goo, T. and Chen, N.J. (1987) Growth and compositional changes during development of lanzone fruit. *HortScience* 22, 1252–1253.

Watson, B.J. (1984a) Santol – *Sandoricum koetjape* (Burm. f) Merrill. In: Page, P.E. (compiler) *Tropical Tree Fruits for Australia.* Queensland Department of Primary Industries, Brisbane, Australia, pp. 92–95.

MANGOSTEEN

Rukayah, A. and Zabedah, M. (1992) Studies on early growth of mangosteen (*Garcinia mangostana* L.). *Acta Horticulturae* 292, 93–100.

Tongdee, S.C. and Suwanagul, A. (1989) Postharvest mechanical damage in mangosteens. *ASEAN Food Journal* 4, 151–155.

Verheij, E.W.M. (1992) *Garcinia mangostana* L. In: Verheij, E.W.M. and Coronel, R.E. (eds) *Plant Resources in South East Asia* No. 2. *Edible Fruits and Nuts.* Prosea, Bogor, Indonesia, pp. 177–181.

WAX APPLE

Horng, D.-T. and Peng, C.-H. (1983) Studies on package, transportation, and storage of wax-apple fruit (*Syzygium samarangenese*). *National Chung Hsing University Horticulture Journal* 8, 31–39.

Shu, Z.H., Wang, D.N. and Sheen, T.F. (1990) *Technique of Producing Off-season Wax-apple.* Food and Fertilizer Technology Center Extension Bulletin No. 321, Taipei, Taiwan.

Shu, Z.-H., Wang, D.-N. and Sheen, T.-F. (1996) Wax apple as a potential economic fruit crop for the world. In: *Proceedings of an International Conference on Tropical Fruits, 23–26 July, 1996.* MARDI, Kuala Lumpur, pp. 69–73.

14

OTHER AMERICAN TROPICAL FRUIT

ACEROLA

BOTANY

Introduction

The *Malpighiaceae* family is dicotyledonous, with about 60 genera and 850 species of trees, shrubs and vines. The family is mostly confined to the American tropics, with a few native to the Old World tropics. Some species produce narcotics, while others are grown for their fruit or as an ornamental. A common ornamental from the Caribbean used in Hawaii is Singapore holly (*Malpighia coccigera* L.), with holly-like leaves.

The genus *Malpighia* has about 30 species of evergreen shrubs and trees native to the American tropics, through Mexico, Central America and the Caribbean islands. *Malpighia emarginata* DC (*Malpighia glabra* L. or *Malpighia punicifolia* L.), acerola, Barbados cherry, chercese or the West Indian cherry, is well-known for its commercial potential because of its unusually high amounts of fruit vitamin C – from 10 to 33 g kg^{-1} of edible pulp. Other species tested for ascorbic acid have generally shown lower amounts, except for two species, *Malpighia souzae* and *Malpighia shaferi*, with 20 and 5 g kg^{-1} of vitamin C, respectively (Asenjo, 1980).

Origin and Distribution

Acerola is presumed to be a native of the Caribbean islands, Central America or northern South America (Moscoso, 1956). It is postulated that the early Indian people of the Caribbean islands transported seeds or plants during their migrations between the islands during pre-Columbian times and it was well established and widely used by the Indians by the time of the arrival of the early Spanish explorers. Because of the resemblance of the acerola fruit to the true cherry (*Prunus avium*), the Spanish explorers called the fruit 'cereza', Spanish for cherry. Acerola or the Barbados cherry was probably introduced into Florida from Cuba in the 1880s. The primary source of acerola plants found in areas other than the Caribbean and Latin America has been Puerto Rico, Florida and Hawaii, where cultivar development was a primary objective.

ECOLOGY

The acerola is a tropical plant, thriving in warm, lowland climates. Acerola is best adapted to sea level, although it does fairly well at all elevations in Puerto Rico (Arostegui and Pennock, 1955) and in Hawaii to 150 m.

Soil

A wide variety of soil types, provided they have good drainage, are tolerated. Growth is generally poor in heavy soils with poor drainage. A suitable soil pH appears to be in the range of 5–6.5. Application of 4.5 Mt ha^{-1} of lime to soil with a pH of 5.4 increases yields by 400% in Puerto Rico (Landrau and Samuels, 1956), indicating that a soil pH closer to 6.0 is more desirable. Acerola is grown in Florida on the alkaline rocky land of South Dade County and in the acid soils of central Florida.

Environment

Rainfall

Adequate moisture throughout most of the year is necessary for good production, due to the repeated flowering cycles. The number of cycles depends upon temperature and the rainfall pattern. A well distributed average rainfall of about 1750 mm year^{-1} is satisfactory for growth and fruiting. Under high-rainfall conditions (2500–3125 mm) *Cercospora* leaf spot (*Cercospora bunchosiae*) is a potential problem and can defoliate trees. When rainfall is seasonal, irrigation is required for successful commercial production.

Temperature

Growth continues throughout the year at 15–32°C, while a night temperature of 10–15°C completely prevents growth. There is no tolerance to frost or temperatures lower than 7°C for extended periods, but acerola does grow in Okinawa, Taiwan and South Florida, where winter night temperatures occasionally drop to less than 5°C for short periods.

Radiation and photoperiod

Flowering and fruiting throughout the year in Hawaii suggest a lack of response to photoperiod. Cessation of flowering during winter in subtropical areas, such as Florida, Okinawa and Taiwan, is a function of temperature rather than a response to photoperiod. Shading leads to increased leaf size and reduced ascorbic acid level in the fruit (Fig. 14.1).

Windbreaks

Acerola height can be readily controlled by pruning, but it still benefits from the use of windbreaks. Trees growing from seedlings are generally deep-rooted, due to the presence of tap roots, while trees established from rooted cuttings are entirely dependent on a shallow, fibrous root system and can be blown down under moderately strong gusts of 64 km h^{-1}.

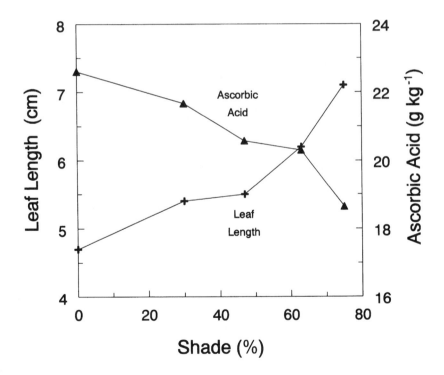

Fig. 14.1. Influence of shading on acerola leaf length and fruit ascorbic acid content for the sour clone 269–2 (after Nakasone *et al.*, 1966).

GENERAL CHARACTERISTICS

Tree

The acerola is a shrub that may be pruned and trained to a single trunk to form a small tree about 4.6 m tall. The root system is shallow, with the majority 6–10 cm below the surface, and equal in diameter to the canopy. Among seedling populations, growth habit varies from semiprostrate to compact or open types with upright, spreading or reclining branches. It has been observed among the experimental population grown in Hawaii that trees producing acid fruit are generally more upright and compact, while trees producing sweet fruit are generally alike, with rank growth and multiple leaders widely spread and rather easily torn by winds. In both types branches are woody, with numerous lenticels, characterized by short, lateral spurs (Nakasone *et al.*, 1968).

Leaves are simple, entire and oppositely arranged, and the shape ranges from elliptical to oval and ovate with apiculate apex (Fig. 14.2). Leaf size varies and ranges from 2.5 to 3.8 cm wide and from 5.0 to 6.4 cm long, with shaded leaves being much larger. Leaves of the sweet type are somewhat undulated. Young leaves and stems are slightly pubescent.

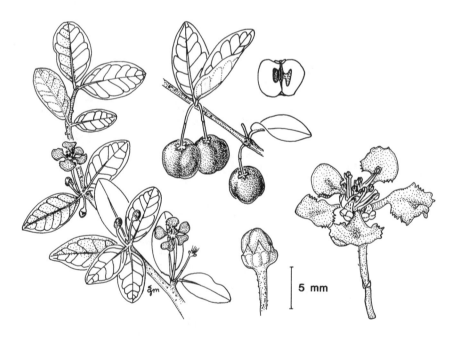

Fig. 14.2. Leaves, flower and fruit of acerola.

Flower

Flowers are 2.0 to 2.5 cm wide, with five petals ranging in colour from white to various shades of pink, depending upon the clone (Fig. 14.2). There are ten stamens, three styles and three carpels fused into a superior ovary. Flowers are produced in clusters from leaf axils on new terminals and on lateral spurs. Floral initiation occurs 8–10 days before bud emergence, anthesis occurring 7 days later.

Pollination and Fruit Set

In Hawaii, flowering occurs in cycles approximately a month apart, commencing in March or April and extending into November or later (Fig. 14.3). Mean fruit set from open pollination ranged from 1.3 to 11.5%; controlled self-pollinations increase fruit set from 16 to 55%. Cross-pollinations significantly increases fruit set further, from 32 to 74%. Reciprocal crosses sometimes show different fruit-set percentages, suggesting a degree of incompatibility.

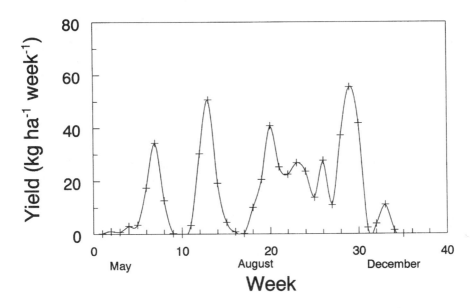

Fig. 14.3. Acerola monthly production in Pahea, Hawaii, of 3-year-old tree showing natural fruiting cycles.

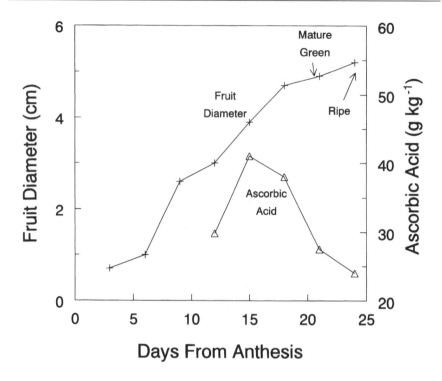

Fig. 14.4. Acerola fruit growth of clone 269–2 and ascorbic acid content of clone 5A-28 at different stages of development (after Miyashita *et al.*, 1964; Nakasone *et al.*, 1966).

Pollen-dissemination studies indicated poor wind distribution; flowers picked during the midafternoon showed only 1.24 pollen grains per stigma. Insects are considered to be the major pollinating agent; however, observations in Hawaii have not shown extensive insect activities. Honey-bees (*Apis mellifera*) and syrphid flies (*Eristalis arrorum*) are the only insects readily visible and they are found in small numbers. Honey-bee hives placed in the orchard did not increase insect activity and fruit set did not improve (Yamane and Nakasone, 1961).

The acerola also has extremely low viable seed production. Studies on reproductive morphology of the flowers found that seedlessness in acerola may be the result of: (i) failure of fertilization, followed by parthenocarpic development (fruit development is achieved with emasculated flowers bagged to prevent pollination); (ii) successful fertilization, with developing-embryo abortion, caused by failure of endosperm development; and (iii) ovule sterility, caused by absence of division or delayed divisions of the megaspore mother cells, disintegration of nucellar cells and abortion of embryo sac before anthesis (Miyashita *et al.*, 1964).

Fruit

The fruit resembles the true cherry but is a three-carpellate drupe. On the surface, the carpels are faintly discernible as three shallow lobes (Fig. 14.2). Inside the fleshy mesocarp are three triangular ridged stones or pyrenes, each capable of containing a seed. In Hawaii, the fruit matures 21–22 days after floral anthesis (Fig. 14.4). Three days after pollination, there is a period of abscission of the peduncle, with a second period 16–18 days after pollination. The embryo is morphologically mature *c*. 15 days after anthesis. Fruit size ranges from approximately 3–4 g to as large as 10 g. The colour of mature fruit ranges from orange-red or red to deep purplish red. The skin is delicate and easily bruised (Nakasone *et al.*, 1968).

Green fruit 12 days after floral anthesis has ascorbic acid levels from 27 to 36 g kg^{-1} of pulp. This increases to a maximum, at slightly over 40 g kg^{-1}, 15 days after anthesis (Fig. 14.4). After attaining the maximum, ascorbic acid content rapidly declined to 24–26 g kg^{-1} of pulp by day 24, when fruit is fully ripe (Nakasone *et al.*, 1966).

CULTIVAR DEVELOPMENT

Cytogenetics and Breeding

Information on the cytogenetics and genetics of acerola is scarce. Miyashita (1963) found a diploid number of 2n = 20 for a clone of acerola.

From 1946 to the 1960s, the breeding objectives were to select desirable cultivars, to develop cultural practices and processing methods and to understand the biochemical aspects of the fruit. The approach used was the planting of large seedling populations having wide diversity. Selection criteria were similar in Puerto Rico, Florida and Hawaii, with emphasis on tree type, fruit yield, ascorbic acid content, fruit size, juice yield, firmness and ease of propagation from cuttings. In Florida, cold tolerance was also considered.

Cultivars

The pH of the juice from acid types ranges from 3.1 to 3.3, while that of the sweet types ranges from 3.4 to 3.6 (Table 14.1). Acid and sweet types contain different ranges of ascorbic acid content; sweet types having less than 10–20 g kg^{-1} and acid types up to 29 g kg^{-1}. Seedlings have been grouped in terms of fruit taste, namely, sour, semisweet and sweet (Ledin, 1958). Malic acid makes up 25–50% of the total acids of tart clones and is possibly absent from the semisweet clones (Santini, 1952).

The acid types are generally better yielders, with more desirable tree forms. Vitamin C levels in excess of 10 g kg^{-1} of pulp are considered acceptable

Table 14.1. Characteristics of sweet and acid acerola cultivars in Hawaii (Nakasone *et al.*, 1968). The yields are for 2-year-old trees.

Cultivar	Fruit mass (g)	Juice pH	Total soluble solids (%)	Ascorbic acid (mg 100 g^{-1})	Fruit yield tree^{-1} (kg)
Sweet					
Manoa Sweet	6.2	3.6	10.0	1537	7.8
Tropical Ruby	7.2	3.5	9.0	1437	5.3
Hawaiian Queen	6.1	3.5	9.6	1577	5.1
Acid					
Maunawili	4.3	3.1	8.9	2010	4.0
J.H. Beaumont	6.9	3.3	11.0	2820	23.5
C.F. Rehnborg	6.2	3.3	8.3	1717	32.0
F. Haley	4.5	3.2	8.2	2663	21.2
Red Jumbo	9.3	3.1	8.7	2183	14.1

in Florida where fruit is used fresh or blended to fortify other juices low in vitamin C. The sweet cultivars were developed for general home planting and for potential use in baby foods and products requiring minimum acidity. Puerto Rico's selection 'B-17', an acid-fruited type, produces a fruit with an average weight of 9–12 g and yields of 8.1–12.1 Mt ha^{-1} from 4-year-old trees. Juice yield is as high as 73% by fruit mass, with a vitamin C content of from 13.3 to 22.5 g L^{-1} juice (Arostegui *et al.*, 1954). A Florida selection, 'Florida Sweet', has upright, open-type growth and some cold tolerance and is readily propagated by cuttings, with large, firm 14 g fruit. Ascorbic acid content ranges from 15 to 20 g kg^{-1} of pulp and fruit yield per plant over a 4-year period averaged about 45.4 kg tree^{-1}, with an estimated fruit yield of 33.6 Mt ha^{-1} (Ledin, 1958). Hawaiian selections ranged in ascorbic acid from 15 to a high of 27 g kg^{-1} purée and fruit yields under natural pollination produced from 5 to 32 kg tree^{-1} (Table 14.1).

Severe *Cercospora* leaf spot can occur during periods of heavy rainfall. Plants from clones nos 8 'Maunawili', and 20–26 show a relatively good level of tolerance. 'Florida Sweet' and other sweet clones showed moderate tolerance to leaf spot (Holtzmann and Aragaki, 1966).

CULTURAL PRACTICES

Propagation

Seed propagation is used primarily to produce seedlings for selection purposes or for rootstocks, if grafting is to be utilized for clonal propagation.

Due to the high heterozygosity, seedlings are not recommended for production purposes. Large quantities of pyrenes are necessary, because as many as 50% may be devoid of viable seeds. The seeds are washed, dried and sown in a well-draining medium, such as vermiculite, perlite, sand or various mixed media, covered up to around 1 cm and left in a shade house. Germination usually occurs within 10–12 days. Seedlings 4–5 cm tall are transplanted into individual containers, 7–8 cm in diameter, for further growth. Six- to eight-month-old seedlings, well-hardened under direct sunlight, may be field-transplanted.

Grafting on seedling rootstocks is recommended only if rootstocks can provide special advantages, such as disease or nematode resistance; otherwise, it is costly and time-consuming. Propagation by rooting cuttings allows large quantities of clonal plants to be obtained within a relatively short time. Cuttings 20–25 cm long and 0.64–1.3 cm in diameter, with some leaves retained, are treated at the base with a root-inducing hormone and stuck in a medium such as vermiculite, sand or some mixtures of porous material under mist. Cuttings rooted in 6 weeks, with rooting ability differing among clones (G.M. Yamane and H.Y. Nakasone, 1957, unpublished data).

Field Preparation

Field preparation follows conventional practices, subsoiling, ploughing and discing thoroughly.

Transplanting and Plant Spacing

Field-transplanting may be done at any time of the year when irrigation is available. In rain-fed areas, transplanting may be restricted to the beginning of or during the rainy season. Approximately 15 g of a complete fertilizer, such as triple-16, is placed at the bottom of each hole, which is refilled with soil up to 2.5–3 cm above the original soil line on the stem. A postplanting application of 7–10 g of fertilizer is spread and mixed with the surface soil and irrigated.

Spacing depends upon the growth habits of the clones selected for planting. Clones with compact, upright growth may be planted in a hedgerow system with about 2.44 m within rows and 4.6 m between rows. Conventional orchard spacing of 3.7 m × 4.6 m to 5.5 m × 5.5 m, with tree density of 478 and 332 trees ha^{-1}, respectively, has been suggested (Ledin, 1958).

Irrigation

Where rainfall is well distributed throughout the year, as in Puna, Hawaii, irrigation may not be necessary . However, in most tropical areas, with a well-defined dry season, irrigation is necessary to maintain production and fruit size.

Pruning

Pruning of acerola begins at the nursery stage, in order to train young plants into a manageable tree type. If left to grow naturally, cultivars will tend to become shrubby, making harvesting difficult. Plants trained to a single trunk system, with scaffold branches produced about 60–90 cm above ground, appear to be a desirable type in terms of weed control and harvesting.

Bearing trees benefit from light pruning when fruiting is completed. Some cultivars have rank growth and long, widely spread, multiple leaders which should be cut back to encourage more axillary branching. Others produce thick, compact growth, which may require some thinning each year. When pruning is done in the non-flowering period, there is a long period between pruning of primary branches and production of flowers on secondary branches, compared with pruning in the flowering period. Flower-bud production appears to occur between 15 and 18 days after pruning (Miyashita et al., 1964).

Fertilization

Nitrogen (N) deficiency shows the strongest influence on growth and yields (Landrau and Hernandez-Medina, 1959). Deficiencies in phosphorus (P), boron (B), sulphur (S) and iron (Fe) resulted in some growth depression, with significant reductions in fruit yields. Omission of potassium (K) reduces tree growth, with no effect on yield in the reported test; deficiencies in magnesium (Mg) and manganese (Mn) are also without apparent effect. The flowering cycles of acerola trees occur almost monthly to bimonthly, so several applications of fertilizer per year are necessary.

Pest Management

Diseases

The acerola is relatively free from serious fungal diseases. In Hawaii, *Cercospora* leaf spot (*C. bunchosiae*) occasionally poses a serious problem of defoliation. Incidence of this disease is low where rainfall is low to medium. Root-knot nematode (*Meloidogyne incognita*) is one of the limiting factors in acerola production in Florida and Puerto Rico, but not in Hawaii. It is especially severe in sandy, acid soils, compared with alkaline, dry soils. Where nematodes pose a problem, soil fumigation may be a solution, though costly and only temporary.

Insects and other pests

Approximately 23 different pests on acerola have been listed by Pollard and Alleyne (1986) in the Caribbean region, but none have been found to be serious. Sucking insects, particularly the scale insects, pose a threat. In Florida, the southern green stink bug (*Nezara viridula*) and the leaf-footed plant bug (*Leptoglossus phyllopus*) damage fruit (Ledin, 1958). Fruit flies are a constant threat if fresh fruit is the preferred use. Some species of birds can cause serious fruit loss.

Weed Management

A considerable amount of weed control can be achieved by heavy mulching with organic material or by the use of polyethylene sheets. Mulching also conserves moisture, regulates soil temperature and reduces damage by nematodes. Mowing between rows is an option.

HARVESTING AND POSTHARVEST HANDLING

Commercial-size yields can be expected from 2-year-old trees, some clones in Hawaii producing as much as 23–32 kg of fruit tree^{-1} year^{-1} (Nakasone *et al.*, 1968). In Puerto Rico, Arostegui and Pennock (1955) reported single-tree yields of 4-year-old trees of from 14 to 28 kg for one season. Older trees give higher yields and continue to do so for 15–20 years. An orchard of 200 trees should produce 2722–4082 kg year^{-1} (Moscoso, 1956).

Harvesting

The acerola fruit is thin-skinned, delicate and highly susceptible to bruising. Whether or not the fruit is to be eaten out of hand or used for processing or for extraction of vitamin C, fruit should not be left on the tree, as they abscise and fall upon ripening. Half- to fully-ripe fruit can be utilized for any purpose. Mature green fruit are suitable for vitamin C extraction inasmuch as these fruit contain a much higher content of vitamin C (Fig. 14.4). It has also been shown that there is little or no increase in fruit size after 18 days from anthesis (Fig. 14.4). Since acerola fruit mature and ripen in 22–25 days, harvesting should be done every other day. If all ripe fruit and those just turning colour are picked, the harvesting cycle may be stretched to every 3 days without loss of fruit.

Acerola is harvested by hand. In Hawaii during the 1960s, an experimental shaker harvester removed all ripe and most of the green fruit, demonstrating the feasibility of a machine being adjusted for acerola.

Postharvest Handling

Harvested fruit, packed in boxes or baskets, should be stored in the shade, as fruit left in the field exposed to sun can lose as much as 25% of the vitamin C over an 8-h period (Nakasone *et al.*, 1966). Fruit should be processed without delay or refrigerated immediately (Alves *et al.*, 1995). For vitamin C extraction, fruit should be immediately frozen or processed upon arrival at the plant.

UTILIZATION

The acerola has not attained the level of commercial production anticipated after the discovery of the high content of vitamin C. Plantings occur in Puerto Rico, Barbados and Brazil and in some of the east Caribbean Windward Islands. The acerola fruit is delicate and highly perishable; for this reason, it appears unlikely to become important as an item for fresh consumption except in village markets. There is considerable potential in the form of processed products and also for the food and pharmaceutical industries, if cost of production and processing can be minimized.

Depending on the size of the fruit and concentration of ascorbic acid, one or two fruit can provide the recommended daily allowance of vitamin C. Acerola, like most fruit, is considered a fair source of provitamin A (408 IU vitamin A), but is low in the three B vitamins, thiamine, riboflavin and niacin (Table 14.2). The minerals calcium, phosphorus and iron, are low.

The acerola fruit can be used in a variety of ways for home consumption. Besides eating out of hand, the fruit can be juiced for punch, for jellies and in the preparation of gelatin desserts, salads and sherbets, or for fortifying other juices low in vitamin C, particularly as a supplement to baby fruit juices. Miller *et al.* (1961) provide numerous recipes and instructions on the preparation of these products.

Cold- or hot-pressed juice may be frozen as cubes and used in fruit punch or added to other juices. Hot-pressed acerola juice may be boiled with added sugar and poured into sterilized bottles or jars for later use. Both hot- and cold-pressed frozen juice retain about 85% of the original vitamin C after 8 months of freezer storage (Miller *et al.*, 1961).

Acerola makes excellent jelly. It is a regular practice at the senior author's home to prepare jelly from fruit accumulated from two home-grown trees. Mustard (1946) reported relatively high retention of vitamin C ($5–7.6 \text{ g kg}^{-1}$ jelly), even after boiling in preparation of the jelly. Acerola juice can be used to prevent oxidation of fruit used in salads and fruit cups, simultaneously enhancing the vitamin C content.

Table 14.2. Composition of acerola, chiku and abiu per 100 g edible portion (Leung and Flores, 1961; Wenkam, 1990).

		Acerola	Chiku	Abiu
Proximate				
Energy	kJ	163	393	586
Protein	g	1.8	0.5	1.8
Fat	g	1.0	1.1	0.4
Carbohydrate	g	6.8	23	36.3
Fibre	g	1.2	1.6	0.9
Ash	g	0.2	0.4	0.9
Moisture	g	90	75	61
Minerals				
Calcium	mg	12	24	22
Phosphorus	mg	11	10	41
Iron	mg	0.2	1	1
Vitamins				
Thiamine	mg	0.03	0.01	0.02
Riboflavin	mg	0.05	0.01	0.02
Niacin	mg	0.6	0.02	34
Vitamin C	mg	1790	15	49
Vitamin A	IU	–	10	78

CHIKU AND ABIU

BOTANY

Sapote is believed to be derived from the Aztec name for all soft sweet fruits – 'tzapotl'. The black sapote (*Diospyros digyra* Jacq.) belongs to the family *Ebenaceae* and is therefore not a true sapote. The family *Sapotaceae* has a number of species in different genera with edible fruit: *Manilkara zapota* (L) von Royen – chiku, zapote (Martin and Malo, 1978; Scholefield, 1983); *Chrysophyllum cainito* L. – star apple (Campbell, 1974; Pratt and Mendoza, 1980); *Pouteria sapota* (Jacq.) H.E. Moore and Stearn – mammey sapote (Almeyda and Martin, 1976); *Pouteria campechiana* (HBK) Beahni – canistel; *Pouteria obovata* HBK – luemo; *Pouteria caimito* (Ruiz and Pav.) Radlk – abiu; *Calocarpum viride* Pittier – green sapote (Whitman, 1965). Only two of these will be discussed in this section, chiku and the lesser known abiu, regarded by some as the better fruit.

   M. *zapota* (L.) von Royen (syn. *Achras zapota* L., *Pouteria mammosa* (L) Cronquist, *Manilkara achras* (Mill) Fosberg) is known by a number of names, including chico (Philippine), sapodilla (English), sanomanila

(Indonesian), lamut (Thai.), chicle (Mexico), chico zapote (Guatemala, Mexico, Venezuela) and naseberry (Jamaica). It is native to Yucatán, southern Mexico, and nearby Central American countries. It has been cultivated since ancient times to extract chicle for chewing-gum and for the fruit. It was introduced long ago to the Caribbean Islands, South America and Florida, taken by the Spaniards to the Philippines and then spread to the Old World tropics (Coronel, 1992). It is widely grown in coastal India and Sri Lanka.

Pouteria caimito (syn. Achras caimito Banth, Guapeda caimito Pierre, Labatia caimito Mart, Laucuma caimito Roem. & Sch.) is known as abiu in Brazil and in English, cauje (Ecuador), caimo (Colombia), abieiro and abio. This is a native fruit of the Amazon headwaters area (Clement, 1989). The fruit was as well known in the pre-Columbian civilizations of South and Central America. It is well distributed throughout the tropics, though not grown to any large extent commercially.

ECOLOGY

Soil

Well-drained soils, from sand to heavy clays, with acid to neutral pH have been successfully used for both crops (Campbell et al., 1967). Calcareous soils (pH 6–8) also give good chiku growth, while abiu is tolerant of a range of pH levels but is sensitive to saline conditions.

Climate

Both species are drought resistant. Chiku can withstand salt spray but probably not saline soils (Patil and Patil, 1983). Chiku is grown in the drier areas with 100 cm rainfall and wetter humid areas with 300 cm. A short dry spell is favoured by chiku for flowering but not required by abiu. Supplemental irrigation is required for good yields and fruit size in drier areas. Unlike most tropical species, mature trees of both species can withstand −2°C with little damage. Best growth occurs within 12° of the equator and up to 1000 m elevation and in warmer areas to within 25° of the equator. Seedlings may require some shade during establishment, with the mature trees growing in full sun (Marshall, 1991). Sun scorching of abiu leaves can occur under the hot and dry conditions of Northern Territory, Australia (Lim and Ramsay, 1992). The species are apparently not photoperiodic.

GENERAL CHARACTERISTICS

Tree

An evergreen, upright to spreading, low-branching tree, chiku (5–20 m) is somewhat irregular in shape, while abiu is pyramidal in shape and slightly shorter (5–15 m). Grafted chiku trees are shorter (up to 15 m). The alternate, glossy, entire leaves (chiku 5–15 cm long; abiu 10–15 cm long), pointed at both ends, tend to cluster at the end of the shoot, with short internodes (Figs 14.5, 14.6). The acute apex of abiu leaves is often curved.

Flowers

The bell-like, hermaphrodite chiku flowers are small (8–12 mm), with three outer and three inner sepals, enclosing a pale green to white, tubular corolla and six stamens. The chiku flowers on short stalks from the leaf-base cluster at the end of the small branches (Fig. 14.5). Chiku flowering peaks occur in the summer in Australia, early in the Philippine rainy season (April–June), February–April and October–December in Mexico and December–March in India. Reducing vegetative growth tends to induce flowering (Gonzalez and Feliciano, 1953).

The almost sessile abiu flowers occur along branches between the shoot tip and the main trunk (Fig. 14.6) and are enclosed by two bracts and four overlapping, light green sepals. The abiu's cylindrical, white, corolla tube (5 mm long), which never opens completely, encloses four free stamens, alternating with four staminodes. The white abiu style arises from a superior ovary and has a four-lobed stigma, which pushes through the petals at anthesis and is 2–3 mm above the corolla tube before the stigma becomes receptive (Lim and Ramsay, 1992). The abiu ovary is 10–12-celled. Abiu flowering can occur throughout the year and it is not uncommon to find fruit of various ages on the same tree and sometimes on the same branches.

Pollination and Fruit Set

Full chiku petal opening for anthesis occurs at night, 45–60 days after initiation. The stigma is receptive 1 day before and 3 days after flower opening. On the day of opening, a strong scent is produced and the chiku stigma is covered with a sticky fluid. Cross-pollination by insects is apparently

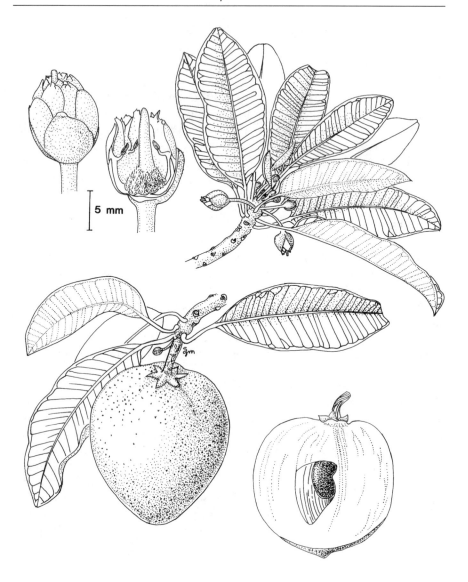

Fig. 14.5. Chiku: leaves, flower and fruit.

necessary for low-yielding chiku cultivars, which produce little pollen, with hand-pollination used to increase fruit set. Other chiku varieties appear to be self-fertile. Abiu is insect-pollinated and has abundant fruit set during the warmer months. Abiu can have up to three crops a year, with flowering overlapping the previous crop.

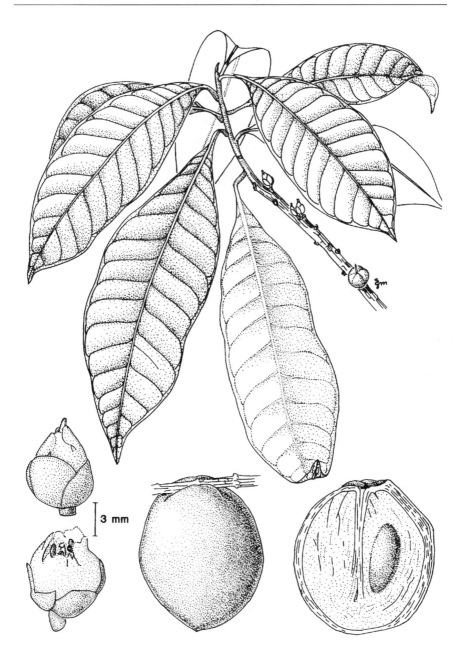

Fig. 14.6. Abiu: leaves, flower and fruit.

Fruit

The pendulous chiku berry can be round, conical to oval, 5–10 cm in width and up to 12 cm long, weighing from 100 to 400 g (some varieties up to 1 kg). The skin is thin and smooth, covered with a light brown scurf (Fig. 14.5). The flesh ranges from yellowish to light brown, with copious latex and astringent when cut immature, and with no latex, a smooth, very juicy, sometimes grainy texture and sweet, low-acid taste when ripe (Lakshminarayana, 1980). Shiny black seeds (0–12) are oblong and up to 2 cm long.

The skin colour observed when the brown scurf is scrubbed off changes from dark green when the fruit is immature to a light yellowish green when mature. Seed colour changes from white and soft to black and hard at the same time. The chiku takes from 168 to 240 days to mature (Fig. 14.7) and in India occurs in three stages: in the first stage of 112 days, the fruit increases in diameter, followed by a 28-day transitory stage, and then during the last 63 days the fruit increases in length to achieve its characteristic shape (Selvaraj and Pal, 1984). Chiku starch content during the last 4 months declines with maturity, along with the tannin content (Abdul-Karim *et al.*, 1987). This coincides with

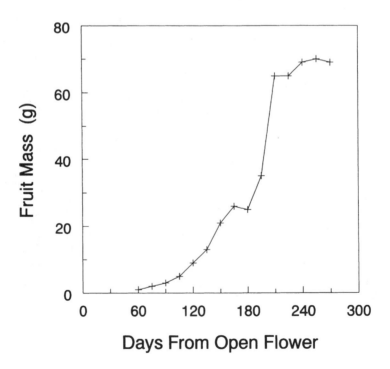

Fig. 14.7. Growth of chiku cv. 'Subang' fruit from flower opening (after Ali and Lin, 1996).

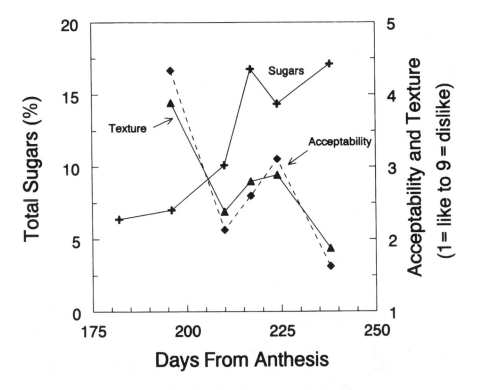

Fig. 14.8. Change in total sugars, fruit texture, aroma, and acceptability of chiku cv. Subang, after the major fruit growth period (after Ali and Lin, 1996). Fruit harvested 195 days (6.5 months) after flower opening did not ripen properly.

the increase in sugar, fruit softening and eating acceptability (Fig. 14.8).

The abiu fruit is round to elliptic and 5–12 cm long and normally weighs from 50 to 500 g (Fig. 14.6). Large fruit – up to 1000 g – have been reported in a semidomesticated race (Clement, 1983). The skin of immature fruit is green, changing to light yellow when ripe. The leathery (5 mm thick) skin encloses a sweet, translucent, jelly-like, white flesh and one to five large oblong seeds at maturity. Fruit development takes about 3 months (Scholefield, 1984). When immature, all parts of the fruit contain a sticky white latex and the cut surface rapidly browns. As fruit matures, the latex disappears from the pulp and remains only in the peel. In some races, the latex is also missing from the peel.

CULTIVAR DEVELOPMENT

There are a few genetic studies of chiku (2n = 26). Hybridization studies were started in India in the 1950s, but new cultivars have not been produced. There is considerable variation of seedling population in plant and fruit characteristics (Fig. 14.9). Some seedlings never bear fruit, due to pollen sterility. Large fruit size, reliable high yields, good eating quality and seedlessness are the major objectives of selection programmes.

Numerous named chiku varieties exist, many probably being synonyms. The most popular varieties in Indonesia include 'Sawo Manila' with oval fruit and 'Sawo Kulan' and 'Sawo Betawi' with round fruit, in the Philippines the high-yielding 'Pineras', the large-fruited, poor-pollen 'Ponderosa' and 'Sao Manila', and in Thailand the small-fruited 'Krasuey', the large-fruited 'Kai Hahn' and the medium-sized 'Makok'. India has many varieties, including the small-fruited 'Kali Patti', the very large-fruited and crisp 'Cricket Ball' and 'Dwarapudi' with its large sweet fruit (Morton, 1987). The most high-yielding Florida variety is the consistently early-bearing 'Prolific'; others include 'Russel' and 'Brown Sugar' (Campbell *et al.*, 1967).

Abiu occurs in the wild and in cultivation, showing great variation in morphology. Named varieties in Queensland, Australia, include 'Inca Gold'

Fig. 14.9. Variation in chiku fruit shape and size. Fruit can vary from round to oval and weigh from 100 to 1000 g.

and 'Cape Oasis', with other selected seedlings (Parker, 1986). Selection criteria include early and regular bearing, round fruit (> 180 g), few seeds, low skin latex, firm, sweet, translucent flesh (13–18% total soluble solids) and good shelf-life.

CULTURAL PRACTICES

Propagation and Nursery Management

Dry chiku seeds germinate readily (2–4 weeks) after several years' storage and trees begin to bear after 6–10 years. Abiu seeds are short-lived and seedlings take 2–5 years to fruit, with great variation in tree form, yield and fruit quality. Vegetative propagation is used in both species to achieve uniform planting material and avoid initial slow growth of seedling trees. Cleft or wedge grafting or marcotting in the wet season are used on the seedling trees of chiku or the related more vigorous species *Manilkara hexandra* (Roxb.) Dubard (India) and *Manilkara kauki* (L.) Dubard (Indonesia) (Gonzalez and Fabella, 1952). Greater grafting success is achieved in the cooler, drier periods. Mist propagation of leafy chiku-stem cuttings treated with a root-promoting powder were also successful (Rowe-Dutton, 1976). Vegetatively propagated chiku trees should bear in 2–3 years and abiu in about 2 years.

Field Preparation, Transplanting and Spacing

No special field preparation is needed. Trees are planted out when 1–2 years old. If planted in exposed situations, the trees are staked and provided with shade for the first 12 months. Spacing of 6–14 m is recommended for both species; mature trees may need to be thinned. Abiu is reported to thrive well on a 2–3 m inter-row spacing.

Irrigation Practices

Young trees require irrigation to become established. High yield and larger fruit are obtained if mature trees receive supplemental irrigation during the dry season. Flower and small fruit abscise under moisture stress.

Pruning

A tree with a central leader and whorls of laterals requires little pruning. Grafted trees may require some initial trimming to stimulate lateral growth. Pruning of mature trees is limited to removal of dead and thin branches and branches of the lower whorl that bend towards the ground.

Fertilization

Mature chiku trees require at least 1.5 kg N, 0.5 kg P_2O_5, and 0.5 kg K_2O tree^{-1} year^{-1}, applications being split into two or three per year, with one before the wet season and one just before the end of or after each harvest (Marshall, 1991). Manure is also used. Liming and other nutrients may be needed. A similar schedule has been suggested for abiu, with applications after harvest (Lim and Ramsay, 1992).

Pest Management

A canker that kills affected chiku branches (*Corticium salmonicolor*, pink disease) and leaf spots caused by *Phaeophleospora indica* in India and *Septoria* sp. in Florida have been reported. Leaf rust (*Uredo sapotae*) can occasionally cause serious damage to chiku. Some leaf-eating larvae, twig borers, various scales, aphids and mealy bugs attack various parts of the tree of both species. The chiku and abiu fruit become hosts for fruit flies as they ripen. These fruit flies are frequently quarantine pests and limit the fruit's export potential. Fruit bats often mean that the tree has to be netted for protection as the fruit ripen.

Weed Management

The most common practice for controlling weeds is the use of mulch around the base of the tree. Care is needed to avoid the mulch touching the trunk. Weeds in the general area are controlled by mowing.

Orchard Protection

Although chiku is tolerant of strong winds, higher yields of less wind-damaged fruit are obtained in sheltered conditions. Staking may be necessary for young trees in strong winds.

HARVESTING AND POSTHARVEST HANDLING

Maturity for a particular chiku variety is judged by size and appearance. Rubbing the brown powdery scruff bloom off the skin to determine if there has been a change in the underlying green of immature fruit to light greenish-yellow of mature fruit is not always successful as a maturity index. Waiting for the first ripe fruit to fall and then harvesting the largest fruit is more reliable. Little or no latex coming from the skin when scratched is another indicator; alternatively, with experience, the changes during abscission can be judged. Some varieties shed the scruffy coat when mature.

The chiku fruit are carefully handled, as they can be readily damaged by abrasion and impact. After harvest, the stalk is pruned and the latex allowed to drain into water. Harvested immature fruit shrivel as they soften, without the full sweet taste. Mature fruit ripen in 3–7 days at 25°C and can be stored at 15°C for about 14 days (Broughton and Wong, 1979). Lower storage temperatures lead to chilling injury and a failure to soften. Ripening can be accelerated with ethylene (Shanmuganelu *et al.*, 1971). In some South-east Asian markets, the fruit is dyed brownish red to improve eye appeal. Limited postharvest life, difficulty of grading for similar maturity and the difficulty of using low temperature for storage have limited the fruit's market potential. The fruit shape lends itself to shipping in cartons, with cell packs of fruit graded to the same size.

Abiu fruit are harvested when bright yellow and they continue to ripen after harvest. Full ripening occurs in 1–5 days, when the fruit pulp does not have the sticky latex. The translucent flesh becomes jelly-like, with a pleasant, somewhat caramel-flavoured pulp. Fruit can be stored for about 1 week at 12°C. The tough leathery skin can be easily bruised.

UTILIZATION

The chiku fruit is very popular in southern and South-east Asia. Cultivation is greatest in India and then Mexico. Commercial planting is also found in Sri Lanka, Malaysia, the Philippines, Thailand, Central America, Venezuela, Florida and some Caribbean islands. Annual yields of 20–30 Mt ha^{-1} (Florida), 20–25 Mt ha^{-1} (the Philippines) and 20–80 Mt ha^{-1} (India) have been reported. Mature trees can yield 1000–2500 fruit annually. Mature abiu trees can yield up to 200 kg tree^{-1} and 6-year-old trees 60–85 kg tree^{-1} (Lim and Ramsay, 1992).

Chiku is normally served fresh, frequently chilled, cut in half and the flesh eaten with a spoon or out of the hand. Some chiku fruit have a gritty taste and these varieties are avoided. About 84% of the chiku fruit is edible, containing 69–75% water (Table 14.2). The chiku fruit is low in vitamin A and C. The acidity of chiku is due to malate. Abiu is an excellent source of vitamin C and a good source of vitamin A and niacin (Table 14.2).

The ripe chiku and abiu pulp can be used dried or in sherbets, jams, fruit salads and yoghurt. It is difficult to retain the characteristic taste and aroma of chiku during processing. The flavour volatiles imparting the distinctive odour are methyl benzoate and methyl salicylate (MacLeod and de Troconis, 1982). The delicate flavour is due to the small quantity of volatiles produced compared with other fruits. The latex is collected from the tree and used to produced chicle (15% rubber, 38% resin), a component of chewing gum. Abiu is always used fresh, sometimes in salads with other fruit.

FURTHER READING

ACEROLA

Asenjo, C.F. (1980) Acerola. In: Nagy, S. and Shaw, P.E. (eds) *Tropical and Subtropical Fruits.* AVI Publishing, Westport, Connecticut, pp. 341–374.
Yamane, G.M. and Nakasone, H.Y. (1961) Pollination and fruit set of acerola, *Malpighia glabra* L. in Hawaii. *Proceedings of the American Society for Horticultural Science* 78, 141–148.

CHIKU AND ABIU

Clement, C.R. (1989) A center of crop genetic diversity in Western Amazonia. *BioScience* 39, 624–631.
Lim, T.K. and Ramsay, G. (1992) Abiu – a new fruit with potential for the Northern Territory. *Acta Horticulturae* 321, 99–105.

REFERENCES

CHAPTER 1. THE TROPICS AND ITS SOILS

Anon. (1956) *Weather for Aircrew Trainees*. US Air Force Manual 105–5, US Air Force, Washington, DC.

Bluthgen, J. (1966) *Allgemeine Klimageografie*. Walter de Gruyter, Berlin.

Gray, W.M. (1968) Global views of the origin of tropical disturbances and storms. *Monthly Weather Review* 96, 669–700.

Henderson–Sellers, A. and Robinson, P.J. (1986) *Contemporary Climatology*. Longman Group, London, John Wiley & Sons, New York.

List, R.J. (1966) *Smithsonian Meterological Tables*. Smithsonian Institution, Washington, DC.

Nieuwolt, S. (1977) *Tropical Climatology – An Introduction to the Climates of the Low Latitudes*. John Wiley & Sons, London.

Oliver, J.E. and Hidore, J.J. (1984) *Climatology – an Introduction*. C.E. Merrill, Columbus, Ohio.

Sanchez, P.A. (1976) *Properties and Management of Soils in the Tropics*. John Wiley & Sons, New York, 618 pp.

Sanchez, P.A. and Salinas, J.G. (1981) Low impact management technology for managing Oxisols and Ultisols in tropical America. *Advances in Agronomy* 34, 279–406.

Sanchez, P.A. and Uehara, G. (1980) Management considerations for acid soils with high phosphorus fixation. In: Khaswaneh, F.E., Sample, E. and Kamprath, E.J. (eds) *The Role of Phosphorus in Agriculture*. American Agronomy Society, Madison, Wisconsin, pp. 471–514.

Soil Conservation Service, USDA (1971) *Soils of the Tropics*. US Department of Agriculture, Washington, DC.

Soil Survey Staff (1990) *Keys to Soil Taxonomy*, 4th edn. Technical Monograph No. 19, SMSS, Blacksburg, Virginia.

Thornthwaite, C.W. (1948) An approach toward a rational classification of climate. *Geographical Review* 38, 55–94.

Walter, H. (1973) *Vegetation of the Earth in Relation to Climate and the Eco–Physiological Conditions* (translated from the second German edition by J. Wieser.). Springer-Verlag, New York.

CHAPTER 2. CULTIVATION AND POSTHARVEST HANDLING

Chen, N.M. and Paull, R.E. (1986) Development and prevention of chilling injury in papaya fruit. *Journal of the American Society for Horticultural Science* 114, 639–643.

Choudhury, J.K. (1963) Early bearing of guava by photoperiodic induction. *Science and Culture* 29, 213–214.

Jones, J.B., Wolf, B. and Mills, H.A. (1991) *Plant Analysis Handbook.* Micro–Macro Publishing, Athens, Georgia.

O'Hare, T.J., McLaughlan, R.L. and Turner, C.D. (1993) Effect of co-shipment of exotic tropical fruit on existing fruit transported in bulk shipment. In: *Proceedings of the Australasia Postharvest Conference. 1993 September 20–24.* University of Queensland, Lawes, Australia, pp. 245–249.

Paull, R.E. and Armstrong, J.W. (1994) Introduction. In. Paull, R.E. and Armstrong, J.W. (eds) *Insect Pests and Fresh Horticultural Products, Treatments and Responses.* CAB International, Wallingford, UK, pp. 1–36.

Reuter, D.J. and Robinson, J.B. (eds) (1986) *Plant Analysis: an Interpretative Manual.* Inkata Press, Melbourne.

Tomlinson, P.B. (1987) Architecture of tropical plants. *Annual Review of Ecology Systemics* 18, 1–21.

Verheij, E.W.M. (1986) Towards a classification of tropical fruit trees. *Acta Horticulturae* 175, 137–150.

Watson, B.J. and Moncur, M. (1985) *Guideline Criteria for Determining Survival, Commercial, and Best Mean Minimum July Temperature for Various Tropical Fruit in Australia.* West Tropical Regional Publication, Queensland Department of Primary Industry, Brisbane, 3 pp.

CHAPTER 3. ANNONAS

Alvarez-Garcia, L.A. (1949) Anthracnose of the Annonaceae in Puerto Rico. *University of Puerto Rico, Journal of Agriculture* 33, 27–43.

Campbell, C.W. (1985) Cultivation of fruits of the Annonaceae in Florida. *Proceedings of the American Society for Horticultural Science of the Tropical Region* 29, 68–70.

Cresswell, G. and Sanewski, G. (1991) Diagnosing nutrient disorders. In. Sanewski, G. (ed.) *Custard Apples – Cultivation and Crop Protection.* Information Series QI90031, Queensland Department of Primary Industry, Brisbane.

Dhingra, J., Mehrotra, R.S. and Aneja, I.R. (1980) A new postharvest disease of *Annona squamosa* L. *Current Science* 49, 477–478.

Duarte, O., Villagarcia, J. and Franciosi, R. (1974) The effect of different treatments on propagation of cherimoya. *Proceedings of the American Society for Horticultural Science of the Tropical Region* 18, 41–48 (in Spanish).

Ellstrand, N.C. and Lee, J.M. (1987) Cultivar identification of cherimoya (*Annona cherimola* Mill.) using isozyme markers. *Scientia Horticulturae* 32, 25–31.

Gazit, S. and Eistenstein, D. (1985) Floral biology of *Annona squamosa* and *Annona cherimola* in relation to spontaneous appearance of atemoya in Israel. *Proceedings of the American Society for Horticultural Science of the Tropical Region* 29, 66–67.

George, A.P. and Nissen, R.J. (1986a) Effect of pruning and defoliation on precocity of bearing of custard apple (*Annona atemoya* Hort.) var. African Pride. *Acta Horticulturae* 175, 237–241.

George, A.P. and Nissen, R.J. (1986b) The effects of root temperature on growth and dry matter production of *Annona* species. *Scientia Horticulturae* 31, 95–99.

George, A.P. and Nissen, R.J. (1987a) Effects of cincturing, defoliation and summer pruning of vegetative growth and flowering of custard apple (*Annona cherimola* × *Annona squamosa*) in subtropical Queensland. *Australian Journal of Experimental Agriculture* 27, 915–918.

George, A.P. and Nissen, R.J. (1987b) The effects of day/night temperatures on growth and dry matter production of custard apple. *Scientia Horticulturae* 31, 269–274.

George, A.P. and Nissen, R.J. (1988) The effects of temperature, vapor pressure deficit and soil moisture stress on growth, flowering and fruit set of custard apple (atemoya) 'African Pride'. *Scientia Horticulturae* 34, 183–191.

George, A.P., Nissen, R.J. and Carseldine, M.L. (1989) Effect of season (vegetative flushing) and leaf position on the leaf nutrient composition of *Annona* spp. hybrid cv. Pink's Mammoth in south–eastern Queensland. *Australian Journal of Experimental Agriculture* 29, 587–595.

George, A.P., Nissen, R.J. and Campbell, J.A. (1992) Pollination and selection in *Annona* species (cherimoya, atemoya, and sugar apple). *Acta Horticulturae* 32, 178–185.

Kshirsaga, S.V., Shinde, N.N., Rane, D.A. and Borikar, S.T. (1976) Studies on the floral biology in atemoya (*Annona atemoya* Hort.). *South Indian Horticulture* 24, 6–10.

Kumar, R, Hoda, M.N. and Singh, D.K. (1977) Studies on the floral biology of custard apple (*Annona squamosa* Linn.). *Indian Journal of Horticulture* 34, 252–256.

Lo, S.S. (1987) Pruning technique, use of sprouting chemical and flower initiation in sugar apple (*Annona squamosa*). In: Chang, L.R. (ed.) *Forcing Culture of Horticultural Crops*. Special Publication No. 10, Taichung District Agricultural Improvement Station, Taichung, Taiwan (in Chinese, English summary), pp. 147–150.

Nakasone, H.Y. (1972) Production feasibility for soursop. *University of Hawaii, Hawaii Farm Science* 21, 10–11.

Pascual, L., Perfectti, F., Gutierres, M. and Vargas, A.M. (1993) Characterizing isozymes of Spanish cherimoya cultivars. *HortScience* 28, 845–847.

Paull, R.E. (1982) Postharvest variation in composition of soursop (*Annona muricata* L.) fruit in relation to respiration and ethylene production. *Journal of the American Society for Horticultural Science* 107, 582–585.

Paull, R.E. (1983) Changes in organic acids, sugars, and headspace volatiles during fruit ripening of soursop (*Annona muricata* L.). *Journal of the American Society for Horticultural Science* 108, 931–934.

Saavedra, E. (1979) Set and growth of *Annona cherimola* Mill. fruit obtained by hand-pollination and chemical treatment. *Journal of the American Society for Horticultural Science* 104, 668–673.

Samuel, R., Pineker, W., Balasubramaman, S. and Morawetz, W. (1991) Allozyme diversity and systematics in Annonaceae – a pilot project. *Plant System Evolution* 178, 125–134.

Smith, D. (1991) Insect pests. In: Sanewski, G. (ed.) *Custard Apples – Cultivation and Crop Protection*. Information Series QI90031, Queensland Department of Primary Industry, Brisbane, pp. 73–79.

Thakur, D.R. and Singh, R.N. (1964) Studies on pollen morphology, pollination and fruit set in some annonas. *Indian Journal of Horticulture* 22, 10–17.

Thakur, D.R. and Singh, R.N. (1965) Studies on floral biology of Annonas. *Indian Journal of Horticulture* 23, 238–252.

Wenkam, N.S. (1990) *Foods of Hawaii and the Pacific Basin. Fruits and Fruit Products, Raw, Processed, and Prepared*, Vol. 4, *Composition*. Research Extension series 110, HITAHR, College of Tropical Agriculture and Human Resources, Honolulu.

Worrell, D.B., Carrington, C.M.S. and Huber, D.J. (1994) Growth, maturation, and ripening of soursop (*Annona muricata* L.) fruit. *Scientia Horticulturae* 57, 7–15.

Yang, C.S. (1987) Production of sugar-apple fruits in the winter. In: Chang, L.R. (ed.) *Forcing Culture of Horticultural Crops. Proceeding of a Symposium*. Special Publication No. 10, Taichung District Agricultural Improvement Station, Taichung, Taiwan, pp. 129–140.

Yang, C.S. (1988) Application of plant growth regulators on *Annona* culture. In: Lin, H.S., Chang, L.R. and Lin, J.H. (eds) *The Application of Plant Growth Regulators on Horticultural Crops. Symposium Proceedings*. Special Publication No. 12, Taichung District Agricultural Improvement Station, Changhua, Taiwan (Chinese, English summary), pp. 305–320.

CHAPTER 4. AVOCADO

Ahmed, E.M. and Barmore, C.R. (1980) Avocado. In: Nagy, S. and Shaw, P.E. (eds) *Tropical and Subtropical Fruits*. AVI Publishing, Westport, Connecticut, pp. 121–156.

Anon. (1995) *International Standardisation of Fruits and Vegetables, Avocado*. Organization for Economic Cooperation and Development, Paris, France, 73 pp.

Arpaia, M.L., Bender, G.S. and Witney, G.W. (1992) Avocado clonal rootstock production trial. In: *Proceedings of the 2nd World Avocado Congress*, Vol. I, pp. 305–310, Los Angeles.

Barmore, C.R. (1976) Avocado fruit maturity. In: Sauls, J.W., Phillips, R.L. and Jackson, L.K. (eds) *The Avocado: Proceedings of the 1st International Tropical Fruit Short Course*. Fruit Crops Department, University of Florida, Gainesville, Florida, pp. 103–109.

Bergh, B.O. (1969) Avocado. In: Ferweda, F.P. and Wit, F. (eds) *Outline of Perennial Crop Breeding in the Tropics*, Landbouwhogeschool miscellaneous paper no. 4, Wageningen, The Netherlands, pp. 23–51.

Bergh, B.O. (1975) Avocados. In: Janick, J. and Moore, J.N. (eds) *Advances in Fruit Breeding*. Purdue University Press, West Lafayette, Indiana, pp. 541–567.

Bergh, B.O. (1976) Avocado breeding and selection. In: Sauls, J.W., Phillips, R.L. and Jackson, L.K. (eds) *The Avocado. Proceedings of the 1st International Tropical Fruit Short Course*. Fruit Crops Department, University of Florida, Gainesville, Florida, pp. 24–33.

Bergh, B. (1990a) The avocado and human nutrition, I. Some human health aspects of the avocado. In: *Proceedings of the 2nd World Avocado Congress*, Vol. I, pp. 25–35. Los Angeles.

Bergh, B. (1990b) The avocado and human nutrition II. Avocado and your heart. In: *Proceedings of the 2nd World Avocado Congress*, Vol. I, pp. 37–47. Los Angeles.

Bergh, B. and Ellstrand, N. (1986) Taxonomy of the avocado. *California Avocado Society Yearbook* 70, 135–145.

Bower, J.P. (1981a) *Climatic Requirements of Avocado*. B.1, Farming in South Africa, Department of Agriculture and Water Supply, Pretoria.

Bower, J.P. (1981b) *Layout of an Avocado Orchard*. D.3, Farming in South Africa. Department of Agriculture and Water Supply, Pretoria.

Bower, J.P. (1981c) *Avocado Pollination and Pollinator Interplanting*. G.4, Farming in South Africa, Department of Agriculture and Water Supply, Pretoria (reprinted 1986).

Broadley, R.H. (ed.) (1991) *Avocado Pests and Disorders*. Information Series QI90013, Queensland Department of Primary Industry, Brisbane, Australia, 74 pp.

Campbell, C.W. and Malo, S.E. (1976) A survey of avocado cultivars. In: Sauls, J.W., Phillips, R.L. and Jackson, L.K. (eds) *The Avocado. Proceedings of the 1st International Tropical Fruit Short Course*. Fruit Crops Department, University of Florida, Gainesville, Florida, pp. 20–24.

Chaikiattiyos, S., Menzel, C.M. and Rasmussen, T.S. (1994) Floral induction in tropical fruit trees, effects of temperature and water supply. *Journal of Horticultural Science* 69, 397–415.

Coffey, M.D. (1987) Phytophthora root rot of avocado – an integrated approach to control in California. *Plant Disease* 71, 1046–1052.

Crane, J.H., Schaffer, B., Davenport, T.L. and Baberdi, C. (1992) Rejuvenation of a mature, non-productive 'Lola' and 'Booth 8' avocado grove by topping and tree removal. *Proceedings of the Florida State Horticulture Society* 105, 282–285.

Currier, W. (1992) New variety introduction under US conditions. In: *Proceedings of the 2nd World Avocado Congress*, Vol. II, pp. 609–613. Los Angeles.

Cutting, J.G.M., Cocker, B. and Wolstenholme, B.N. (1994) Time and type of pruning cut affect shoot growth in avocado (*Persea americana* (Mill.)). *Journal of Horticultural Science* 69, 75–80.

Davenport, T.L., Parnitzki, P., Fricke, S. and Hughes, M.S. (1994) Evidence and significance of self-pollination of avocado in Florida. *Journal of the American Society for Horticultural Science* 119, 1200–1207.

Diaz-Avelar, J. (1979) *The Cultivation of Avocado*. XXV Anniversary, FIRA, Bano de Mexico (in Spanish).

Finazzo, S.A., Davenport, T.L. and Schaffer, B. (1994) Partitioning of photoassimilates in avocado (*Persea americana* Mill.) during flowering and fruit set. *Tree Physiology* 14, 153–164.

Furnier, G.P., Cummings, P.M. and Clegg, M.T. (1990) Evolution of the avocado as revealed by DNA restriction fragment variation. *Journal of Heredity* 81, 183–188.

Gabor, B.K., Guillemet, F.B. and Coffey, M.D. (1990) Comparison of field resistance to *Phytophthora cinnamoni* in twelve avocado rootstocks. *HortScience* 25, 1655–1656.

Garcia, V.A. (1975) Cytogenetic studies in the genus *Persea* (Lauraceae). I. Karyology of seven species. *Canadian Journal of Genetic Cytology* 17, 170–180.

Garcia, V.A. and Tsunewaki, K. (1977) Cytogenetical studies in the genus *Persea* (Lauraceae). III. Electrophoretical studies on peroxidase isoenzymes. *Japanese Journal of Genetics* 52, 379–386.

Gazit, S. (1976) Pollination and fruit set of avocado. In: Sauls, J.W., Phillips, R.L. and Jackson, L.K. (eds) *The Avocado. Proceedings of the 1st International Tropical Fruit Short Course*. Fruit Crops Department, University of Florida, Gainesville, Florida, pp. 88–92.

Gomez, R.E., Soule, J. and Malo, S.E. (1973) Anatomical aspects of avocado stems with reference to rooting. *Proceedings of the American Society for Horticultural Science of Tropical Region* 17, 23–28.

Gustafson, D. (1976) Avocado water relations. In: Sauls, J.W., Phillips, R.L. and Jackson, L.K. (eds) *The Avocado. Proceedings of the 1st International Tropical Fruit Short Course.* Fruit Crops Department, University of Florida, Gainesville, Florida, pp. 47–53.

Hatton, T.T., Jr and Campbell, C.W. (1959) Evaluation indices for Florida avocado maturity. *Proceedings of the Florida State Horticulture Society* 72, 349–353.

Hatton, T.T. and Reeder, W.F. (1972) Relationship of bloom date to the size and oil content of 'Booth 8' avocados. *Citrus Industry* 53, 20–21.

Kadman, A. and Ben-Ya'acov, A. (1970a) Avocado. selection of rootstocks and other work related to salinity and time. In: *1960–1969 Report.* Division of Subtropical Horticulture, Volcanic Institute of Agricultural Research, Bet Dayan, Israel, pp. 23–39.

Kadman, A. and Ben-Ya'acov, A. (1970b) Vegetative propagation, A. Rooting of leaf-bearing cuttings. In: *1960–1969 Report.* Division of Subtropical Horticulture, Volcanic Institute of Agricultural Research, Bet Dayan, Israel, pp. 47–50.

Kaiser, C. and Wolstenholme, B.N. (1994) Aspects of delayed harvest of 'Hass' avocado (*Persea americana* Mill.) fruit in a cool subtropical climate. II. Fruit size, yield, phenology and whole tree starch cycling. *Journal of Horticultural Science* 69, 447–457.

Kawano, Y., Hylin, J.W. and Hamilton, R.A. (1976) *Plant Products of Economic Potential in Hawaii, III. Quantity and Quality of Oil Obtained from Hawaii-grown Avocado Varieties.* Research Report 211, Hawaii Agricultural Experimental Station, Honolulu.

Knight, R.J., Jr (1976) Breeding avocados for cold hardiness. In: Sauls, J.W., Phillips, R.L. and Jackson, L.K. (eds) *The Avocado. Proceedings of the 1st International Tropical Fruit Short Course.* Fruit Crops Department, University of Florida, Gainesville, Florida, pp. 33–36.

Lavi, U., Lahav, E., Genizi, A., Degani, D., Gazit, S. and Hillel, J. (1991) Quantitative genetic analysis of traits in avocado cultivars. *Plant Breeding* 106, 149–160.

Lavi, U., Lahav, E., Degani, C. and Gazit, S. (1992) The genetics of the juvenile phase in avocado and its application for breeding. *Journal of the American Society for Horticultural Science* 117, 981–984.

Lavi, U., Lahav, E., Degani, C. and Gazit, S. (1993a) Genetics of skin color, flowering group and anise scent in avocado. *Journal of Heredity* 84, 82–84.

Lavi, U., Lahav, E., Degani, C., Gazit, S. and Hillel, J. (1993b) Genetic variance components and heritabilities of several avocado traits. *Journal of the American Society for Horticultural Science* 118, 400–404.

Lutz, J.M. and Hardenburg, P.E. (1968) *The Commercial Storage of Fruits, Vegetables and Florist and Nursery Stocks.* ARS Handbook No. 66, USDA, Washington, DC.

McKellar, M.A., Buchanan, D.W., Ingram, D.L. and Campbell, C.W. (1992) Freezing tolerance of avocado leaves. *HortScience* 27, 341–343.

McMillan, R.T., Jr (1976) Diseases of avocado. In: Sauls, J.W., Phillips, R.L. and Jackson, L.K. (eds) *The Avocado. Proceedings of the 1st International Tropical Fruit Short Course.* Fruit Crops Department, University of Florida, Gainesville, Florida, pp. 66–70.

Malo, S.E. (1976) Mineral nutrition of avocados. In: Sauls, J.W., Phillips, R.L. and

Jackson, L.K. (eds) *The Avocado. Proceedings of the 1st International Tropical Fruit Short Course.* Fruit Crops Department University of Florida, Gainesville, Florida, pp. 42–46.

Miller, C.D., Bazore, K. and Bartow, M. (1965) *Fruits of Hawaii. Description, Nutritive Value, and Recipes.* University of Hawaii Press, Honolulu, Hawaii.

Nishimoto, R.K. and Yee, W.Y.J. (1980) *A Guide to Chemical Weed Control in Tropical and Subtropical Fruit and Nut Crops in Hawaii.* CES Circular 423 (revised), University of Hawaii, Honolulu.

Ploetz, R.C., Ramos, J.L., Parrado, J.L. and Shepard, E.S. (1991) Shoot and root growth cycles of avocado in South Florida. *Proceedings of the Florida State Horticulture Society* 104, 21–24.

Popenoe, W. (1920) *Manual of Tropical and Subtropical Fruits.* MacMillan, New York.

Rainey, C.A., Gilette, G., Brydon, A., McIntyre, S., Rivers, O., Vasquez, C.A. and Wilson, E. (1992) Physiological maturity and percent dry matter of California avocado. In: *Proceedings of the 2nd World Avocado Congress,* Vol. II, pp. 379–385.

Rainey, C., Afflick, M., Bretschger, K. and Alfin-Slater, R.B. (1994) The California avocado. *Nutrition Today* 29(3), 23–27.

Schaffer, B., Pena, J., Lara, S.P. and Buisson, D. (1986) Net photosynthesis, transpiration, and stomatal conductance of avocado leaves infested by avocado red mites. *Proceedings of the Interamerican Society for Tropical Horticulture* 30, 73–82.

Scora, R.W. and Bergh, B.O. (1992) Origin of and taxonomic relationship within the genus *Persea.* In: *Proceedings of the 2nd World Avocado Congress,* Vol. II, pp. 505–514. Los Angeles.

Sedgley, M. and Grant, W.J.R. (1983) Effect of low temperature during flowering on floral cycle and pollen tube growth in nine avocado cultivars. *Scientia Horticulturae* 18, 207–213.

Sedgley, M., Scholefield, P.B. and Alexander, McE.D. (1985) Inhibition of flowering of Mexican- and Guatemalan-type avocados under tropical conditions. *Scientia Horticulturae* 25, 21–30.

Spalding, D.H. (1976) Storage of avocados. In: Sauls, J.W., Phillips, R.L. and Jackson, L.K. (eds) *The Avocado. Proceedings of the 1st International Tropical Fruit Crop Short Course.* Fruit Crops Department, University of Florida, Gainesville, Florida, pp. 109–113.

Valmayer, R.V. (1967) Cellular development of the avocado from blossom to maturity. *The Philippine Agriculturalist* 50, 907–976.

Wenkam, N.S. (1990) *Foods of Hawaii and the Pacific Basin, Fruits and Fruit Products: Raw, Processed, and Prepared,* Vol. 4, *Composition,* Research Extension Series 110, College of Tropical Agriculture and Human Resources, University of Hawaii, Honolulu.

Whiley, A.W. (1984) Lauraceae. In: Page, P.E. (compiler) *Tropical Tree Fruit for Australia.* Information Series Q183018, Queensland Department of Primary Industry, Brisbane, pp. 70–77.

Whiley, A.W., Chapman, K.R. and Saranah, J.B. (1988b) Water loss by floral structures of avocado (*Persea americana* cv. Fuerte) during flowering. *Australian Journal of Agricultural Research* 39, 457–467.

Whiley, A.W., Saranah, J.B. and Rasmussen, T.S. (1992) Effect of time of harvest on fruit size, yield and trunk starch concentrations of 'Fuerte' avocado. *Proceedings of the 2nd World Avocado Congress,* Vol. I, pp. 155–159. Los Angeles.

Whiley, A.W., Hargreaves, P.A., Pegg, K.G., Doogan, V.J., Ruddle, L.J., Saranah, J.B. and Langdon, P.W. (1995) Changing sink strengths influence translocation of phosphorate in avocado (*Persea americana* Mill.) tree. *Australian Journal of Agricultural Research* 46, 1079–1090.

Williams, L.O. (1976) The botany of the avocado and its relatives. In: Sauls, J.W., Phillips, R.L. and Jackson, L.K. (eds) *The Avocado. Proceedings of the 1st International Tropical Fruit Crop Short Course*. Fruit Crops Department, University of Florida, Gainesville, Florida, pp. 9–15.

Wolstenholme, B.N. and Whiley, A.W. (1992) Requirements for improved fruiting efficiency in the avocado tree. *Proceedings of the 2nd World Avocado Congress*, Vol. I, 161–167. Los Angeles.

Yee, W. (1978) *Producing Avocado in Hawaii*. CES Circular 382 (revised), University of Hawaii, Honolulu.

Zentmyer, G.A. (1976) Soil-borne pathogens of avocado. In: Sauls, J.W., Phillips, R.L. and Jackson, L.K. (eds) *The Avocado. Proceedings of the 1st International Tropical Fruit Crop Short Course*. Fruit Crops Department, University of Florida, Gainesville, Florida, pp. 75–82.

CHAPTER 5. BANANA

Acedo, A.L. and Bautista, O.K. (1991) Enhancing ripening of 'Saba' banana (*Musa*, BBB group) fruits with *Gliricidia* leaves as ethylene source. *The Philippine Agriculturalist* 71, 351–365.

Baldry, J., Coursey, D.G. and Howard, G.E. (1981) The comparative consumer acceptability of triploid and tetraploid banana fruit. *Tropical Science* 23, 33–66.

Daniells, J.W. and O'Farrell, P.J. (1987) Effect of cutting height of the parent pseudostem on yield and time of production of the following sucker in banana. *Scientia Horticulturae* 31, 89–94.

Espino, R.R.C., Jamalualdin, S.H., Silayoi, B. and Nasution, R.E. (1992) *Musa* L. (edible cultivars). In: Verheij, E.W.M. and Coronel, R.E. (eds) *Plant Resources in South East Asia*, No. 2, *Edible Fruits and Nuts*. Prosea, Bogor, Indonesia.

Galan–Sauco, V. (1992) *Los Fruitakes Tropicales en los Subtropicas. II. Platano (banano)*. Mundiprensa, Madrid, 173 pp. (in Spanish).

Israeli, Y. and Lahav, E. (1986) Banana. In: Monselise, S.P. (ed.) *CRC Handbook of Fruit Set and Development*. CRC Press, Boca Raton, Florida, pp. 45–73.

Lahav, E. and Turner, D.W. (1983) *Fertilizing for High Yield-Banana*. Bulletin No. 7, International Potash Institute, Bern, 62 pp.

Lebot, V., Aradhya, K.M., Manshardt, R. and Meilleur, B. (1993) Genetic relationships among cultivated bananas and plantains from Asia and the Pacific. *Euphytica* 67, 163–175.

Lodh, S.B., Ravel, P., Selvaraj, V. and Kohli, R.R. (1971) Biochemical changes with growth and development of Dwarf Cavendish banana. *Indian Journal of Horticulture* 28, 38–45.

Loyola Santos, J.L., Shephard, K. and Alves, E.J. (1986) Propagacao rapida da bananeira. *Informe Agropecuario Belo Horizonte* 12(133), 33–38.

Martin–Prevel, P. (1990) Past, present, and future of tropical fruit nutrition with special reference to banana. *Acta Horticulturae* 275, 523–534.

Murray, D.B. (1961) Shade and fertilizer relations in the banana. *Tropical Agriculture* 38, 123–132.

Novak, F.J. (1992) *Musa* (banana and plantains). In: Hammerschlag, F.A. and Ritz, R.E. (eds) *Biotechnology of Perennial Fruit Crops*. CAB International, Wallingford, UK, pp. 449–488.

Reynolds, P.K. (1951) Earliest evidence of banana culture. *Journal of American Oriental Society* 71(4) (Suppl.), 28 pp.

Robinson, J.C. and Human, N.B. (1988) Forecasting of banana harvest ('Williams') in the subtropics using seasonal variation in bunch development rate and bunch mass. *Scientia Horticulturae* 34, 249–263.

Robinson, J.C. and Nel, D.J. (1985) Comparative morphology, phenology and production potential of banana cultivars 'Dwarf Cavendish' and 'Williams' in the Eastern Transvaal Lowveld. *Scientia Horticulturae* 25, 149–161.

Robinson, J.C. and Nel, D.J. (1990) Competitive inhibition of yield potential in a 'Williams' banana plantation due to excessive sucker growth. *Scientia Horticulturae* 43, 225–236.

Rowe, P. and Rosales, F.E. (1996) Bananas and plantains. In: Janick, J. and Moore, J.N. (eds) *Fruit Breeding*, Vol. I, *Tree and Tropical Fruits*. John Wiley & Sons, New York, pp. 167–211.

Smith, M.K. (1988) A review of factors influencing the genetic stability of micropropagated bananas. *Fruits* 43, 219–223.

Stover, R.H. and Simmonds, N.W. (1987) *Bananas*. Longmans, London.

Turner, D.W. (1970) Daily variation in banana leaf growth. *Australian Journal of Experimental Agriculture and Animal Husbandry* 10, 231–234.

Vuylsteke, D., Ortiz, R., Pasberg-Gauhl, C., Gauhl, F., Gold, C., Ferris, S. and Speizer, P. (1993) Plantain and banana research at the International Institute of Tropical Agriculture. *Hortscience* 28, 873–874, 970–971.

Wainwright, H. (1992) Improving the utilization of cooking bananas and plantains. *Outlook on Agriculture* 21, 177–181.

Wenkam, N.S. (1990) *Foods of Hawaii and the Pacific Basin, Fruits and Fruit Products, Raw, Processed, and Prepared*, Vol. 4, *Composition*. Research Extension Series 110, College of Tropical Agriculture and Human Resources, University of Hawaii, Honolulu.

CHAPTER 6. CARAMBOLA

Bookeri, M. (1996) Effect of irrigation on carambola (*Averrhoa carambola* L.) production in drought prone area. In: *Proceedings of the International Conference on Tropical Fruit, Kuala Lumpur, Malaysia, 1996 July 23–26*. Vol. I, pp. 317–320.

Campbell, C.A. (1971) Commercial production of minor tropical fruit crops in Florida. *Proceedings of the Florida State Horticultural Society* 84, 320–323.

Campbell, C.A. and Koch, K.E. (1989) Sugar/acid composition and development of sweet and tart carambola fruit. *Journal of American Society of Horticultural Science* 114, 455–457.

Darshana, N. (1970) Flowering and bearing behaviour of carambola (*Averrhoa carambola* Linn.). *Indian Journal of Horticulture* 27, 147–152.

Dave, Y.S., Patel, N.D. and Rupera, P.J. (1975) Structural studies in the pistil and developing fruit of *Averrhoa carambola* L. *Flora* (*Alemania*) 164, 479–486.

Galan Sauco, V. (1993) *Carambola Cultivation*. Plant Production and Protection Paper No. 108, FAO, Rome, 74 pp.

Galan Sauco, V., Hernandez Delgado, P.M. and Fernandez Galvan, D. (1989) Preliminary observations on carambola in the Canary Islands. *Proceedings of the Inter-American Society for Tropical Horticulture* 33, 55–58.

Ismail, M.R., Yusaf, M.K. and Masturi, A. (1996) Growth and flowering of water stressed starfruit plants and responses to ameliorated water stress. In: *Proceedings of the International Conference on Tropical Fruit, Kuala Lumpur, Malaysia, 1996 July 23–26*, Vol. II, pp. 97–106.

Knight, R.J. (1965) Heterostyly and pollination in carambola. *Proceedings of the Florida State Horticultural Society* 78, 375–378.

Knight, R.J. (1982) Partial loss of self-incompatibility in 'Golden Star' carambola. *Hortscience* 17, 72.

Knight, R.J. Jr (1983) Tropical fruits of Asia with potential for expanded world production. *Proceedings of the American Society for Horticultural Science in Tropical Regions* 27, 71–95.

Matthews, R.F. (1989) Processing of carambola. *Proceedings of the Inter-American Society for Tropical Horticulture* 33, 83–90.

Salakpetch, S. (1987) Flower ontogeny and pollen germination in carambola (*Averrhoa carambola* cv. Fwang Tung). MSc (Agric.) University of Western Australia, Perth, Australia, 82 pp.

Salakpetch, S., Turner, D.W. and Dell, B. (1990) Flowering in carambola (*Averrhoa carambola*). *Scientia Horticulturae* 43, 83–94.

Schnell, R.J. and Knight, R.J. Jr (1989) Variation in isozyme alleles in *Averrhoa carambola* L. *Proceedings of the Inter-American Society for Tropical Horticulture* 33, 127–132.

Sedgley, M. (1983) Oxalidaceae. In: Page, P.E. (ed.) *Tropical Tree Fruits for Australia*. Queensland Department of Primary Industries, Brisbane, pp. 125–128.

Shiesh, C.C., Lin, T.S., Wang, U.C. and Tsai, P.L. (1985) Flower morphology and flowering habit of carambola. *Chinese Horticulture* 31, 157–163.

Wang, W.C. (1994) Effect of fruit setting positions on yield and quality of carambola. *Journal of Agricultural Research in China* 43, 330–335 (in Chinese).

Watson, B.J., George, A.P., Nissen, R.J. and Brown, B.I. (1988) Carambola: a star on the horizon. *Queensland Agricultural Journal* 114, 45–51.

Wenkam, N. (1990) *Foods of Hawaii and the Pacific Basin Fruits and Fruit Products: Raw, Processed, and Prepared*, Vol. 4, *Composition*. Research Extension Series 110, College of Tropical Agriculture and Human Resources, University of Hawaii, Honolulu.

Yang, S.H. and Wang, W.C. (1993) Storage quality of Chang–Chwei carambola fruit. *Journal of Agricultural Research in China* 42, 387–395 (in Chinese).

CHAPTER 7. GUAVA

Batten, D.J. (1984) Guava (*Psidium guajava*). In: Page, P.E. (compiler) *Tropical Tree Fruits for Australia*. Information Series QI83018, Queensland Department of Primary Industry, Brisbane, Australia, pp. 113–120.

Chan, H.T., Jr and Cavaletto, C.G. (1982) Aseptically packaged papaya and guava puree: changes in chemical and sensory quality during processing and storage. *Journal of Food Science* 47, 1164–1174.

Chapman, K.R., Paxton, B., Saranah, J. and Scudamore-Smith, P.D. (1981) Growth, yield and preliminary selection of seedling guavas in Queensland. *Australian Journal of Agriculture and Animal Husbandry* 21, 119–123.

Choudhury, J.K. (1963) Early bearing of guava by photoperiodic induction. *Science and Culture* 29, 213–214.

Foss, S.L. (1980) Evaluation of resistance to *Selenothrips rubrocinctus* Giard. among various *Psidium* genotypes based on some fruit parameters. MS thesis, Department of Horticulture, University of Hawaii, Honolulu.

Hamilton, R.A. and Nakasone, H.Y. (1967) Bud grafting of superior guava cultivars. *University of Hawaii, Hawaii Farm Science* 16(2), 6–8.

Hirano, R.T. and Nakasone, H.Y. (1969a) Chromosome numbers of ten species and clones in the genus *Psidium. Journal of the American Society for Horticultural Science* 94, 83–86.

Hirano, R.T. and Nakasone, H.Y. (1969b) Pollen germination and compatibility studies of some *Psidium* species. *Journal of the American Society for Horticultural Science* 94, 287–289.

Huang, P.C. (1961) The investigations on the flowering and fruiting habits of guava tree. *Chinese Horticulture* 7(3), 27–36.

Ito, P.J. and Nakasone, H.Y. (1968) Compatibility and the inheritance of a seedling character in guava (*Psidium guajava*). *Proceedings of the American Society for Horticultural Science of Tropical Regions* 12, 216–221.

Ito, P.J., Kunimoto, R. and Ko, W.H. (1979) Transmission of *Mucor* rot of guava by three species of fruit flies. *Tropical Agriculture (Trinidad)* 56, 49–52.

Lam, P.F. (1987) Physico–chemical changes and eating quality of bagged and unbagged guava (*Psidium guajava* L. cv. Taiwan) during maturation. *MARDI Research Bulletin* 15, 27–30.

Maggs, D.H. (1984) Myrtaceae. In: Page, P.E. (compiler). *Tropical Tree Fruits of Australia*. Information Series Q 183018, Queensland Department of Primary Industry, Brisbane, Australia, pp. 108–112.

Malo, S.E. and Campbell, C.W. (1968) *The Guava*. Fruit Crops Fact Sheet No. 4, University of Florida, Homestead, Florida.

Mitchell, W.C. (1973) Insect and mite pests of guava. In: *CTA Statewide Guava Industry Seminar*. Miscellaneous Publication No. III, Cooperate Extension Service and Hawaii Agricultural Experiment Station, University of Hawaii, Honolulu.

Mohammed, S., Wilson, L.A. and Prendergast, N. (1984) Guava meadow orchard: effect of ultra high density plantings and growth regulators on growth, flowering and fruiting. *Tropical Agriculture (Trinidad)* 61, 297–301.

Nakasone, H.Y., Hamilton, R.A. and Ito, P.J. (1967) Evaluation of introduced cultivars of guavas. *University of Hawaii, Hawaii Farm Science* 16 (2), 4–6.

Nakasone, H.Y., Brekke, J.E. and Cavaletto, C.G. (1976) *Fruit and Yield Evaluation of Ten Clones of Guava (*Psidium guajava *L.)*. HAES Research Report 218, University of Hawaii, Honolulu.

Paull, R.E. and Goo, T. (1983) Relationship of guava (*Psidium guajava* L.) fruit detachment force to the stage of fruit development and chemical composition. *HortScience* 18, 65–67.

Paxton, B., Saranah, J. and Chapman, K.R. (1980) Guava propagation by cuttings. In: *Maroochy Horticultural Research Station Biennial Research Report. No. 2 for 1979–1980*, Nambour, Queensland, Australia, pp. 25–26.

Shanmugvelu, K.G. (1962) A preliminary study on the induction of parthenocarpic guava by gibberellic acid. *Indian Journal of Horticulture* 19, 125–131.

Shigeura, G.T. and Bullock, R.M. (1983) *Guava (*Psidium guajava *L.) in Hawaii – History and Production.* Hawaii Institute of Tropical Agriculture and Human Resources, Research Extension Series 035, University of Hawaii, Honolulu.

Wenkam, N.S. (1990) *Foods of Hawaii and the Pacific Basin Fruits and Fruit Products, Raw, Processed, and Prepared,* Vol. 4, *Composition.* Research Extension Series 110, College of Tropical Agriculture and Human Resources, University of Hawaii, Honolulu.

CHAPTER 8. LITCHI, LONGAN AND RAMBUTAN

Alexander, D.McE., Scholefield, P.B. and Frodsham, A. (1982) Litchi. In: *Some Tree Fruits for Tropical Australia.* Commonwealth Scientific and Industrial Research Organization, Canberra, p. 31.

Allen, B.M. (1967) *Malayan Fruits.* Donald Moore Press, Singapore.

Anon. (1985) *An Album of Guangdong Litchi Varieties in Full Color.* Guangdong Province Scientific Technology Commission, Guangdong, China, 78 pp.

Aradhya, M.K., Zee, F.T. and Manshardt, R.M. (1995) Isozyme variation in lychee (*Litchi chinensis* Sonn.). *Scientia Horticulturae* 63, 21–25.

Campbell, C.W. and Knight, R.J. Jr (1987) Production of litchis. In: *Cultivation and Production of Tropical Fruits. 23rd Congress of NORCOFEL, September 1983.* Tenerife (in Spanish), pp. 215–221.

Chapman, K.R. (1984) Sapindaceae. In: Page, P.E. (compiler) *Tropical Tree Fruits for Australia.* Information Series Q183018, Queensland Department of Primary Industry, Brisbane, pp. 179–191.

Chen, W.S. and Ku, M.L. (1988) Ethephon and kinetin reduce shoot length and increase flower bud formation in lychee. *HortScience* 23, 1078.

Chian, G.Z., Hee, Y.Q., Lin, Z.Y., Liu, M.Y. and Feng, X.B. (1996) *Longan High Production Cultivation Techniques.* Guangxi Science Technology Publishers, Nanning, 90 pp. (in Chinese).

Diczbalis, Y.A., Eamus, D. and Menzel, C.M. (1996) Environmental factors influencing growth and yield of rambutan, grown in the wet/dry tropics of Northern Australia. In: *Proceedings of an International Conference on Tropical Fruit. 23–26 July 1996, Kuala Lumpur, Malaysia,* Vol. II, pp. 15–24. MARDI, Kuala Lumpur.

Galan Sauco, V. (1990) *Los Frutales Tropicales en las Subtropicos. I. Aquacate, Mango, Litchi, Longan.* Mundiprensa, Madrid (in Spanish), 133 pp.

Groff, G.W. (1921) *The Lychee and Lungan.* Orange Judd, New York.

Joubert, A.J. (1986) Litchi. In: Monselise, S.P. (ed.) *CRC Handbook of Fruit Set and Development.* CRC Press, Boca Raton, Florida, pp. 233–246.

Kadman, A. and Gazit, S. (1970) *Flowering and Fruting of Litchi.* Division of Subtropical Horticulture, Volcani Institute of Agricultural Research, Bet Dagan, pp. 120–122.

Ke, G.W. (1990) Development of flesh and seed of longan. In: *Symposium on Litchi and Longan, December 1990.* Guangxi Subtropical Fruit Research and Development Office, Nanning, pp. 176–178.

Khan, I., Misra, R.S. and Srivastava, R.P. (1976) Effects of plant growth regulators on the fruit drop, size and quality of litchi cultivar Rose Scented. *Progress in Horticulture* 8, 61–69.

Lam, P.F. and Kosiyachinda, S. (eds) (1987) *Rambutan, Fruit Development, Postharvest*

Physiology and Marketing in ASEAN. ASEAN Food Handling Bureau, Kuala Lumpur, Malaysia, 82 pp.

Landrigan, M., Morris, S.C. and McGlasson, B.W. (1996) Postharvest browning of rambutan is a consequence of water loss. *Journal of the American Society for Horticultural Science* 121, 730–734.

Lian, T.K. and Chen, L.L. (1965) A preliminary observation on the flower bud initiation of longan in Foochow. *Acta Horticulture Sinica* 4, 13–18.

Limangkura, L. (1966) Studies on the nature of ovule abortion in the Groff variety of lychee (*Litchi chinensis* Sonn.). MS thesis, Department of Horticulture, University of Hawaii, Honolulu.

Lin, T.S., Shau, S.M., Yen, C.R., Lin, T.C. and Chang, J.W. (1991) Effects of fruits, leaves, light intensity and pruning on flowering of litchi (*Litchi chinensis* Sonn.). In: *Second Symposium on Forcing Culture of Horticultural Crops*. Special Publication No. 23, Taichung District Agricultural Improvement Station, Taichung, Taiwan (in Chinese, English summary), pp. 127–136.

Menzel, C.M. (1985) Propagation of lychee, a review. *Scientia Horticulturae* 25, 31–48.

Menzel, C.M. and Paxton, B.F. (1986a) A screening technique for lychee genotypes adapted to warm climates. *Acta Horticulturae* 175, 59–61.

Menzel, C.M. and Paxton, B.F. (1986b) The effect of cincturing on flowering of lychee in subtropical Queensland. *Acta Horticulturae* 175, 233–235.

Menzel, C.M. and Simpson, D.R. (1987) Effect of cincturing on growth and flowering of lychee over several seasons in subtropical Queensland. *Australian Journal of Experimental Agriculture* 27, 733–738.

Menzel, C.M. and Simpson, D.R. (1988) Effect of temperature on growth and flowering of litchi (*Litchi chinensis* Sonn.) cultivars. *Journal of Horticulture Science* 63, 349–360.

Menzel, C.M. and Simpson, D.R. (1990) Performance and improvement of lychee cultivars, a review. *Fruit Varieties Journal* 44, 197–215.

Menzel, C.M. and Simpson, D.R. (1991) A description of lychee cultivars. *Fruit Varieties Journal* 45, 45–56.

Menzel, C.M. and Simpson, D.R. (1994) Lychee. In: Schaffer, B. and Anderson, P.C. (eds) *Handbook of Environmental Physiology of Fruit Crops*, Vol II, *Subtropical and Tropical Crops*. CRC Press, Boca Raton, Florida, pp. 123–144.

Menzel, C.M., Rasmussen, T.S. and Simpson, D.R. (1989) Effect of temperature and leaf water stress on growth and flowering of litchi (*Litchi chinensis* Sonn.). *Journal of Horticultural Science* 64, 739–752.

Mo, B.C. (ed.) (1992) *Litchi High Production Cultivation and Technology*. Guangxi Science and Technology Publishers, Nanning, 97 pp. (in Chinese).

Morton, J.F. (1987) *Fruits of Warm Climates*. Creative Resource Systems, Winterville, North Carolina.

Mustard, M.J. (1954) Fundamentals of panicle differentiation. *Proceedings of the Florida Lychee Growers' Association* 1, 7–8.

Nakata, S. (1953) *Girdling as a Means of Inducing Flower Bud Initiation in Litchi*. Progress Note 95, Hawaii Agricultural Experiment Station, University of Hawaii, Honolulu.

Nakata, S. and Suehisa, R. (1969) Growth and development of *Litchi chinensis* as affected by soil-moisture stress. *American Journal of Botany* 56, 1121–1126.

Nakata, S. and Watanabe, Y. (1966) Effects of photoperiod and night temperature on the flowering of *Litchi chinensis*. *Botanical Gazette* 127, 146–152.

O'Hare, T.J. (1995) Postharvest physiology and storage of rambutan. *Postharvest Biology and Technology* 6, 189–199.

Pandey, R.S. and Yadava, R.P.S. (1970) Pollination of litchi (*Litchi chinensis* Sonn.) by insects with special reference to honey bees. *Journal of Apiculture Research* 9, 103–105.

Paull, R.E. and Chen, N.J. (1987) Effect of storage temperature and wrapping on quality characteristics of litchi fruit. *Scientia Horticulturae* 33, 223–236.

Paull, R.E., Chen, N.J., Deputy, J., Huang, H.B., Cheng, G.W. and Gao, F.F. (1984) Litchi growth and compositional changes during fruit development. *Journal of the American Society for Horticultural Science* 109, 817–821.

Paull, R.E., Reyes, M.E.Q. and Reyes, M.U. (1995) Litchi and rambutan insect disinfestation, treatments to minimize induced pericarp browning. *Postharvest Biology and Technology* 6, 139–148.

Prasad, A. and Jauhari, C.S. (1963) Effect of 2,4,5-T and alpha NAA on 'drop stop' and size of litchi fruits. *Madras Agriculture Journal* 50, 28–29.

Salma, I. (1986) *Rambutan (Nephelium lappaceum L.) Clones and their Classification.* Report No. 107, Malaysian Agricultural Research and Development Institute, Kuala Lumpur, 33 pp.

Singh, L.B. and Singh, U.P. (1954) *The Litchi* (Litchi chinensis *Sonn.*) Superintendent of Printing and Stationery, Lucknow, Uttar Pradesh, India.

Singh, U.S. and Lal, R.K. (1980) Influence of growth regulators on setting, retention and weights of fruit in two cultivars of litchi. *Scientia Horticulturae* 12, 321–326.

Storey, W.B. (1973) The lychee. In: *California Avocado Society Yearbook 1972–1973*, pp. 75–86.

Storey, W.B., Hamilton, R.A. and Nakasone, H.Y. (1953) *Groff – A New Variety of Lychee.* Circular 39, Hawaii Agricultural Experiment Station, University of Hawaii, Honolulu.

Wanichkul, K. (1980) A study on fruit development, harvesting index and postharvest changes of rambutan (*Nephelium lappaceum* L.) var. Seechompo. MSc thesis, Horticulture, Kasetsart University, Thailand, 90 pp.

Wenkam, N.S. (1990) *Foods of Hawaii and the Pacific Basin. Fruits and Fruit Products, Raw, Processed and Prepared*, Vol. 4, *Composition.* Research Extension Series No. 110, College of Tropical Agriculture and Human Resources, Honolulu, 96 pp.

Wills, R.B.H., Lim, J.S.K. and Greenfield, H. (1986) Composition of Australian foods. 31. Tropical and subtropical fruit. *Food Technology Australia* 38, 118–123.

Yen, C.R. (1995) Litchi. In: *Taiwan Agricultural Encyclopedia, Crop Edition*, Vol. 2. Agricultural Publisher Councils, Taipei, Taiwan, pp. 35–42 (in Chinese).

Yuan, R.C. and Huang, H.B. (1988) Litchi fruit abscission, its patterns, effect of shading, and relation to endogenous abscisic acid. *Scientia Horticulturae* 36, 281–292.

CHAPTER 9. MANGO

Adlan, H.A. (1965) Floral morphology as related to the fruit drop in mango (*Mangifera indica* L.). MSc thesis, Department of Horticulture, University of Hawaii, Honolulu, Hawaii.

Akamine, E.K. (1976) Problems in shipping fresh Hawaiian tropical and subtropical fruits. *Acta Horticulturae* 57, 157–161.

Azzouz, S., Said, G.A. and Abdella, M.Y. (1989) Studies on the control of malformation in some mango cultivars. *Zagazig Journal of Agricultural Research* 16, 43–47.

Bondad, N.D. and Apostol, C.J. (1979) Induction of flowering and fruiting in immature mango shoots with KNO₃. *Current Science* 48, 591–593.

Campbell, C.W. (1961) Comparison of yields of polyembryonic and monoembryonic mangos. *Proceedings of the Florida State Horticultural Society* 74, 363–365.

Campbell, C.W. (1988) Progress in mango cultivation. *Proceedings of the Interamerican Society Tropical Horticulture* 32, 8–19.

Campbell, R.J. and Campbell, C.W. (1992) Commercial Florida mango cultivars. *Acta Horticulturae* 341, 55–59.

Chang, M.T. and Lion, M.F. (1987) Delay fruit harvest by pinching mango inflorescence. In: Chang, L.R. (ed.) *Proceedings of the Symposium on Forcing Culture of Horticultural Crops*. Special Publication No. 10, Taichung District Agricultural Improvement Station, Taichung, Taiwan, pp. 119–128.

Crane, J.H. and Campbell, C.W. (1991) *The Mango*. Fruit Crops Fact Sheet FC-2, Florida Cooperative Extension Service, Gainesville, FL.

Degani, C., El–Batsri, R. and Gazit, S. (1990) Enzyme polymorphism in mango. *Journal of the American Society for Horticultural Science* 115, 844–847.

Galan Sauco, V. (1979) Actual situation, problems and prospects of mango development in the Canary Islands. *Acta Horticulturae* 102, 7–14.

Galan Sauco, V. (1996) Mango world production (outside Israel, Egypt and India). 5th International Mango Symposium. Tel Aviv, Israel, 1–6 September 1996. *Acta Horticulturae* (in press).

Gazit, S. and Kadman, A. (1980) 13–1 mango root stock selection. *HortScience* 15, 669.

Hamilton, R.A., Chia, C.L. and Evans, D.O. (1992) *Mango Cultivars in Hawaii*. Information Text Series 042, HITAHR, University of Hawaii, Honolulu.

Hayes, W.B. (1966) *Fruit Growing in India*, 3rd revised edn. Indian Universities Press, Allahabad, India.

Issarakraisila, M., Considine, J.A. and Turner, D.W. (1993) Effects of temperature on pollen viability in mango cv. Kensington. *Acta Horticulturae* 341, 112–124.

Ito, P.J., Hamilton, R.A. and Rapoza, H. (1992) *'Exel', a High Quality Dessert Mango*. Commodity Fact Sheet 169 MAN–3(B) Fruit, HITAHR, University of Hawaii, Manoa.

Joel, D.M. (1978) The secretory ducts of mango fruits, a defense system effective against Mediterranean fruit fly. *Israel Journal of Botany* 27, 44–45.

Knight, R.J., Jr (1985) Criteria for evaluating important characters in mango (*Mangifera indica* L.) germplasm. In: *Technology for Agricultural Development: Joint Proceedings of the 21st Caribbean Food Crops Society and the 32nd Annual Meeting of the American Society of Horticultural Science, Tropical Region*. ASHSTR, Port of Spain, pp. 57–60.

Knight, R.J. and Schnell, R.J. (1993) Mango (*Mangifera indica* L.) introduction and evaluation in Florida and its impact on the world industry. *Acta Horticulturae* 341, 125–135.

Lakshminarayana, S., Gomez-Cruz, A. and Martinez-Romero, S. (1985) Preliminary study of a new preharvest stem-end rot and associated microflora in mango. *HortScience* 20, 947–948.

Lovey, B.R., Robinson, S.P. Brophy, J.J. and Chacko, E.K. (1992) Mango sapburn, components of fruit sap and their role in causing skin damage. *Australian Journal of Plant Physiology* 19, 449–457.

Malo, S.E. (1976) Recent advances and possibility of mango culture in tropical

America with emphasis on the Florida situation. *Acta Horticulturae* 57, 47–52.

Malo, S.E. and Campbell, C.W. (1978) Studies on mango fruit breakdown in Florida. *Proceedings of the American Society for Horticultural Science of Tropical Regions* 22, 1–15.

Mukherjee, S.K. (1950) Cytological investigation of the mango and the allied Indian species. *Proceedings of the National Institute of Science India* 16, 281–303.

Nagao, M.A. and Nishina, M.S. (1993) Use of potasium nitrate on mango flowering. In: *Proceedings of a Conference on Mango in Hawaii.* College of Tropical Agriculture and Human Resources, University of Hawaii, Manoa, Honolulu, Hawaii, pp. 61–66.

Nakasone, H.Y. (1979) *Observations and Comments on Tropical Fruit Cultivation.* FIRA, Bank of Mexico, Mexico City (in Spanish).

Nakasone, H.Y., Bowers, F.A.I. and Beaumont, J.H. (1955) Terminal growth and flowering behavior of the 'Pirie' mango (*Mangifera indica* L.) in Hawaii. *Proceedings of the American Society of Horticultural Science* 66, 183–191.

Oosthuyse, S.A. and Jacobs, G. (1995) Relationship between branching frequency, and growth, cropping, and structural strength of 2 year old mango trees. *Scientia Horticulturae* 64, 85–93.

Pennock, W., Torres-Sepulveda, A, Lopez-Garcia, J., Reyes-Soto, I., Valle-Lamboy, S., Cedeno-Maldonado, A. and Jackson, G. (1972) Yield and fruit size comparison in the first size crops of 16 mango varieties. *Journal of Agriculture Puerto Rico* 56, 343–365.

Pinto, A.C.Q. and Sharma, D.K. (1984) Studies on growth behavior and sex expression in some exotic introductions of mango (*Mangifera indica* L.) in Brazil. *Proceedings of the American Society for Horticultural Science, Tropical Region* 28, 25–27.

Pollard, G.V. and Alleyne, E.H. (1986) Insect pests as constraints to the production of fruits in the Caribbean. In: Brathwaite, C.W.D., Marte, R. and Porsche, E. (eds) *Pests and Diseases as Constraints in the Production and Marketing of Fruits in the Caribbean.* Technical Events Series A2/TT-86-001, IICA, Bridgetown, Barbados.

Pope, W.T. (1929) *Mango Culture in Hawaii.* Hawaii Agricultural Experiment Station Bulletin 58, Honolulu.

Ram, S. and Sirohi, S.C. (1991) Feasibility of high density orcharding mango in Dashehari mango. *Acta Horticulturae* 291, 207–212.

Shawky, I., Zidan, Z., El-Tomi, A. and Dashan, D. (1980) Flowering malformation in relation to vegetative growth of 'Taimour' mangoes. *Egypt Journal of Horticulture* 7, 1–8.

Singh, L.B. (1960) *The Mango – Botany, Cultivation and Utilization.* Interscience Publishers, New York.

Tandon, D.K. and Kalra, S.K. (1983) Changes in sugars, starch, and amylase activity during development of mango fruit cv. Dashehari. *Journal of Horticultural Science* 58, 449–453.

Turnball, C.G.N., Anderson, K.L. and Winston, E.C. (1996) Influence of gibberellin treatment on flowering and fruiting patterns in mango. *Australian Journal of Experimental Agriculture* 36, 603–611.

Wenkam, N.S. (1990) *Foods of Hawaii and the Pacific Basin. Fruits and Fruit Products, Raw, Processed, and Prepared,* Vol. 4, *Composition.* Research Extension Series 110. College of Tropical Agriculture and Human Resources, University of Hawaii, Honolulu.

Whiley, A.W. and Saranah, J. (1981) Mango evaluation. In: *Biennial Research Report*

No. 2 for 1979–80. Maroochy Horticultural Research Station, Queensland Department of Primary Industries and Forestry, Brisbane, p. 47.

Whiley A.W., Mayers, P.E., Bartley, J.B. and Saranah, J.B. (1993) Breeding mangoes for Australian conditions. *Acta Horticulturae* 341, 136–145.

Yee, W. (1958) *The Mango in Hawaii.* Agriculture Extension Service Circular No. 388, University of Hawaii, Honolulu.

Young, T.W. and Koo, R.C.J. (1974) Increasing yield of 'Parvin' and 'Kent' mangos on Lakewood sand by increasing nitrogen and potassium fertilizer. *Proceedings of the Florida State Horticulture Society* 87, 380–384.

CHAPTER 10. PAPAYA

Adsuar, J. (1946) Transmission of papaya bunchy top by a leafhopper of the genus *Empoasca. Science* 103, 316.

Arkle, T.D., Jr (1973) Reproductive morphology and genetics of *Carica papaya* L. PhD dissertation, University of Hawaii.

Arkle, T.D., Jr and Nakasone, H.Y. (1984) Floral differentiation in the hermaphroditic papaya. *HortScience* 79, 832–834.

Awada, M. (1958) *Relationship of Minimum Temperature and Growth Rate with Sex Expression of Papaya Plants* (Carica papaya L.). Hawaii Agricultural Experiment Station Technical Bulletin 38, University of Hawaii, Honolulu.

Awada, M. (1977) Relations of nitrogen, phosphorus, and potassium fertilization to nutrient composition of the petiole and growth of papaya. *Journal of the American Society for Horticultural Science* 102, 413–418.

Awada, M. and Ikeda, W. (1957) *Effects of Water and Nitrogen Application on Composition, Growth, Sugars, in Fruit Yield, and Sex Expression of the Papaya Plants* (Carica papaya L). Hawaii Agricultural Experiment Station Technical Bulletin 33, University of Hawaii, Honolulu.

Awada, M., Suehisa, R.H. and Kanehiro, Y. (1975) Effects of lime and phosphorus on yield, growth, and petiole composition of papaya. *Journal of the American Society for Horticultural Science* 100, 294–298.

Awada, M., Wu, J.P., Suehisa, R.H. and Padgett, M.M. (1979) *Effects of Drip Irrigation and Nitrogen Fertilization on Vegetative Growth, Fruit Yield, and Mineral Composition of the Petioles and Fruits of Papaya.* Hawaii Agricultural Experiment Station Technical Bulletin No. 103, University of Hawaii, 20 pp. Honolulu.

Awada, M., De La Pena, R. and Suehisa, R.H. (1986) *Effects of Nitrogen and Potassium Fertilization on Growth, Fruiting and Petiole Composition of Bearing Plants.* Research Series 043, Hawaii Institute of Tropical Agriculture and Human Resources, University of Hawaii, Honolulu.

Badillo, V.M. (1993) *Caricaceae Segundo Esquema.* Alcance 43, Facultad de Agronoma, Universidad Central de Venezuela, 111 pp. (in Spanish).

Chan, H.T., Jr, Hibbard, K.L., Goo, T. and Akamine, E.K. (1979) Sugar composition of papaya during fruit development. *HortScience* 14, 140–141.

Chan, Y.K. (1985) Evaluation of the performance and stability of papaya varieties bred at MARDI. *MARDI Research Bulletin* 13, 1–7.

Chittiraichelvan, R. and Shanmugavelu, K.G. (1979) Studies of the growth and development of the fruit of CO. 2 papaya (*Carica papaya* L.) II. Changes in biochemical constituents. *Indian Journal of Horticulture* 36, 42–48.

Clemente, H. and Marler, T.E. (1996) Drought stress influences gas-exchange responses of papaya leaves to rapid changes in irradiance. *Journal of the American Society for Horticultural Science* 121, 292–295.

Conover, R.A., Litz, R.E. and Malo, S.E. (1986) 'Cariflora', a papaya for South Florida and the Caribbean. *HortScience* 21, 1072.

Davis, M.Y., Kramer, J.B., Ferwerdo, F.H. and Brunner, B.R. (1996) Association of a bacterium and not a phytoplasm with papaya bunchy top disease. *Phytopathology* 86, 102–109.

Gonzalves, D. and Ishii, M. (1980) Purification and serology of papaya virus. *Phytopathology* 70, 1028–1032.

Hamilton, R.A. (1954) *A Quantitative Study of Growth and Fruiting in Inbred and Crossbred Progenies from Two Solo Papaya Strains*. Hawaii Agricultural Experiment Station Technical Bulletin 20, Universitiy of Hawaii, Honolulu.

Higgins, J.S. and Holt, V.S. (1914) *The Papaya in Hawaii*. Hawaii Agricultural Experiment Station Bulletin 32, University of Hawaii, Honolulu.

Hiranpradit, H. (1983) Research on production of papaya in Thailand. In: Schirmer, A. (ed.) *Promoting Research on Tropical Fruits*. German Foundation for International Development, pp. 137–143.

Hofmeyer, J.D.J. (1938) *Genetical Studies of* Carica papaya L. Science Bulletin 187, Department of Agriculture and Forestry, Pretoria.

Hofmeyer, J.D.J. (1941) Metaxenia in *Carica papaya* L. *South African Journal of Science* 38, 130–132.

Ito, P.J. (1976) The effect of leaf pruning on yield and quality of 'Solo' papayas in Hawaii. *Proceedings of the American Society for Horticultural Science of Tropical Regions* 20, 46–50.

Kuhne, F.A. and Allan, P. (1970) Seasonal variation in fruit growth of *Carica papaya* L. *Agroplantae* 2, 99–104.

Lange, A.H. (1961) The effects of temperature and photoperiod on the growth of *Carica papaya*. *Ecology* 42, 481–486.

Marler, T. and Discekici, H.M. (1996) Root and stem extension of young papaya plants. *HortScience* 13, 685 (abstract).

Mekako, H.U. and Nakasone, H.Y. (1975) Interspecific hybridization among 6 *Carica* species. *Journal of the American Society for Horticultural Science* 100, 237–242.

Morshidi, M. (1996) Genetic variability in *Carica papaya* and related species. PhD dissertation, University of Hawaii at Manoa.

Mosqueda-Vazquez, R., Aragaki, M. and Nakasone, H.Y. (1981) Screening of *Carica papaya* L. seedlings for resistance to root rot caused by *Phytophthora palmivora* Butl. *Journal of the American Society for Horticultural Science* 106, 484–497.

Nakasone, H.Y. and Aragaki, M. (1973) Tolerance to *Phytophthora* fruit and root rot in *Carica papaya* L. *Proceedings of the American Society for Horticultural Science of Tropical Regions* 17, 176–185.

Nakasone, H.Y. and Storey, W.B. (1955) Studies on the inheritance of fruiting height of *Carica papaya* L. *Proceedings of the American Society for Horticultural Science* 66, 168–182.

Namba, R. (1985) Control of papaya ringspot virus in Hawaii. In: *Proceedings of the 20th Annual Hawaii Papaya Industry Association Conference, 1984*. College of Tropical Agriculture and Human Resources, University of Hawaii, Honolulu, pp. 86–87.

Paull, R.E. (1993) Pineapple and papaya. In: Seymour, G., Taylor, J. and Tucker, G.

(eds) *Biochemistry of Fruit Ripening*. Chapman & Hall, London, pp. 291–323.

Quintana, M.E.C. and Paull, R.E. (1993) Mechanical injury during postharvest handling of 'Solo' papaya fruit. *Journal of the American Society for Horticultural Science* 118, 618–622.

Roth, I. and Clausnitzer, I. (1972) Development and anatomy of the fruit and seed of *Carica papaya* L. *Acta Botonica Venezuela* 7, 187–206 (in Spanish).

Seo, S.T., Farias, G.J. and Harris, E.J. (1982) Oriental fruit fly, ripening of fruit and its effect on index of infestation of Hawaiian papayas. *Journal of Economic Entomology* 75, 173–178.

Shoji, K., Nakamura, M. and Matsumura, M. (1958) *Growth and Yield of Papaya in Relation to Fertilizer Application*. Progress Notes No. 118, College of Agriculture, University of Hawaii, Honolulu.

Storey, W.B. (1938) Segregations of sex types in Solo papaya and their application to the selection of seed. *Proceedings of the American Society for Horticultural Science* 35, 83–85.

Storey, W.B. (1941) *Papaya Production in the Hawaiian Islands*. Part I. *The Botany and Sex Relationships of the Papaya*. Hawaii Agricultural Experiment Station Bulletin 87, University of Hawaii, Honolulu.

Storey, W.B. (1953) Genetics of the papaya. *Journal of Heredity* 44, 70–78.

Storey, W.B. (1958) Modifications of sex expression in papaya. *Horticultural Advances* 2, 49–60.

Wenkam, N.S. (1990) *Foods of Hawaii and Pacific Basin Fruits and Fruit Products, Raw, Processed, and Prepared*, Vol. 4, *Composition*. Research Extension Series 110, College of Tropical Agriculture and Human Resources, Honolulu, 96 pp.

Yee, W. (1970) (ed.) *Papayas in Hawaii*. Cooperative Extension Service Circular 436, University of Hawaii, Honolulu.

Zee, F.T.P. (1985) Breeding for papaya ringspot virus tolerance in Solo papaya, *Carica papaya* L. PhD dissertation, Department of Horticulture, University of Hawaii.

CHAPTER 11. PASSION-FRUIT

Abanto, A.M. and Muller, L.E. (1976) Alterations produced in passion flower plants (*Passiflora edulis* Sims) by deficiencies of nitrogen, phosphorus and potassium. *Turrialba* 26, 331–336.

Abanto, A.M. and Muller, L.E. (1977a) Alterations produced in passion flower plants (*Passiflora edulis* Sims) by deficiencies in manganese, iron, boron and zinc. *Turrialba* 27, 163–168.

Abanto, A.M. and Muller, L.E. (1977b) Alterations produced in passion flower plants (*Passiflora edulis* Sims) by deficiencies of magnesium, calcium and sulfur. *Turrialba* 27, 221–225.

Akamine, E.K. and Girolami, G. (1959) *Pollination and Fruit Set in the Yellow Passion Fruit*. Hawaii Agricultural Experiment Station Bulletin 39, University of Hawaii, Honolulu.

Anon. (1972) *Passion Fruit Culture in Hawaii*. Cooperative Extension Service Circular 345 (revised), University of Hawaii, Honolulu.

Anon. (1984) *Passionfruit Culture*, 7th edn. District Crop Survey, North Moreton Region, Queensland Department of Primary Industry, Brisbane.

Arjona, H.E., Matta, F.B. and Garner, J.O. (1992) Temperature and storage time affect quality of yellow passionfruit. *HortScience* 27, 809–810.

Beal, P.R. and Farlow, P.J. (1984) Passifloraceae. In: *Tropical Tree Fruits for Australia*. Information Series Q183018, Queensland Department of Primary Industry, Brisbane.

Bester, C.W.J. (1980) Growing purple granadilla on yellow granadilla rootstocks. *Deciduous Fruit Grower* 30, 324–327.

Bowers, F.A.I. and Dedolph, R.R. (1959) A preliminary report on pruning of passionfruit. *Hawaii Farm Science* 7(4), 6,8.

Campbell, C.W. and Knight, R.J. (1987) Production of granadilla. In. *Cultivation and Production of Tropical Fruit: 13th NORCOFEL Congress*. Tenerife, Canary Islands, pp. 225–231 (in Spanish).

Chai, T.B. (1979) *Passion Fruit Culture in Malaysia*. Fruit Research Branch, MARDI, Sungaibaging.

Chapman, K.R., George, A.P. and Cull, B.W. (1978) A mechanical harvester for passionfruit. In: *Biennial Research Report No. 1 for 1977–78*. Maroochy Horticultural Research Station, Queensland Department of Primary Industry, Brisbane, pp. 71–75.

Farlow, P.J., Winks, C.W., Lanham,T.E. and Mayers, P.E. (1984) Genetic improvement in passionfruit. In: *Research Report 3 for 1981–1983*. Maroochy Horticultural Research Station, Queensland Department of Primary Industry, Brisbane, pp. 96–98.

Grech, N.M. and Frean, R.T. (1986) Preliminary comparisons as to tolerance of three *Passiflora* species to *Fusarium oxysporum* PV. *passiflorae* and *Phytophthora parasitica*. In: *CSFRI Symposium – Research into Citrus and Subtropical Crops*, Citrus and Subtropical Fruit Research Institute, Nelspruit, p. 49 (abstract).

Haddad, G.O. and Figueroa, R.M. (1972) Studies on flowering and fruiting in *Passiflora quadrangularis*. *Agronomie Tropicale (Venezuela)* 22, 483–496.

Hardin, L.C. (1986) Floral biology and breeding system of the yellow passionfruit, *Passiflora edulis* f. *flavicarpa*. *Proceedings of the Interamerican Society for Tropical Horticulture* 30, 35–44.

Ito, P.J. (1978) 'Noel's Special' passionfruit. *HortScience*. 13, 197.

Jan, F.J. and Yeh, S.D. (1995) Purification, *in situ* localization, and comparative serological properties of passionfruit woodeners virus-encoded amorphous inclusion protein and two other virus proteins. *Phytopathology* 85, 64–71.

Knight, R.J., Jr (1980) Origin and world importance of tropical and subtropical fruit crops. In: Nagy, S. and Shaw, P.W. (eds) *Tropical and Subtropical Fruits*. AVI Publishing, Westport, Connecticut, pp. 1–120.

McCarthy, G.J.P. (compiler) (1982) Passionfruit. In: *A Handbook of Plant Diseases in Color*, Vol. 1 (2nd edn). Information Publication Q182011, Queensland Department of Primary Industry, Brisbane.

Manders, G., Otani, W.C., d'Utra Vaz, F.B., Blackhall, N.W., Powers, J.B. and Davey, M.R. (1994) Transformation of passionfruit (*Passiflora edulis* f. *flavicarpa* Degener.) using *Agrobacterium tumefaciens*. *Plant Cell Reports* 13, 697–702.

Manica, I., Ritzinger, R., Koller, O.C., Riboldi, J., Ramos, R.M. and Rodrigues, A.E.C. (1985) Effect of six thicknesses of planting on the production of passion fruit (*Passiflora edulis* f. *flavicarpa* Deg.) during its third year, in Viamao, Rio Grande do Sul, Brazil. *Fruits* 40, 265–270.

Menzel, C.M. and Simpson, D.R. (1988) Effects of continuous shading on growth, flowering, and nutrient uptake of passionfruit. *Scientia Horticulturae* 35, 77–88.

Menzel, C.M., Simpson, D.R. and Dowling, A.J. (1986) Water relations in passion fruit: effect of moisture stress on growth, flowering and nutrient uptake. *Scientia Horticulturae.* 29, 239–249.

Menzel, C.M., Simpson, D.R. and Winks, C.W. (1987) Effect of temperature on growth, flowering, and nutrient uptake of three passion fruit cultivars under low irradiance. *Scientia Horticulturae.* 31, 259–268.

Menzel, C.M., Hayden, G.F., Doogan, V.J. and Simpson, D.R. (1993) New standard leaf nutrient concentrations for passion fruit based upon seasonal phenology and leaf composition. *Journal of Horticulture Science* 68, 215–229.

Nakasone, H.Y., Hirano, R. and Ito, P.J. (1967) *Preliminary Observations on the Inheritance of Several Factors in the Passionfruit* (Passiflora edulis Sims and forma flavicarpa *Deg.*). Hawaii Agricultural Experiment Station Progress Report 161, University of Hawaii, Honolulu.

Nakasone, H.Y., Aragaki, M. and Ito, P. (1973) *Alternaria* brown spot tolerance in passionfruit. *Proceedings of the American Society for Horticulture Science* 17, 159–165.

Ruggiero, C., Lam–Sanchez, A. and Banzatto, D.A. (1976) Studies on natural and controlled pollination in yellow passion fruit (*Passiflora edulis* f. *flavicarpa* Deg.). *Acta Horticulturae* 57, 121–124.

Seale, P.E. and Sherman, G.D. (1960) *Commercial Passionfruit Processing in Hawaii.* Circular 58, Hawaii Agricultural Experiment Station, University of Hawaii, Honolulu.

Shiomi, S., Wanoeho, L.S. and Agong, S.G. (1996) Ripening characteristics of purple passionfruit on and off the vine. *Postharvest Biology and Technology.* 7, 161–170.

Shyy, H.T., Fang, T.T. and Chen, H.E. (1987) Studies on picking maturity of passionfruit. I. Effect on the general quality of fruit juice. *Journal of the Chinese Society for Horticultural Science* 33, 51–67. II. Effect on the carbohydrate content of juice. *Journal of the Chinese Society for Horticultural Science* 33, 68–81.

Staveley, G.W. and Wolstenholme, B.N. (1990) Effects of water stress on growth and flowering of *Passiflora edulis* (Sims) grafted to *P. caerulea* L. *Acta Horticulturae* 275, 551–558.

Storey, W.B. (1950) Chromosome numbers of some species of *Passiflora* occurring in Hawaii. *Pacific Science* 4, 37–42.

Watson, D.P. and Bowers, F.A.I. (1965) Long days produce flowers on passion fruit. *Hawaii Farm Science* 14(2), 3–5.

Wenkam, N.S. (1990) *Food of Hawaii and the Pacific Basin, Fruits and Fruit Products, Raw, Processed, and Prepared,* Vol. 4, *Composition.* Research Extension Series No. 110, Hawaii Agricultural Experiment Station, College of Tropical Agriculture and Human Resources, University of Hawaii, Honolulu.

CHAPTER 12. PINEAPPLE

Aldrich, W.W. and Nakasone, H.Y. (1975) Day versus night application of calcium carbide for flower induction in pineapple. *Journal of the American Society for Horticultural Science* 100, 410–413.

Aradhya, M.K., Zee, F. and Manshardt, R.M. (1994) Isozyme variation in cultivated and wild pineapple. *Euphytica* 79, 87–99.

Armstrong, J.W. and Vargas, R.I. (1982) Resistance of pineapple variety 59-650 to field populations of Oriental fruit flies and melon flies. (Diptera, Tephritidae). *Journal of Economic Entomology* 75, 781–782.

Bartholomew, D.P. (1977) Inflorescence development of pineapple (*Ananas comosus* [L.] Merr). induced to flower with ethephon. *Botanical Gazette* 138, 312–320.

Bartholomew, D.P. and Criley, R.A. (1983) Tropical fruit and beverage crops. In: Nichells L.G. (ed.) *Handbook of Plant Growth Regulating Chemicals*, Vol. II. CRC Press, Boca Raton, Florida, pp. 1–34

Bartholomew, D.P. and Kadzimin, S.B. (1977) Pineapple. In: Alvin, P.T. and Kozlowski, T.T. (eds) *Ecophysiology of Tropical Crops*. Academic Press, New York, pp. 113–156.

Bartholomew, D.P. and Paull, R.E. (1986) Pineapple. In: Monselise, S.P. (ed.) *CRC Handbook of Fruit Set and Development*. CRC Press, Boca Raton, Florida, pp. 371–385.

Brewbaker, J.L. and Gorrez, D.D. (1967) Genetics of incompatibility in monocotyledonous genera *Ananas* (pineapple) and *Gasteria*. *American Journal of Botany* 54, 611–616.

Broadley, R.H., Wassman, R.C., III and Sinclair, E. (1993) *Pineapple Pests and Disorders*. Information Series QI92033, Queensland Department of Primary Industry, Brisbane, Australia.

Chan, Y.K. and Lee, C.K. (1985) The hybrid 1 pineapple, a new canning variety developed at MARDI. *Teknologi Buah-buahan Jil.* 1(1) (Mac), 24–30.

Collins, J.L. (1933) Morphological and cytological characteristics of triploid pineapple. *Cytologia* 4, 248–256.

Collins, J.L. (1948) Pineapples in ancient America. *Science Monthly* 65, 372–377.

Collins, J.L. (1949) History, taxonomy and culture of the pineapple. *Economic Botany* 3, 335–359.

Collins, J.L. (1960) *The Pineapple: Botany, Cultivation and Utilization*. World Crops Books, InterScience Publishers, New York.

Collins, J.L. and Kerns, K.R. (1938) Mutations in the pineapple: a study of thirty inherited abnormalities in the Cayenne variety. *Journal of Heredity* 29, 162–172.

Drew, R.A. (1980) Pineapple tissue culture unequalled for rapid multiplication. *Queensland Agriculture Journal* 106, 447–451.

Evans, D.O., Stanford, W.G. and Bartholomew, D.P. (1988) *Pineapple*. Commodity Fact Sheet PIN-3(A), College of Tropical Agriculture and Human Resources, Hawaii Institute of Tropical Agriculture and Human Resources, Honolulu.

Friend, D.C. and Lydon, J. (1978) Effect of daylength on flowering, growth and CAM of pineapple (*A. comosus* (L.)). *Botanical Gazette* 140, 280–283.

Glennie, J.D. (1981) The effects of temperature on the induction of pineapples with ethephon. In: *Biennial Report 2 for 1978–80*. Maroochy Horticultural Research Station, Queensland Department of Primary Industry, Brisbane, p. 69.

Gowing, D.P. (1961) Experiments on the photoperiodic response in pineapple. *American Journal of Botany* 48, 16–21.

Grazia, M., Antoni, S. and Leal, F. (1980) Key to the identification of commercial varieties of pineapple (*Ananas comosus*). *Proceedings of the American Society for Horticultural Science of Tropical Regions* 24, 107–112 (in Spanish).

Krauss, B.H. (1948) Anatomy of the vegetative organs of the pineapple, *Ananas comosus* (L) Merr. I. Introduction, organography, the stem, and the lateral branches or axillary buds. *Botanical Gazette* 110, 159–217.

Krauss, B.H. (1949) Anatomy of the vegetative organs of the pineapple, *Ananas comosus* (L) Merr. II. The leaf. *Botanical Gazette* 110, 333–404.

Lacoeuilhe, J.J. (1975) Etudes sur le contrôlle du cycle de l'ananas en Côte d'Ivoire. *Fruits* 30, 307–312.

Leal, F.J. and Antoni, M.G. (1980) Species of the genus *Ananas*: origin and geographic distribution. *Proceedings of the American Society for Horticultural Science of Tropical Regions* 24, 103–106. (in Spanish).

Leal, F.J. and Soule, J. (1977) 'Maipure', a new spineless group of pineapple cultivars. *HortScience* 12, 301–305.

Lim, W.H. (1985) *Diseases and Disorders of Pineapples in Peninsular Malaysia*. MARDI Report No. 97, Kuala Lumpur, Malaysia.

Loison–Cabot, C. (1992) Origin, phylogeny and evolution of pineapple species. *Fruits* 47, 25–32.

McCall, W.W. (1975) *Soil Classification in Hawaii*. Circular 426, Cooperative Extension Service, University of Hawaii, Manoa.

Malezieux, E., Zhang, J.B., Sinclair, E.R. and Bartholomew, D.P. (1994) Predicting pineapple harvest date in different environments, Amiga computer simulation models. *Agronomy Journal* 86, 609–617.

Mapes, M.O. (1973) Tissue culture of bromeliads. *International Plant Propagators Society* 23, 47–55.

Moreau, B. and Moreuil, C. (1976) L'ananas dans la région de Tamatave (Côte est de Madagascar): contribution à la connaissance de sa végétation en condition naturelle et dirigée. *Fruits* 31, 21–30.

Nightingale, G.T. (1942) Nitrate and carbohydrate reserves in relation to nitrogen nutrition of pineapple. *Botanical Gazette* 103, 409–456.

Okimoto, M.C. (1948) Anatomy and histology of the pineapple inflorescence and fruit. *Botanical Gazette* 110, 217–230.

Paull, R.E. (1993) Pineapple and papaya. In: Seymour, G., Taylor, J. and Tucker, G. (eds) *Biochemistry of Fruit Ripening*. Chapman & Hall, London, pp. 291–323.

Paull, R.E. and Reyes, M.E.Q. (1996) Preharvest weather conditions and pineapple fruit translucency. *Scientia Horticulturae* 66, 59–67.

Paull, R.E. and Rohrbach, K.G. (1985) Symptom development of chilling injury in pineapple fruit. *Journal of the American Society for Horticultural Science* 110, 100–105.

Rodriguez, A.G. (1932) Influence of smoke and ethylene on the flowering of pineapple (*Ananas sativas* Shult). *Journal of the Department of Agriculture Puerto Rico* 26, 5–18.

Rohrbach, K.G. (undated) *Pineapple: the Plant and Its Culture*. Leaflet, Hawaii Institute of Tropical Agriculture and Human Resources, Honolulu.

Rohrbach, K.G. (1989) Unusual tropical fruit diseases with extended latent periods. *Plant Disease* 73, 607–609.

Rohrbach, K.G. and Apt, W.J. (1986) Nematode and disease problems of pineapple. *Plant Disease* 70, 81–87.

Rohrbach, K.C. and Schmitt, D.P. (1994) Pineapple. In : Ploetz, R.C., Zentmyer, G.A., Nishijima, W.T., Rohrbach, K.G. and Ohr, H.D. (eds) *Compendium of Tropical Fruit Diseases*. American Phytopathological Society, St Paul, Minnesota, pp. 45–55.

Rohrbach, K.G., Beardsley, J.W., German, T.L., Reiner, N.J. and Sanford, W.G. (1988) Mealybug wilt, mealybugs, and ants on pineapple. *Plant Disease* 72, 558–565.

Sanford, W.G. (1962) Pineapple crop log – concept and development. *Better Crops Plant Food* 46, 32–43.

Sanford, W.G. (1964) Factors influencing the interpretation of the crop log. 4. In:

Bould, C., Prevot, P. and Magness, J.R. (eds) *Plant Analysis and Fertilizer Problems.* W.F. Humphrey Press, Geneva, New York, pp. 255–267.

Sideris, C.P. and Krauss, B.H. (1936) The classification and nomenclature of groups of pineapple leaves, sections of leaves, and sections of stem based on morphology and anatomical differences. *Pineapple Quarterly* 6, 135–147.

Soler, A. (1992) Métabolisme de l'éthéphon dans l'épiderme de l'ananas (*Ananas comosus*, (L.) Merr.). *Fruits* 47, 471–477.

Teisson, C. (1979a) Le brunissement interne de l'ananas. I. Historique. II. Matériel et méthodes. *Fruits* 34, 245–261.

Teisson, C. (1979b) Le brunissement interne de l'ananas. III. Symptomatologie. IV. Approche biochémique du phénomène. *Fruits* 34, 315–339.

Wassman, R.C. (1990) Effects of seasonal temperature variation on pineapple scheduling for canning in Queensland. *Acta Horticulturae* 275, 131–138.

Wenkam, N.S. (1990) *Foods of Hawaii and the Pacific Basin Fruits and Fruit Products, Raw Processed, and Prepared, Vol. 4. Composition.* Research Extension Series No. 110, College of Tropical Agriculture and Human Resources, Honolulu, Hawaii.

Winks, C.W. and Glennie, J.D. (1981) Pineapple breeding. In: *Biennial Research Report 2 for 1979–80.* Maroochy Horticultural Research Station, Queensland Department of Primary Industry, Brisbane, pp. 69–70.

CHAPTER 13. OTHER ASIAN TROPICAL FRUIT

Breadfruit

Atchley, J. and Cox, P.A. (1985) Breadfruit fermentation in Micronesia. *Economic Botany* 39, 326–335.

Bates, R.P., Graham, H.D., Matthews, R.F. and Clos, L.R. (1991) Breadfruit chips, preparation, stability, and acceptability. *Journal of Food Science* 56, 1608–1610.

Bennett, F.D. and Nozzolillo, C. (1987) How many seeds in a seeded breadfruit? *Economic Botany* 41, 370–374.

Dignan, C.A., Burlingame, B.A., Arthur, J.M., Quigley, R.J. and Milligan, G.C. (1994) *The Pacific Islands Food Composition Tables.* South Pacific Commission, Noumea, New Caledonia.

Fosberg, F.R. (1960) Introgression in *Artocarpus* (Moraceae) in Micronesia. *Brittonia* 12, 101–113.

Graham, H.D. and Negron, E. (1981) Composition of breadfruit. *Journal of Food Science* 46, 535–539.

Hamilton, R.A., Criley, R.A. and Chia, C.L. (1982) Rooting of stem cuttings of breadfruit (*Artocarpus altilis* [Parkins.] Fosb.) under intermittent mist. *International Plant Propagator's Society Proceedings* 32, 347–350.

Hasan, S.M.Z. and Razak, A.R. (1992) Parthenocarpy in seedless breadfruit (*Arthocarpus incircus* (Thumb) L.). *Acta Horticulturae* 321, 648–652.

Ragone, D. (1988) *Breadfruit Varieties in the Pacific Atolls.* Office of Project Services, United Nations Development Programme, 45 pp.

Rajendran, R. (1992) *Artocarpus altilis* (Parkinson) Fosberg. In, Verheij, E.W.M. and Coronel, R.E. (eds) *Plant Resources of South East Asia, No. 2, Edible Fruits and Nuts.* Prosea, Bogor, Indonesia, pp. 83–86.

Rowe-Dutton, P. (1976) *Artocarpus altilis* – breadfruit. In: Garner, R.J. and Chaudhri, S.A. (eds) *The Propagation of Tropical Trees*. Horticultural Review No. 4, Commonwealth Agricultural Bureau, pp. 248–268.

Trujillo, E.E. (1971) *The Breadfruit Diseases of the Pacific Basin*. Information Document 27, South Pacific Commission, Noumea, New Caledonia, 28 pp.

Wenkam, N.S. (1990) *Foods of Hawaii and the Pacific Basin. Fruits and Fruit Products, Raw, Processed, and Prepared*, Vol. 4, *Composition*. Research Extension Series No. 110, College of Tropical Agriculture and Human Resources, University of Hawaii, 96 pp.

Whitney, P.J. (1988) The microbiology of breadfruit and cassava preservation by pit fermentation. *Tropical Science* 28, 43–50.

Wooten, M. and Tumoalii, F. (1984) Breadfruit production, utilization, and composition – a review. *Food Technology Australia* 36, 464.

Jackfruit and Chempedak

Bhutani, D.K. (1978) Pests and diseases of jackfruit in India and their control. *Fruits* 33, 352–357.

Chatterjee, B.K. and Mukherjee, S.K. (1980) Effect of different media on rooting of cuttings of jackfruit (*Artocarpus heterophyllus* Lam). *Indian Journal of Horticulture* 37, 360–363.

Jarrett, F.M. (1959) Studies in *Artocarpus* and related genera III. A revision of *Artocarpus* subgenus *Artocarpus*. *Journal of Arnold Arboretum Harvard University* 40, 329–334.

McMillan, R.T. (1974) *Rhizopus artocarpi* rot of jackfruit (*Artocarpus heterophyllus*). *Proceedings of Florida State Horticulture Society* 87, 392–393.

Moncur, M.W. (1985) Floral ontogeny of the jackfruit, *Artocarpus heterophyllus* Lam. (Moraceae). *Australian Journal of Botany* 33, 585–593.

Muda, P., Othman, A. and Noor, H.M. (1996) Physico-chemical changes during development and maturation of NS1 jackfruit. In: *Proceedings of an International Conference on Tropical Fruits, Kuala Lumpur, Malaysia, 23–26 July*. Vol. 1, pp. 381–385. Kuala Lumpur.

Mukherjee, S.K. and Chatterjee, B.K. (1979) Effects of forcing, etoliation, and indole butyric acid on rooting of cuttings of *Artocarpus heterophyllus* Lam. *Scientia Horticulturae* 10, 295–300.

Roy, S.K., Rahman, S.L. and Majumdar, R. (1990) *In vitro* propagation of jackfruit (*Artocarpus heterophyllus* Lam.) *Journal of Horticulture Science* 65, 355–358.

Sambamurty, K. and Ramalingam, V. (1954) Preliminary studies in blossom biology of the jack (*Artocarpus heterophyllus* Lam) and pollination effects. *Indian Journal of Horticulture* 11, 24–29.

Soepadmo, E. (1992) *Artocarpus heterophyllus* Lamk. In: Verheij, E.W.M. and Coronel, R.E. (eds) *Plant Resources of South East Asia*, No. 2, *Edible Fruits and Nuts*. Prosea, Bogor, Indonesia, pp. 86–91.

Thomas, C.A. (1980) Jackfruit, *Artocarpus heterophyllus* (Moraceae), a source of food and income. *Economic Botany* 34, 154–159.

Durian

Baldry, J., Dougan, J. and Howard, G.E. (1972) Volatile flavoring constituents of durian. *Phytochemistry* 11, 2081–2084.

Brooncherm, P. and Siriphanich, J. (1991) Postharvest physiology of durian pulp and husk. *Kasetsart Journal (Natural Science)* 25, 119–125.

Chandraparnik, S., Hiranpradit, H., Punnachit, U. and Salakpetch, S. (1992a) Paclobutrazol influences flower induction in durian *Durio zibethinus* Murr. *Acta Horticulturae* 321, 282–290.

Chandraparnik, S., Hiranpradit, H., Salakpetch, S. and Punnachit, U. (1992) Influence of thiourea on flower bud burst in durian, *Durio zibethnius* Murr. *Acta Horticulturae* 321, 636–640.

Chua, S.E. and Teoh, T.S. (1973) Propagation by approach grafting and woody cutting of some tropical fruit and ornamental trees. *Singapore Journal of Primary Industry* 1, 87–95.

Chua, S.E. and Young, S.K. (1978) The use of approach, bud, and wedge grafting technique to propagation durian (*Durio zibethinus* Murr.), rambutan (*Nephelium lappaceum*, L.) and mango (*Mangifera indica*, L.). *Singapore Journal of Primary Industry* 6, 94–101.

Hasan, B.M. and Yaacob, O. (1986) The growth and productivity of selected durian clones under the plantation system at Serdang, Malaysia. *Acta Horticulturae* 175, 55–58.

Hiranpradit, H., Someri, S., Chandraparnik, S. and Detpittayanan, V. (1992a) Clonal selection of *Durio zibethinus* Murr. *Acta Horticulturae* 321, 164–172.

Hiranpradit, H., Lee–Ungulasatian, N., Chandraparnik, S. and Jantigoo, S. (1992b) Quality standardization of Thai durian, *Durio zibethinus* Murr. *Acta Horticulturae* 321, 695–704.

Ketsa, S. and Pangkool, S. (1994) The effect of humidity on ripening of durians. *Postharvest Biology and Technology* 4, 159–165.

Lim, T.K. (1990) *Durian – Diseases and Disorders*. Tropical Press SDN, BHD, Kuala Lumpur, Malaysia, pp. 60–72.

Malo, S.E. and Martin, F.W. (1979) *Cultivation of Neglected Tropical Fruits with Promise, Part 7, The Durian*. US Department of Agricultural Science and Education Administration, New Orleans.

Mamat, A.S. and Wahab, A.A. (1992) Gibberellins in the developing flower and fruit of *Durio zibethinus* Murr. *Acta Horticulturae* 292, 101–106.

Ng, S.K. and Thamboo, S. (1967) Nutrient removal studies on Malaysian fruits, durian and rambutan. *Malaysian Agriculture Journal* 46, 164–182.

Punnachit, U., Kwangthong, C. and Chandraparnik, S. (1992) Effect of plant growth regulators and fertilizers on leaf flushing and quality of durian. *Acta Horticulturae* 321, 343–347.

Salakpetch, S., Chandraparnik, S. and Hiranpradit, H. (1992a) Pollen grains and pollination in durian, *Durio zibethinus* Murr. *Acta Horticulturae* 321, 636–640.

Soegeng-Reksodihardjo, W. (1962) The species of *Durio* with edible fruits. *Economic Botany* 16, 270–282.

Sriyook, S., Siriatiwat, S. and Siriphanich, J. (1994) Durian fruit dehiscence – water status and ethylene. *HortScience* 29, 1195–1198.

Tai, L.H. (1973) Susceptibility of durian clones to patch canker disease. *MARDI Research Bulletin* 1, 5–9.

Tongdee, S.C., Chayasombat, A. and Neamprem, S. (1987a) Effects of harvest maturity on respiration, ethylene production, and composition of internal atmospheres of durian (*Durio zibethinus*, Murray). *Proceedings of a Durian Conference 25–26 February 1987*. Thailand Institute of Science and Technology, Bangkok, pp. 31–36.

Tongdee, S.C., Neamprem, S. and Chayasombat, A. (1987b) Control of postharvest infection of *Phytophthora* fruit rot in durian with Fosetyl–Al and residue levels in fruit. In: *Proceedings of a Durian Conference, 25–26 February 1987*. Thailand Institute of Science and Technology, Bangkok, pp. 55–66.

Watson, B.J. (1983) Durian (*Durio zibethinus* Murr.). In: Page, P.E. (compiler) *Tropical Tree Fruits for Australia*. Queensland Department of Primary Industries, Brisbane, Queensland, pp. 45–50.

Langsat, Duku and Santol

Anon. (1984) *Tips on Handling Lanzones*. ASEAN Postharvest Horticultural Trainings and Research Center, University of the Philippines at Los Baños, College, Laguna, Philippines, 2 pp.

Blackler, M.H. (1976) *Lansium domesticum* – langsat. In: T*he Propagation of Tropical Fruit Trees*. Horticultural Review 4, Commun. Ber. Horticulture Plantation Crops, Commonwealth Agricultural Bureau, East Malling, UK, pp. 376–385.

Salma, I. and Razali, B. (1987) The reproductive biology of Duku langsat, *Lansium domesticum* Corr. (Meliaceae) in peninsular Malaysia. *MARDI Research Bulletin* 15, 141–150.

Sotto, R.C. (1992) *Sandoricum koetjape* (Burm. f) Merr. In: Verheij, E.W.M. and Coronel, R.E. (eds) *Plant Resources of South East Asia*, No. 2, *Edible Fruits and Nuts*. Prosea, Bogor, Indonesia, pp. 284–287.

Yaacob, O. and Bamroongrigsa, N. (1992) *Lansium domesticum* Correa. In: Verheij, E.W.M. and Coronel, R.E. (eds) *Plant Resources of South East Asia*, No. 2, *Edible Fruits and Nuts*. Prosea, Bogor, Indonesia, pp. 186–190.

Mangosteen

Achmad, S., Mohamed, Z.A., Teck, C.S., Hamidah, W. and Hussein, W. (1983) Past, present and suggested future research on mangosteen with example of research and production in Malaysia. In: *International Workshop for Promoting Research on Tropical Fruits, Jakarta, 30 May – 6 June*. Jakarta, 18 pp.

Alexander, D.McE. (1983) Guttiferae. In: Page, P.E. (compiler) *Tropical Tree Fruits for Australia*. Queensland Department of Primary Industry, Brisbane, Australia, pp. 66–68.

Almeyda, N. and Martin, F.W. (1976) *Cultivation of Neglected Tropical Fruits with Promise, Part 1. Mangosteen*. ARS–S–155, US Department of Agriculture, New Orleans.

Augustin, M.A. and Azudin, M.N. (1986) Storage of mangosteen (*Garcinia mangostana* L.). *ASEAN Food Journal* 2, 78–80.

Campbell, C.W. (1967) Growing the mangosteen in Southern Florida. *Proceedings of the Florida State Horticultural Society* 79, 399–400.

Gonzalez, L.G. and Anoos, Q.A. (1951) The growth behavior of mangosteen and its graft affinity with relatives. *Philippine Agriculturists* 35, 379–395.

Hume, E.P. and Cobin, M. (1948) Relative of seed size to germination and early growth of mangosteen. *Proceedings of the American Society for Horticultural Science* 48, 289–302.

Krishnamurthi, S., Rao, V.N.M. and Ravoof, N.A. (1964) A note on the flowers and floral biology in mangosteen (*Garcinia mangostana* L.). *South Indian Horticulture* 12, 99–101.

MacLeod, A.J. and Pieris, W.M. (1982) Volatile flavor components of mangosteen *Garcinia mangostana. Phytochemistry* 21, 117–119.

Marshall, J.R. and Marshall, J. (1983) *Mangosteen (Purple).* Fact Sheet No. 3, Rare Fruit Council of Australia, Cairns, Queensland.

Poonnachit, U., Salakpetch, S., Chandraparnik, S. and Hiranpradit, H. (1992) Integrated technology to improve mangosteen production. Chanthaburi Horticultural Research Center, Thailand (in Thai). (Communicated by S. Salakpetch, 1996.)

Richards, N.J. (1990) Studies in *Garcinia,* dioecious tropical fruit trees, the origin of the mangosteen (*Garcinia mangostana* L.) *Botanical Journal of the Linnean Society* 103, 301–308.

Salakpetch, S. (1996) *Technology to Improve Mangosteen Production. Orchard Management Workshop 22–27 July 1996.* Chanthaburi Horticultural Research Center, Chanthaburi, Thailand.

Siong, T.E., Noor, M.I., Azudin, M.N. and Idris, K. (1988) *Nutrient Composition of Malaysian Foods.* ASEAN Food Habits Project, Kuala Lumpur, Malaysia.

Tongdee, S.C. (1985) *Mangosteen.* Second Progress Report ACIAR Project No. 8356. Thailand Institute of Scientific and Technological Research, Bangkok, Thailand.

Weibel, J., Eamms, D., Chacko, E.K., Downton, W.J.S. and Ludders, P. (1993) Gas exchange characteristics of mangosteen (*Garcinia mangostana* L.) leaves. *Tree Physiology* 13, 55–69.

Wax Apple

Leung, W-T.W. and Flores, M. (1961) *Food Composition Table for Use in Latin America.* National Institute of Health, Bethesda, Maryland.

Panggabean, G. (1992) *Syzygium aqueum* (Burm f.) Aeston, *Syzygium malaccense* (L.) Merr. & Perry. *Syzygium samarangense* (Blume) Merr. & Perry. In: Verheij, E.W.M. and Coronel, R.E. (eds) *Plant Resources of South East Asia,* No. 2, *Edible Fruits and Nuts.* Prosea, Bogor, Indonesia, pp. 292–298.

Shu, Z.H., Wang, D.N., Wong, R.H., Lee, K.C. and Lin, H.L. (1994) Studies on the relationship between flowering and leaf color as well as leaf and soil nutrient status of wax apple. In: Lin, H.S. and Chang, L.R. (eds) *Proceedings of a Symposium on the Practical Aspects of Some Economical Fruit Trees in Taiwan, Pingtung, Taiwan.* Taichung District Agricultural Improvement Station Special Publication No. 33, Taichung, Taiwan, pp. 5–11.

Wang, D.N. (1989) Nutrition and fertilization of wax apple. In: Chang, L.R. (ed.) *Fruit Tree Nutrition and Orchard Soil Management.* Taichung District Agricultural Improvement Station Special Publication No. 20, Taichung, Taiwan, pp. 119–132 (Chinese, English abstract).

Wang D.-N., Shu, Z.-H. and Sheen, T.-F. (1994) Wax apple production in Taiwan. *Chronica Horticulturae* 35(4), 11–12.

Yang, R.M., Lin, T.S., Wang, D.N. and Lee, C.L. (1991) Studies on bud morphogenesis in wax apple (*Syzygium samarangense* Merr. & Perry). In: Lin, H.S. and Chang, L.R. (eds) *Proceedings of the 2nd Symposium on Forcing Culture of Horticultural Crops.* Nan Ton Taiwan, Taichung District Agricultural Improvement Station Special Publication No. 23, Taichung, Taiwan, pp. 137–151 (in Chinese, English abstract).

CHAPTER 14. OTHER AMERICAN TROPICAL FRUIT

Acerola

Alves, R.E., Chitarra, A.B. and Chitarra, M.I.F. (1995) Postharvest physiology of acerola (*Malpighia emarginata* D.C.) fruits, maturation changes, respiratory activity, and refrigerated storage at ambient and modified atmospheres. *Acta Horticulturae* 370, 223–229.

Arostegui, F. and Pennock, W. (1955) *The Acerola*. Miscellaneous Publication 15, Agricultural Experiment Station, University of Puerto Rico, Rio Piedras, Puerto Rico.

Arostegui, F., Asenjo, C.F., Muniz, A.I. and Alemany, L. (1954) Studies on the West Indian cherry, *Malphigia punicifolia* L., observation and data on a promising selection. *Proceedings of the Florida State Horticulture Society* 67, 250–255.

Holtzmann, O.V. and Aragaki, M. (1966) Susceptibility of acerola to *Cercospora* leaf spot. *Phytopathology* 56, 1114–1115.

Landrau, P., Jr and Hernandez Medina, F. (1959) Effects of major and minor elements, lime, and soil amendments on the yield and ascorbic acid content of acerola (*Malphigia punicifolia* L.). *University of Puerto Rico Journal of Agriculture* 43, 19–33.

Landrau, P., Jr and Samuels, G. (1956) Results of lime and minor element fertilizer research in Puerto Rico, 1949–1950. *University of Puerto Rico Journal of Agriculture* 40, 224–234.

Ledin, R.B. (1958) *The Barbados or West Indian Cherry*. Bulletin 594, Agricultural Experiment Station, University of Florida, Gainesville.

Miller, C.D., Wenkam, N.S. and Fitting, K.O. (1961) *Acerola – Nutritive Value and Home Use*. Circular 59, Hawaii Agricultural Experiment Station, University of Hawaii, Honolulu.

Miyashita, R.K. (1963). Reproductive morphology of acerola (*Malphigia glabra* L.). MS thesis, Department of Horticulture, University of Hawaii, Honolulu.

Miyashita, R.K., Nakasone, H.Y. and Lamoureux, C.H. (1964) Reproductive morphology of acerola (*Malphigia glabra* L.). Technical Bulletin 63, Hawaii Agricultural Experiment Station, University of Hawaii.

Moscoso, C.G. (1956) West Indian cherry – richest known source of natural vitamin C. *Economic Botany* 10, 280–294.

Mustard, M.J. (1946) The ascorbic acid content of some *Malpighia* fruits and jellies. *Science* 104, 230–231.

Nakasone, H.Y., Miyashita, R.K. and Yamane, G.M. (1966) Factors affecting ascorbic acid content of the acerola (*Malpighia glaba* L.) *Proceedings of the American Society for Horticultural Science* 89, 161–166.

Nakasone, H.Y., Yamane, G.M. and Miyashita, R.K. (1968) Selection, evaluation, and naming of acerola (*Malpighia glabra* L.) cultivars. Circular 65, Hawaii Agricultural Experiment Station, University of Hawaii, Honolulu.

Pollard, G.V. and Alleyne, E.H. (1986) Insect pests as constraints to the production of fruits in the Caribbean. In: Brathwaite, C.W.D., Marte, R. and Porsche, E. (eds) *Proceedings of a Seminar on Pests and Diseases as Constraints in the Production and Marketing of Fruits in the Caribbean, Barbados*. Technical Events Series A2/TT-86-001, IICA, Barbados.

Santini, R., Jr (1952). Identification and determination of polybasic organic acids present in West Indian cherries (*Malpighia punicifolia* L.) and in three varieties of guava (*Psidium guajava* L.). *University of Puerto Rico Journal of Agriculture* 37, 195–198.

Wenkam, N.S. (1990). *Foods of Hawaii and the Pacific Basin, Fruits and Fruit Products, Raw, Processed, and Prepared*, Vol. 4, *Composition*. Research Extension Series No. 110, College of Tropical Agriculture and Human Resources, University of Hawaii, Honolulu.

Chiku and Abiu

Abdul-Karim, M.N.B., Tarmizi, S.A. and Bakar, A.A. (1987) The physico-chemical changes in Ciku (*Achras sapota* L.) of Jan Tung variety. *Pertanika* 10, 277–282.

Ali, S.H. and Lin, T.S. (1996) Fruit development and maturation of sapodilla cv. Subang. In: *Proceedings of an International Conference on Tropical Fruits, Kuala Lumpur, Malaysia, 23–26 July*. Vol .1, pp. 397–402. MARDI, Kuala Lumpur.

Almeyda, N. and Martin, F.W. (1976) *Cultivation of Neglected Tropical Fruits with Promise. 2. The Mammey Sapote*. ARS-S-156, United States Department of Agriculture, New Orleans.

Broughton, W.J. and Wong, H.C. (1979) Storage conditions and ripening of chiku fruits *Achras sapota* L. *Scientia Horticulturae* 10, 377–385.

Campbell, C.W. (1974) Research on the cainito (*Chrysophyllum cainito* L.) *Proceedings of the American Society for Horticultural Science of Tropical Regions* 18, 123–127.

Campbell, C.W., Malo, S.E. and Goldweber, S. (1967) *The Sapodilla*. Fruit Crop Fact Sheet No. 1, Cooperative Extension, University of Florida, Gainesville, Florida.

Clement, C.R. (1983) Underexploited Amazonian fruits. *Proceedings of the American Society for Horticultural Science of Tropical Regions* 27(A), 117–141.

Coronel, R.E. (1992) *Manilkara zapota* (L.) P. van Royen. In: Verheij, E.W.M. and Coronel, R.E. (eds) *Plant Resources in South-East Asia*, No. 2, *Edible Fruits and Nuts*. Prosea, Bogor Indonesia, pp. 220–223.

Gonzalez, L.G. and Fabella, E.L. (1952) Intergeneric graft-affinity of the chico. *The Philippine Agriculturalist* 35, 402–407.

Gonzalez, L.G. and Feliciano, P.A. (1953) The blooming and fruiting habits of the Ponderosa Chico. *Philippine Agriculturist* 37, 384–398.

Lakshminarayana, S. (1980) Sapodilla and prickly pear. In, Nagy, S. and Shano, P.E. (eds) *Tropical and Subtropical Fruits, Composition, Properties, and Uses*. AVI Publishing, Westpoint, Connecticut, pp. 415–441.

Leung, W-T.W. and Flores, M (1961) *Food Composition Tables for Use in Latin America.* Joint Research Project, Institute of Nutrition of Central America and Panama, Guatemala City, and National Institute of Health, Bethesda, Maryland.

Macleod, A.J., de Troconis, N.G. (1982) Volatile flavor components of sapodilla fruit (*Achras sapota* L.). *Journal of Agriculture and Food Chemistry* 30, 515–517.

Marshall, J.R. (1991) *Sapodilla (Chico).* Fact Sheet No. 4, Rare Fruit Council of Australia, Cairns, Queensland.

Martin, F.W. and Malo, S.E. (1978) *Cultivation of Neglected Tropical Fruits with Promise,* Part 5, *The Canistel and its Relatives.* PB84–112515, United States Department of Agriculture, New Orleans, 12pp.

Morton, J.F. (1987) Sapotaceae – sapodilla. In: *Fruits for Warm Climates.* Creative Resource Systems, Winterville, North Carolina, pp. 393–398.

Parker, G.H. (1986) *Abiu.* Fact Sheet No. 13, Rare Fruit Council of Australia, Cairns, Queensland.

Patil, P.K. and Patil, V.K. (1983) Studies on soil salinity tolerance of sapota. *South Indian Horticulture* 31, 3–6.

Pratt, H.K. and Mendoza, D.B. (1980) Fruit development and ripening of the star apple (*Chrysophyllum cainito* L.). *HortScience* 15, 721–722.

Rowe-Dutton, P. (1976) *Manilkara achras* – sapodilla. In: Garner, R.J. and Chandri, S.A. (eds) *The Propagation of Tropical Fruit Trees. CAB Horticultural Review* 4, 475–512.

Scholefield, P.B. (1983) Sapotaceae – Sapodilla. In: P.E. Page. (Compiler) *Tropical Tree Fruits for Australia.* Information Series QI 83018. Queensland Department of Primary Industries. Brisbane, Queensland, pp. 209–217.

Scholefield, P.B. (1984) Abiu. In: Page, P.E. (ed.) *Tropical Tree Fruits for Australia.* Information Series QI 83013, Queensland Department of Primary Industry, Australia.

Selvaraj, Y. and Pal, D.K. (1984) Changes in the chemical composition and enzyme activity of two sapodilla (*Manilkara zapota*) cultivar during development and ripening. *Journal of Horticultural Science* 59, 275–281.

Shanmuganelu, K., Srinivasan, G. and Rao, V.N.M. (1971) Influence of Ethrel (2-chloroethyl phosphonic acid) on ripening of sapota (*Achras zapota* L). *Horticultural Advances* 8, 33.

Whitman, W.F. (1965) The green sapote, a new fruit of south Florida. *Proceedings of the Florida State Horticulture Society* 78, 330–335.

INDEX

Note: page numbers in *italics* refer to figures and tables, numbers in **bold** refer to plates